PERGAMON INTERNATIONAL LIBRARY
of Science, Technology, Engineering and Social Studies
The 1000-volume original paperback library in aid of education,
industrial training and the enjoyment of leisure
Publisher: Robert Maxwell, M.C.

MECHANICS:
CLASSICAL AND QUANTUM

INTERNATIONAL SERIES IN NATURAL PHILOSOPHY

VOLUME 82

GENERAL EDITOR: D. TER HAAR

(Full list of titles in Natural Philosophy see p. 397)

SOME OTHER TITLES OF INTEREST

ABRIKOSOV, A. A. *et al.*
Quantum Field Theoretical Methods in
Statistical Physics

DAVYDOV, A. S.
Quantum Mechanics

KOTKIN, G. L. & SERBO, V. G.
Collection of Problems in Classical
Mechanics

LANDAU, L. D. & LIFSHITZ, E. M.
A Shorter Course of Theoretical Physics
— Volume 1: Mechanics and Electrodynamics
— Volume 2: Quantum Mechanics

TER HAAR, D.
The Old Quantum Theory

MECHANICS:
CLASSICAL AND QUANTUM

by

T. T. TAYLOR

Professor of Physics, Loyola Marymount University,
Los Angeles, California

PERGAMON PRESS

OXFORD NEW YORK TORONTO
SYDNEY PARIS BRAUNSCHWEIG

U. K.	Pergamon Press Ltd., Headington Hill Hall, Oxford OX3 0BW, England
U. S. A.	Pergamon Press Inc., Maxwell House, Fairview Park, Elmsford, New York 10523, U.S.A.
C A N A D A	Pergamon of Canada, Ltd., 207 Queen's Quay West, Toronto 1, Canada
A U S T R A L I A	Pergamon Press (Aust.) Pty. Ltd., 19a Boundary Street, Rushcutters Bay, N.S.W. 2011, Australia
F R A N C E	Pergamon Press SARL, 24 rue des Ecoles, 75240 Paris, Cedex 05, France
W E S T G E R M A N Y	Pergamon Press GmbH, D.3300 Braunschweig, Burgplatz 1, West Germany

First edition 1976

Library of Congress Cataloging in Publication Data

Taylor, Thomas Tallott.
Mechanics: classical and quantum.

(International series in natural philosophy; v. 82)
Bibliography: p.
1. Quantum theory. 2. Mechanics, Analytic.
I. Title.
QC174.12.T4 1975   530.1'2   75–12794
ISBN 0–08–018063–9 h
    0–08–020522–4 f

Printed in Hungary

CONTENTS

PREFACE

EVER since the appearance in 1950 of Herbert Goldstein's deservedly well-known text on classical mechanics, the once esoteric methods of this branch of physics have become increasingly accessible, not only to graduate students but also to undergraduates. Now it is quite common for junior-year textbooks on mechanics to include chapters on Lagrangian and Hamiltonian theory and it is generally agreed that this material, if competently presented, is within the capabilities of the students who use these books. The trend exemplified here is part of an ever-operative process in the teaching of physics whereby advanced topics are simplified (in the constructive sense of that term) and placed earlier in the curriculum. Thus the efficiency of learning is increased and time is made available for new subject matter generated by research. Having established that the Lagrangian–Hamiltonian approach, also known as analytical mechanics, should be welcome at the undergraduate level, one may discuss the manner in which it may best be introduced. If presented in the context of a few classical applications only, students may miss much of its true relevancy. Would it not be better utilized as a logical prelude to quantum theory? This rhetorical question expresses the author's position and reveals the motif of the present work.

Designed for the junior year, this book is intended to give students a solid grounding in the principles of quantum mechanics after leading them to these principles via the medium of analytical mechanics. In doing this, the use of gratuitous postulates is minimized; quantum theory, as exemplified by Schrödinger's formulation, is introduced through the concept that macroscopic particle motion is to be explained by the motion of a packet of waves. This is possible if the eikonal of the typical wave is made proportional to the classical action function. The Hamilton–Jacobi equation then becomes the eikonal equation and the linear wave equation corresponding thereto is shown to be the Schrödinger equation. Once this derivation is completed, the treatment becomes deductive and the implications of the Schrödinger formulation are explored in the microscopic domain where phenomena cannot be described successfully in classical terms. Techniques appropriate to this domain are introduced and the power and universality of quantum theory are made evident.

Beginning as it does with the Lagrangian formulation and ending with the hydrogenic atom and matrix mechanics, this book is ideally suited to a two-semester sequence. However, in curricula where the Lagrangian–Hamiltonian methodology is introduced in an earlier course on intermediate mechanics and where some familiarity with wave motion and with the historical background of quantum theory can be presumed, this book could be adapted to a one-semester presentation. In such an event, it is unlikely that the two final chapters could be covered; the substance of these chapters would probably be incorporated into a subsequent course, e.g. the senior course on atomic physics. In curricula where time is divided into units smaller than traditional semesters, still other ways of utilizing the material made available here will suggest themselves.

The author wishes to thank Dr. James S. Albertson who first suggested that a course of this type be presented and to express his gratitude for the interest and the reactions of the many students who attended the lectures now incorporated into this book. The author is particularly indebted to Dr. Joseph R. Schwartz who audited the entire sequence of lectures and who participated in many stimulating conversations. Also appreciated are the hospitality of Dr. William G. Spitzer of the University of Southern California and the many profitable discussions held with members, particularly with Dr. Jan Smit, of the faculty of that university during a sabbatical leave in which a significant part of this project was completed. The tangible encouragement of the author's immediate administrator, Dr. James E. Fox-worthy, is also gratefully acknowledged.

In the preparation of the illustrations and typescript, the author is greatly indebted to Mr. John F. McCaffrey, to Mrs. Josephine D. Murphy, and to his wife, Grace Bonaro Taylor, for very helpful assistance.

Los Angeles   THOMAS T. TAYLOR

CHAPTER 1

THE LAGRANGIAN FORMULATION OF MECHANICS

1.01. The Harmonic Oscillator; a New Look at an Old Problem

Every student is familiar with the one-dimensional harmonic oscillator consisting of a given mass M attached to a linear spring (whose own mass is negligible) as in Fig. 1.01. The mass M is constrained to

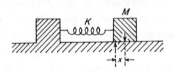

FIG. 1.01. The one-dimensional harmonic oscillator.

move without friction along the line of action of the spring and x denotes its displacement from equilibrium. The force exerted on M by the spring is called F_x and is reckoned positive when it tends to increase x. In terms of the elastic constant K, it is given by:

$$F_x = -Kx, \qquad (1.001)$$

whereupon the equation of motion becomes:[†]

$$M\ddot{x} + Kx = 0. \qquad (1.002)$$

[†] $\dot{x} = \mathrm{d}x/\mathrm{d}t$; $\ddot{x} = \mathrm{d}^2x/\mathrm{d}t^2$; etc.

1

Notice that this is a second-order differential equation since it involves the second derivative of x with respect to t. It has as a general solution:

$$x = A \sin (\Omega t + \theta); \qquad \Omega = (K/M)^{\frac{1}{2}}. \tag{1.003}$$

Here A and θ are the familiar amplitude and phase, respectively. The presence of these two arbitrary and independent parameters indicates that there is a two-fold infinity of motions available to the oscillator; the specification of A and θ determines one particular motion out of this continuum of possible motions. Other pairs of quantities, e.g. the initial displacement x_0 and the initial velocity $\dot{x}_0$, could serve equally well as determinative parameters in this situation and are often used for this purpose. Equations expressing A and θ in terms of x_0 and $\dot{x}_0$ and vice versa can be easily derived.

It might be supposed that nothing could be added to this simple analysis which so quickly produces a complete description of the motion of the oscillator. In a sense, this is true. However, another method, which reduces (1.002) to a first-order differential equation before finally solving it, can also be brought to bear upon this problem. Although obviously not necessary in the present case, this method is very valuable in the analysis of more complicated systems. It employs the concepts of potential energy and kinetic energy and introduces other useful ideas.

The potential energy $V(x)$ is energy stored in the spring when the displacement of M is increased from zero to x; it can be found by integrating the force an external agent would have to apply to the end of the spring to produce the required extension. The force in question, F'_x, is equal to $-F_x$, i.e. to Kx. Thus:

$$V(x) = \int_0^x Kx \, dx = \tfrac{1}{2}Kx^2. \tag{1.004}$$

A potential energy exists and is derivable from the force whenever, as here, the latter can be integrated to yield a single-valued function of position.[†] Conversely, in such cases, the force can be obtained from

---

† A force having this characteristic is called a *conservative* force.

the potential energy by applying the following simple formula:

$$-F_x = F'_x = \frac{\mathrm{d}V}{\mathrm{d}x}; \qquad F_x = -\frac{\mathrm{d}V}{\mathrm{d}x}. \qquad (1.005)$$

From this it is evident that a constant added to the potential energy would not change the force in any way and would therefore have no effect upon the motion of the oscillator. The privilege of adding such a constant to the potential energy (and therefore to the total energy also) is exercised in many situations in physics. In the present example, the potential energy without an additive constant can be plotted as the parabolic curve shown in Fig. 1.02.

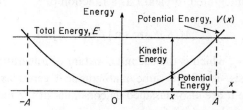

FIG. 1.02. Harmonic oscillator potential energy curve and associated concepts.

The expression for the kinetic energy is the well-known $\frac{1}{2}M\dot{x}^2$; the total energy therefore becomes:

$$E = \tfrac{1}{2}M\dot{x}^2 + \tfrac{1}{2}Kx^2. \qquad (1.006)$$

Differentiation of this with respect to time yields:

$$\dot{E} = M\dot{x}\ddot{x} + Kx\dot{x} = \dot{x}(M\ddot{x} + Kx) = 0. \qquad (1.007)$$

The null result follows because the original differential equation (1.002) asserts that $(M\ddot{x} + Kx)$ is zero. The fact that $\dot{E} = 0$ or that E is constant is, of course, an expected result and E in this context is said to be a *constant of the motion*. The equation expressing this constant in terms of x and $\dot{x}$, namely (1.006), is actually a first-order non-linear differential equation of motion; its discovery is tantamount to an integration of equation (1.002) with E playing the part of a constant of integration.

An expression appearing in this way is often called a *first integral of the motion*.

Equation (1.006) may now be solved for $\dot{x}$ and the variables separated:

$$\dot{x} = \left[\frac{2(E-\frac{1}{2}Kx^2)}{M}\right]^{\frac{1}{2}}. \tag{1.008}$$

On rearrangement, this becomes:

$$dt = \left[\frac{2(E-\frac{1}{2}Kx^2)}{M}\right]^{-\frac{1}{2}} dx. \tag{1.009}$$

This may be integrated to yield t as a function of x:

$$t = (M/K)^{\frac{1}{2}} \text{ arc sin } \left[x(K/2E)^{\frac{1}{2}}\right] + t_e. \tag{1.010}$$

The constant t_e, which is the second constant of integration produced in this analysis, is of the same mathematical genus as E itself and quantities like it are also classified as constants of the motion. Equation (1.010) can be solved for x:

$$x = (2E/K)^{\frac{1}{2}} \sin \Omega(t-t_e). \tag{1.011}$$

This is obviously equivalent to (1.003) with $(2E/K)^{\frac{1}{2}} = A$ and $-\Omega t_e = \theta$. Notice that when $t = t_e$, the mass M passes through its neutral position while moving to the right. A constant of the motion such as t_e, which identifies the instant at which a moving body passes through some specially designated point on its path, is called an *epoch*.[†]

The solution in the form of (1.011) involves, as it must, two arbitrary and independent determinative parameters. These are none other than the constants of the motion or, if one prefers, the constants of integration E and t_e. It follows that E and t_e, A and θ, x_0 and $\dot{x}_0$, are all different forms of the same two entities; any one of these pairs may be obtained from any other. It is clear that two constants of the

---

[†] This term is borrowed from celestial mechanics where it usually refers to the time of periapsidal passage of a planet or satellite.

motion will invariably be associated with a system governed by a second-order differential equation of motion; systems governed by more such equations will involve correspondingly more constants of the motion.

It is not always easy to arrive at an expression like (1.006) which relates a constant of the motion to a function of coordinates, velocities, and perhaps time, i.e. it is not always easy to perform a first integration; in some cases, it may be impossible to do so in terms of recognized functions. Classical analytical mechanics, which begins with the

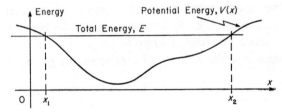

FIG. 1.03. Potential energy curve for a typical anharmonic oscillator.

Lagrangian formulation, is noteworthy for the fact that it facilitates the discovery of some of these first integrals. The influence of this formulation goes beyond the classical domain and affects much of the content of quantum mechanics.

The practical utility of the method just outlined is more apparent in the case of an anharmonic oscillator for which F_x is a non-linear function of x and for which the potential energy $V(x)$, illustrated in Fig. 1.03, is therefore non-parabolic. Once more, a first-order differential equation is obtained and this can be integrated to yield t as a function of x:

$$t = \int \left\{ \frac{2[E - V(x)]}{M} \right\}^{-\frac{1}{2}} dx + t_e. \qquad (1.012)$$

Regardless of the form of $V(x)$, the problem may be solved at least in principle by performing this indicated integration. Even if t in (1.012) cannot be expressed in terms of recognized functions, a result can always be obtained to any desired degree of accuracy using step-by-step integration with the aid of a computer.

Additional physical insight into the role of the energy integral may be gained by superposing a horizontal line of constant ordinate E on the potential energy curve of, for example, Fig. 1.02. Since V cannot be greater than E, the motion must take place between the points of intersection of this straight line and the parabola, i.e. in the region for which $V \leqslant E$. The points of intersection, here $- A$ and A, are called the *turning points of the motion*. Notice that, for given E, there is associated with every value of x a specific amount of kinetic energy just as there is associated a specific amount of potential energy. These two amounts are easily identified by the double-ended arrows in Fig. 1.02 which invariably add up to the total energy E. The direction of motion and the amount of kinetic energy respectively deter-

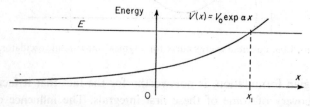

FIG. 1.04. Potential energy curve for a non-oscillatory one-dimensional system.

mine the sign and magnitude of $\dot{x}$; thus $\dot{x}$ may be written in terms of x as was done in equation (1.008). These considerations obviously apply to oscillatory systems with other potential energy functions, e.g. that of Fig. 1.03. In the latter case, the turning points are x_1 and x_2, as shown.

Problems in which the motion is non-oscillatory are equally amenable to the method of treatment outlined here. For example, suppose $V(x) = V_0 \exp \alpha x$ as in Fig. 1.04, and that the moving mass is incident from the left with total energy E. It will stop at the turning point x_1 where $V = E$ and then retrace its motion in the opposite direction, never to return. A few other examples of the infinite variety of possible potential energy curves are shown in Fig. 1.05. The student

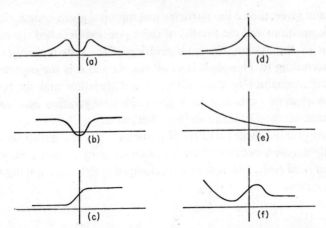

FIG. 1.05. Typical potential energy curves for one-dimensional systems.

should consider, in qualitative fashion, the various types of motion which can occur in each of these cases under various assumed values of total energy.

1.02. A System and its Configuration

As a fundamental approach to classical mechanics, it is convenient to consider a *system* as a collection of *particles*[†] which, in the most general case, have no fixed spatial relationships with one another. Each particle has an intrinsic property called its *mass* (and for this reason is often called a *point mass)* and each may be acted upon by forces. Forces exerted by other members of the collection are called *internal forces*; those arising from sources outside the collection are called *external forces*.

[†] Particles, as the term is used here, are abstractions proper to classical mechanics. They may be regarded simply as pieces of matter so small that their intrinsic structure and motions (e.g. rotation and vibration) are inconsequential in the context of their gross motions with respect to one another.

At any given time t the particles making up a given system occupy specific positions and the totality of these positions is called the *configuration* of the system. The configuration of the system changes with time according to Newton's laws of motion and it is the objective of classical mechanics to determine the configuration and its rate of change at some particular time given the configuration and its rate of change at some other (usually earlier) time.

In many cases, large numbers of particles are held together by theoretically massless bonds to form an extended body of invariable shape called a *rigid body*. Macroscopic mechanical systems are usually com-

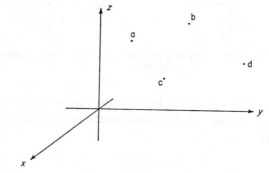

FIG. 1.06. A system of particles with an inertial Cartesian reference frame.

posed exclusively of rigid bodies which, in turn, may be related to one another by additional restrictions such as those which permit rolling without slipping, rotating about a fixed axis, or (as in Fig. 1.01) sliding along a given line without rotating, etc. All such bonds or restrictions are called *constraints* and their presence greatly reduces the diversity of the configurations available to the system. The study of constrained systems will be taken up later after certain fundamental concepts have first been firmly established.

It is of the utmost importance that the configuration of a system be specified in a scientifically acceptable manner. The Cartesian coordinates of the several particles with respect to an inertial frame of reference immediately suggest themselves for this purpose. In Fig. 1.06, for example, a system of N particles (designated a, b, c, ...) is illustrated in conjunction with an inertial coordinate frame x, y, z. Evidently the configuration of this system is completely specified, but not overspecified, by the values of the $3N$ coordinates:

$$\left.\begin{array}{c} x_\text{a}, \ y_\text{a}, \ z_\text{a}; \\ x_\text{b}, \ y_\text{b}, \ z_\text{b}; \\ \cdot \quad \cdot \quad \cdot \end{array}\right\} \tag{1.013}$$

In elementary mechanics, it is convenient to arrange these coordinates in sets of three (as above) and to regard each such set as forming the components of a *vector*. Thus:

$$\boldsymbol{r}_\text{a} = \boldsymbol{e}_x x_\text{a} + \boldsymbol{e}_y y_\text{a} + \boldsymbol{e}_z z_\text{a}, \tag{1.014}$$

where $\boldsymbol{e}_x$, $\boldsymbol{e}_y$, and $\boldsymbol{e}_z$ are *unit vectors*. Although there will still be an occasional need to use vector notation, from now on it will generally be more convenient to consider the totality of the $3N$ inertial Cartesian coordinates as a single set and to designate them by a single symbol with an index which runs from one to $3N$. Thus, one defines:

$$x_1 = x_\text{a}; \quad x_2 = y_\text{a}; \quad x_3 = z_\text{a}; \quad x_4 = x_\text{b}; \text{ etc.} \tag{1.015}$$

The indexed quantities are known collectively as "the x_j", $j = 1, 2, \ldots$, $3N$, and a set of values of these quantities is both necessary and sufficient to specify the configuration of the system.

Instead of regarding the configuration as a constellation of N points in a three-dimensional space, one may regard it as a single point in a $3N$-dimensional space called *configuration space*. This elegant and abstract point of view is often useful and is frequently mentioned in theoretical discussions.

1.03. Generalized Coordinates and Velocities

Although the $3N$ inertial Cartesian coordinates have much to recommend them, they are by no means the only possible set of coordinates or even the most suitable set in particular circumstances. In studies of planetary motion, for example, spherical polar coordinates are more convenient and are, in fact, used. The conventional relationship between the spherical polar coordinates and the Cartesian coordinates of a given point P is illustrated in Fig. 1.07; also illustrated are the mutually orthogonal[†] unit vectors e_r, e_θ, and e_φ. The transformation equations are as follows:

$$r = (x^2+y^2+z^2)^{\frac{1}{2}}; \quad \theta = \arccos\frac{z}{(x^2+y^2+z^2)^{\frac{1}{2}}} ; \quad \varphi = \arctan\frac{y}{x}.$$

(1.016)

$$x = r \sin \theta \cos \varphi; \quad y = r \sin \theta \sin \varphi; \quad z = r \cos \theta. \quad (1.017)$$

The partial derivatives of one set of coordinates with respect to the other are often useful. These derivatives, expressed in terms of r, θ, and φ, are given herewith:

$$\left.\begin{array}{lll}
\dfrac{\partial r}{\partial x} = \sin \theta \cos \varphi. & \dfrac{\partial \theta}{\partial x} = \dfrac{1}{r} \cos \theta \cos \varphi. & \dfrac{\partial \varphi}{\partial x} = \dfrac{-\sin \varphi}{r \sin \theta}. \\[2ex]
\dfrac{\partial r}{\partial y} = \sin \theta \sin \varphi. & \dfrac{\partial \theta}{\partial y} = \dfrac{1}{r} \cos \theta \sin \varphi. & \dfrac{\partial \varphi}{\partial y} = \dfrac{\cos \varphi}{r \sin \theta}. \\[2ex]
\dfrac{\partial r}{\partial z} = \cos \theta. & \dfrac{\partial \theta}{\partial z} = \dfrac{-1}{r} \sin \theta. & \dfrac{\partial \varphi}{\partial z} = 0.
\end{array}\right\} \quad (1.018)$$

$$\left.\begin{array}{lll}
\dfrac{\partial x}{\partial r} = \sin \theta \cos \varphi. & \dfrac{\partial y}{\partial r} = \sin \theta \sin \varphi. & \dfrac{\partial z}{\partial r} = \cos \theta. \\[2ex]
\dfrac{\partial x}{\partial \theta} = r \cos \theta \cos \varphi. & \dfrac{\partial y}{\partial \theta} = r \cos \theta \sin \varphi. & \dfrac{\partial z}{\partial \theta} = -r \sin \theta. \\[2ex]
\dfrac{\partial x}{\partial \varphi} = -r \sin \theta \sin \varphi. & \dfrac{\partial y}{\partial \varphi} = r \sin \theta \cos \varphi. & \dfrac{\partial z}{\partial \varphi} = 0.
\end{array}\right\} \quad (1.019)$$

Using spherical polar coordinates instead of Cartesians in the example of Fig. 1.06, one would write the following as the set of $3N$ coordi-

---

[†] Orthogonal vectors are vectors which meet at right angles.

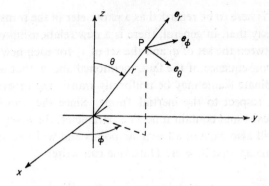

FIG. 1.07. Cartesian coordinates and spherical polar coordinates.

nates for the system:

$$\left.\begin{array}{l} r_a,\ \theta_a,\ \varphi_a; \\ r_b,\ \theta_b,\ \varphi_b; \\ \cdot\ \ \cdot\ \ \cdot \end{array}\right\} \qquad (1.020)$$

Again, the convenient procedure is to place all the coordinates in a sequentially numbered set. For generalized coordinates, in this context, q is the symbol usually chosen. Thus, in the present example:

$$q_1 = r_a; \qquad q_2 = \theta_a; \qquad q_3 = \varphi_a; \qquad q_4 = r_b;\ \text{etc.} \quad (1.021)$$

Generalized coordinates even further removed from inertial Cartesian coordinates are conceivable and useful, so long as the set of coordinate values is necessary and sufficient to specify the configuration of the system. The generalized coordinates need not even belong to an inertial frame of reference so long as the functional relationships connecting them with inertial Cartesians are known. Thus, the following relationships of transformation must exist and must be differentiable to the degree required by subsequent theory:

$$\left.\begin{array}{l} q_1 = q_1(x_1,\ x_2,\ \ldots,\ x_{3N},\ t); \\ q_2 = q_2(x_1,\ x_2,\ \ldots,\ x_{3N},\ t); \\ \cdot\ \ \cdot\ \ \cdot\ \ \cdot\ \ \cdot\ \ \cdot\ \ \cdot\ \ \cdot\ \ \cdot\ \ \cdot \\ q_{3N} = q_{3N}(x_1,\ x_2,\ \ldots,\ x_{3N},\ t). \end{array}\right\} \quad (1.022)$$

The time t is here to be regarded as a parameter of the transformation. This is to say that, in general, there is a new relationship of transformation between the set of q_j and the set of x_k for each new value of t. This is a consequence of the fact, mentioned above, that the generalized coordinate frame may be uniformly translating or even accelerating with respect to the inertial frame. Since the transformations of (1.022) exist and contain time as a parameter, the inverse transformations will also exist at all regular points and will also, in general, contain time as a parameter. Thus, one can write:

$$\left.\begin{aligned} x_1 &= x_1(q_1, q_2, \ldots, q_{3N}, t); \\ x_2 &= x_2(q_1, q_2, \ldots, q_{3N}, t); \\ &\cdot \quad \cdot \quad \cdot \quad \cdot \quad \cdot \quad \cdot \quad \cdot \\ x_{3N} &= x_{3N}(q_1, q_2, \ldots, q_{3N}, t). \end{aligned}\right\} \qquad (1.023)$$

It should be noticed that the generalized coordinates do not necessarily have the dimensions of length. For example, q_2, q_3, q_5, q_6, etc., in (1.021) are angles and are therefore dimensionless.

For any generalized coordinate q_j, there is defined a *generalized velocity*:

$$\dot{q}_j = \frac{dq_j}{dt} . \qquad (1.024)$$

Just as q_j does not necessarily have the dimensions of length, so also $\dot{q}_j$ does not necessarily have the conventional dimensions of a velocity, i.e. length/time. Moreover, the generalized velocity corresponding to an angular coordinate such as $q_2 = \theta_a$ is not synonymous with the θ component of the vector velocity as that term is usually understood; the latter is given by $e_\theta \cdot v_a$ which is equal to $r_a \dot{\theta}_a$, not to $\dot{\theta}_a$. In spite of such departures from usage which the reader may regard as traditional, the utility of the generalized velocities will soon make itself evident.

A few words about the conceptual position of the generalized velocities in the Lagrangian formulation of mechanics are in order. To look upon a particular $\dot{q}_j$ as a quantity which simply duplicates (in another form) the information contained in the function $q_j(t)$, of which it is the derivative, is to take a narrow viewpoint since $q_j(t)$ exists as a particular function of time only in the context of a specific motion of the system. If one considers the totality of all the motions available to the system, however, the $\dot{q}_j$ take on a new sigfinicance. It is clear that a specification of the $3N$ values of the q_j, which determines the instantaneous configuration, is insufficient to determine the future dynamical development of the system; to do the latter, one must specify both the $3N$ values of the q_j and the $3N$ values of the $\dot{q}_j$ as of the initial instant. On this point of view, which is fundamental in the context of the present chapter, all $6N$ of the quantities, q_j and $\dot{q}_j$, are on an equal footing. All must be invoked to specify the dynamical state of the system and the fact that any one of the q_j and its corresponding $\dot{q}_j$ are linked together by (1.024) for the duration of a specific motion of the system is of relatively less importance.

The generalized velocities can be expressed in terms of the inertial Cartesian coordinates, the inertial Cartesian velocities, and the time:

$$\dot{q}_j = \sum_{k=1}^{3N} \frac{\partial q_j}{\partial x_k}\, \dot{x}_k + \frac{\partial q_j}{\partial t} = \dot{q}_j(x_k, \dot{x}_k, t). \qquad (1.025)$$

Similarly, it is possible to express the inertial Cartesian velocities in terms of the generalized coordinates, the generalized velocities, and the time:

$$\dot{x}_k = \sum_{j=1}^{3N} \frac{\partial x_k}{\partial q_j}\, \dot{q}_j + \frac{\partial x_k}{\partial t} = \dot{x}_k(q_j, \dot{q}_j, t). \qquad (1.026)$$

The latter expression, particularly, will be very useful in subsequent sections.

The partial derivatives in (1.025) and (1.026) are partial derivatives of the functions in (1.022) and (1.023) respectively, i.e. of the coordinate transformations. In other contexts, some of the identical partial derivative symbols may have other meanings and values. This is mentioned so that the reader may be prepared for situations which might otherwise cause difficulty.

1.04. Kinetic Energy and the Generalized Momenta

The expression for the kinetic energy in terms of inertial Cartesian velocities is well known:

$$T = \tfrac{1}{2}[m_a(\dot{x}_a^2 + \dot{y}_a^2 + \dot{z}_a^2) + m_b(\dot{x}_b^2 + \dot{y}_b^2 + \dot{z}_b^2) + \ldots]. \qquad (1.027)$$

Mass, of course, is associated with a particle. It is equally possible to associate a mass m_k with each coordinate x_k provided it is done in the following way:

$$\left.\begin{array}{l} m_a = m_1 = m_2 = m_3; \\ m_b = m_4 = m_5 = m_6; \\ \cdots\cdots\cdots\cdots\cdots \end{array}\right\} \qquad (1.028)$$

The justification for this artifice is that it greatly simplifies the writing of many expressions, particularly the expression for kinetic energy. The latter becomes:

$$T = \tfrac{1}{2} \sum_{k=1}^{3N} m_k \dot{x}_k^2. \qquad (1.029)$$

One may now substitute equation (1.026) into the above to obtain a formal expression for T in terms of the generalized coordinates, the generalized velocities, and the time:

$$T = \frac{1}{2} \sum_{k=1}^{2N} m_k \left[\sum_{j=1}^{3N} \frac{\partial x_k}{\partial q_j} \dot{q}_j + \frac{\partial x_k}{\partial t} \right]^2 = T(q_j, \dot{q}_j, t). \qquad (1.030)$$

The final expression above, indicating the functional content of T, is correct because x_k, as expressed by (1.023), is in general a function of the q_j and t and its derivatives $\partial x_k/\partial q_j$ and $\partial x_k/\partial t$ are likewise. It is obvious that T is quadratic in the generalized velocities, which appear explicitly in (1.030). Thus terms containing $\dot{q}_j^2$ or $\dot{q}_j\dot{q}_l$, but none of higher degree, are to be found in this expression. Terms of degree one or zero in the $\dot{q}_j$ are present only if expressions (1.023) for the x_k in terms of the generalized coordinates also involve the time, i.e. only if the $\partial x_k/\partial t$ in (1.030) do not vanish.

It is now possible to define a set of quantities which rivals, and in many contexts surpasses, the set of generalized velocities in importance. The quantities in question are called the *generalized momenta* or *canonical momenta*. Denoted by the symbol p_j, they are defined by:

$$p_j = \frac{\partial T}{\partial \dot{q}_j}.$$ (1.031)

In applying this definition to find p_3, for example, one takes the partial derivative of T with respect to $\dot{q}_3$ holding all other quantities such as q_1, q_2, q_3, q_4, $\cdots$ $\dot{q}_1$, $\dot{q}_2$, $\dot{q}_4$, $\ldots$, and t constant. Thus, following a theme expressed in Section 1.03, a particular one of the $\dot{q}_j$ is treated as formally independent of all the generalized coordinates, all the generalized velocities except itself, and the time. Evidently there are $3N$ p_j just as there are $3N$ q_j; a particular one of these generalized momenta such as p_3 is said to be *conjugate* to the corresponding coordinate, i.e. to q_3. The q_j and the p_j are collectively known as the *dynamical variables* or the *canonical variables* and there are $3N$ conjugate pairs of these variables associated with a system of N unconstrained particles moving in three dimensions.

For the trivial case in which the generalized coordinates are chosen to be synonymous with the inertial Cartesian coordinates,

$$p_j = \frac{\partial T}{\partial \dot{x}_j} = \frac{\partial}{\partial \dot{x}_j} \left[\frac{1}{2} \sum_{k=1}^{3N} m_k \dot{x}_k^2 \right] = m_j \dot{x}_j,$$ (1.032)

and the generalized momenta are seen to be identical with the ordinary Cartesian components of the momentum vectors of the particles. In a one-particle system, for example, $m = m_1 = m_2 = m_3$ and one would have:

$$\left. \begin{array}{l} p_1 = m_1 \dot{x}_1 = m\dot{x} = p_x; \\ p_2 = m_2 \dot{x}_2 = m\dot{y} = p_y; \\ p_3 = m_3 \dot{x}_3 = m\dot{z} = p_z. \end{array} \right\}$$ (1.033)

In cases where the generalized coordinates are not inertial Cartesians, the generalized momenta will usually not be identifiable with any component of conventional momentum and will not necessarily even have the dimensions of mass times velocity. A one-particle system described by spherical polar coordinates is a case in point. The first step, of course, in finding the generalized momenta is to find the expression $T(q_j, \dot{q}_j, t)$ which, in this case, is $T(r, \theta, \varphi, \dot{r}, \dot{\theta}, \dot{\varphi}, t)$. To do this by direct substitution of the partial derivatives (1.019) into (1.030) is tedious; fortunately an easier method is available. This method begins by considering an infinitesimal vector displacement $d\boldsymbol{r}$ of the particle (which may be regarded as taking place at point P of Fig. 1.07) and resolving this in the basis formed by the orthogonal unit vectors $\boldsymbol{e}_r$, $\boldsymbol{e}_\theta$, and $\boldsymbol{e}_\varphi$. One finds that:

$$d\boldsymbol{r} = \boldsymbol{e}_r\, dr + \boldsymbol{e}_\theta r\, d\theta + \boldsymbol{e}_\varphi\, r \sin\theta\, d\varphi. \tag{1.034}$$

Dividing by dt, one obtains the representation of the velocity of the particle in this same basis:

$$\boldsymbol{v} = \boldsymbol{e}_r\dot{r} + \boldsymbol{e}_\theta r\dot{\theta} + \boldsymbol{e}_\varphi r \sin\theta\, \dot{\varphi}. \tag{1.035}$$

Evidently:

$$v^2 = \dot{r}^2 + r^2\dot{\theta}^2 + r^2 \sin^2\theta\, \dot{\varphi}^2, \tag{1.036}$$

and the kinetic energy becomes:

$$\begin{aligned} T &= \tfrac{1}{2}m(\dot{r}^2 + r^2\dot{\theta}^2 + r^2 \sin^2\theta\, \dot{\varphi}^2), \\ &= \tfrac{1}{2}m(\dot{q}_1^2 + q_1^2\dot{q}_2^2 + q_1^2 \sin^2 q_2\, \dot{q}_3^2). \end{aligned} \tag{1.037}$$

It is noteworthy that not every one of the arguments of which $T(q_j, \dot{q}_j, t)$ is in general a function need be present in every case. Thus q_3 (i.e. φ) and t are missing from (1.037). Using (1.031), it is a simple matter to obtain the p_j:

$$\left. \begin{aligned} p_1 &= m\dot{q}_1 = m\dot{r} = p_r\,; \\ p_2 &= mq_1^2\dot{q}_2 = mr^2\dot{\theta} = p_\theta\,; \\ p_3 &= mq_1^2 \sin^2 q_2\, \dot{q}_3 = mr^2 \sin^2\theta\, \dot{\varphi} = p_\varphi. \end{aligned} \right\} \tag{1.038}$$

These may be compared with (1.033). Notice that p_2 and p_3 here have the dimensions of angular momentum and p_3 (or p_φ) in particular is simply L_z, the z component of the total vector angular momentum $\boldsymbol{L} = \boldsymbol{r} \times \boldsymbol{p}$ of the particle about the origin.

1.05. Lagrange's Equations

The Lagrangian formulation of mechanics does not in any way contradict Newton's laws but rather builds upon them. At the very least, it yields the complete set of second-order equations of motion of even a complicated system quickly and surely. In addition, it often permits the replacement of some or all of these second-order equations with first-order equations through the discovery of expressions for the corresponding constants of the motion. Finally, the Lagrangian formulation leads into the Hamiltonian formulation and ultimately, through the latter, into quantum mechanics.

In order to incorporate Newton's second law into the present development, it is necessary to express the generalized momenta in terms of inertial Cartesian quantities. One may do this most easily by beginning with expression (1.030) for T and differentiating it with respect to $\dot{q}_j$. Before so doing, it is advisable to replace the dummy index j with another letter, say l, to avoid confusion. The result is:

$$\frac{\partial T}{\partial \dot{q}_j} = \sum_{k=1}^{3N} m_k \left[\sum_{l=1}^{3N} \frac{\partial x_k}{\partial q_l} \dot{q}_l + \frac{\partial x_k}{\partial t} \right] \frac{\partial x_k}{\partial q_j} . \tag{1.039}$$

Since the content of the square brackets is simply $\dot{x}_k$, one has:

$$p_j = \sum_{k=1}^{3N} m_k \dot{x}_k \frac{\partial x_k}{\partial q_j} . \tag{1.040}$$

The total time derivative of this expression is important in what follows; it is conveniently written as two sums:

$$\dot{p}_j = \sum_{k=1}^{3N} m_k \ddot{x}_k \frac{\partial x_k}{\partial q_j} + \sum_{k=1}^{3N} m_k \dot{x}_k \frac{\mathrm{d}}{\mathrm{d}t} \left(\frac{\partial x_k}{\partial q_j} \right) . \tag{1.041}$$

Pursuant to the introductory remarks of Section 1.01, the present discussion will be limited to the case in which all forces are derivable from a potential energy V. This is quite appropriate when the ultimate goal is the study of quantum mechanics with applications to particle motion on the atomic level. The derivation continues, then, by expressing Newton's second law in the following form for each of the particles of the system:

$$\left.\begin{array}{l} m_a \ddot{\mathbf{r}}_a = -\left(\mathbf{e}_x \dfrac{\partial V}{\partial x_a} + \mathbf{e}_y \dfrac{\partial V}{\partial y_a} + \mathbf{e}_z \dfrac{\partial V}{\partial z_a}\right) = -\nabla_a V; \\[2mm] m_b \ddot{\mathbf{r}}_b = -\nabla_b V; \\[2mm] m_c \ddot{\mathbf{r}}_c = -\nabla_c V; \\[2mm] \cdot \quad \cdot \quad \cdot \quad \cdot \quad \cdot \end{array}\right\} \qquad (1.042)$$

The potential energy depends upon the configuration of the system and may, in addition, depend explicitly upon the time. (The latter situation obtains if any of the force laws operative among the particles are time-dependent in character or, which is more likely, if time-dependent external forces are acting.) Thus $V = V(\mathbf{r}_a, \mathbf{r}_b, \ldots, t)$. In the condensed notation of Section 1.02, $V = V(x_1, x_2, \ldots, x_{3N}, t)$ and the bulky system (1.042) of vector equations reduces to:

$$m_k \ddot{x}_k = -\frac{\partial V}{\partial x_k}. \qquad (1.043)$$

Upon switching from inertial Cartesians to generalized coordinates, one finds that $V = V(q_1, q_2, \ldots, q_{3N}, t)$ and the partial derivatives $\partial V/\partial q_j$ become important. In taking these, time is held constant. Thus:

$$\frac{\partial V}{\partial q_j} = \sum_{k=1}^{3N} \frac{\partial V}{\partial x_k} \frac{\partial x_k}{\partial q_j} = -\sum_{k=1}^{2N} m_k \ddot{x}_k \frac{\partial x_k}{\partial q_j}. \qquad (1.044)$$

This expression reveals a very significant fact, namely that $\partial V/\partial q_j$ is equal to the negative of the first sum in expression (1.041) for $\dot{p}_j$. This result will be utilized shortly.

To simplify the second sum in (1.041), it should be recalled that $\partial x_k / \partial q_j$ is in general a function of the q_j and of t. Therefore:

$$\frac{d}{dt} \left(\frac{\partial x_k}{\partial q_j} \right) = \frac{\partial^2 x_k}{\partial q_j \, \partial q_1} \dot{q}_1 + \frac{\partial^2 x_k}{\partial q_j \, \partial q_2} \dot{q}_2 + \ \dots \ + \frac{\partial^2 x_k}{\partial q_j \, \partial t}. \quad (1.045)$$

However, after writing out the individual terms in (1.026), one finds that:

$$\frac{\partial}{\partial q_j} \dot{x}_k = \frac{\partial^2 x_k}{\partial q_j \, \partial q_1} \dot{q}_1 + \frac{\partial^2 x_k}{\partial q_j \, \partial q_2} \dot{q}_2 + \ \dots \ + \frac{\partial^2 x_k}{\partial q_j \, \partial t}. \quad (1.046)$$

Notice that $\dot{q}_j$ is not affected by the operation $\partial/\partial q_j$. The right-hand sides of the above expressions are identical and therefore:

$$\frac{d}{dt} \frac{\partial x_k}{\partial q_j} = \frac{\partial \dot{x}_k}{\partial q_j}. \quad (1.047)$$

With the aid of this result, one obtains for the second sum in (1.041):

$$\sum_{k=1}^{3N} m_k \dot{x}_k \frac{d}{dt} \left(\frac{\partial x_k}{\partial q_j} \right) = \sum_{k=1}^{3N} m_k \dot{x}_k \frac{\partial \dot{x}_k}{\partial q_j} = \frac{\partial T}{\partial q_j}. \quad (1.048)$$

Using this and (1.044) in (1.041), it is easy to arrive at the important conclusion:

$$\dot{p}_j = -\frac{\partial V}{\partial q_j} + \frac{\partial T}{\partial q_j} = \frac{\partial}{\partial q_j} (T - V). \quad (1.049)$$

This says that the rate of change of the generalized momentum p_j is not merely the q_j derivative of $-V$ alone as would be the case if one were dealing with inertial Cartesian coordinates; rather it is the q_j derivative of a new quantity with the dimensions of energy, namely of $(T - V)$. This new quantity is called the *Lagrangian*, $\mathcal{L}$:

$$\boxed{\mathcal{L}(q_j, \dot{q}_j, t) = T(q_j, \dot{q}_j, t) - V(q_j, t)}. \quad (1.050)$$

It is noteworthy that the Lagrangian is in general a function of the $6N + 1$ quantities q_j, $\dot{q}_j$, and t. This again recalls the theme expressed in Section 1.03. Since V does not contain the $\dot{q}_j$,

$$\frac{\partial \mathscr{L}}{\partial \dot{q}_j} = \frac{\partial T}{\partial \dot{q}_j} = p_j \tag{1.051}$$

and, with appropriate substitutions, the equations of motion of the system expressed by (1.049) may be written very succinctly as follows:

$$\boxed{\frac{d}{dt}\left(\frac{\partial \mathscr{L}}{\partial \dot{q}_j}\right) - \frac{\partial \mathscr{L}}{\partial q_j} = 0} \,. \tag{1.052}$$

These $3N$ equations are called LAGRANGE'S EQUATIONS.

As a simple example of the use of the Lagrangian method to obtain the equations of motion of an unconstrained system, consider the problem of the motion of a planet (the Kepler problem) under the simplifying assumption that the mass of the planet is relatively negligible compared with that of the Sun.[†] Under this assumption, the Sun can be taken at rest at the origin of an inertial frame of reference and the configuration of the system is described by the position of the planet relative to this frame. At present, the objective is merely to find the equations of motion in spherical polar coordinates by the Lagrangian method. A discussion of the actual motion appears in Chapter 2.

The potential energy in this problem is given by:

$$V = -GMmr^{-1}, \tag{1.053}$$

since the negative gradient of this function is the well-known inverse square gravitational force, $-GMmr^{-2}e_r$; G is the constant of universal gravitation, M is the mass of the Sun and m, the mass of the planet. The kinetic energy for this case has already been given in equation (1.037) and the Lagrangian is therefore:

$$\mathscr{L} = \tfrac{1}{2}m(\dot{r}^2 + r^2\dot{\theta}^2 + r^2 \sin^2 \theta \; \dot{\varphi}^2) + GMmr^{-1}. \tag{1.054}$$

[†] An example of a Keplerian problem with a "planet" of non-negligible mass is provided by the hydrogenic atom. This will be studied in a quantum mechanical context in Chapter 12.

As in equations (1.038):

$$p_1 = p_r = \frac{\partial \mathcal{L}}{\partial \dot{r}} = m\dot{r}.$$

$$p_2 = p_\theta = \frac{\partial \mathcal{L}}{\partial \dot\theta} = mr^2\dot\theta. \qquad\qquad (1.055)$$

$$p_3 = p_\varphi = \frac{\partial \mathcal{L}}{\partial \dot\varphi} = mr^2 \sin^2 \theta\, \dot\varphi.$$

The other derivatives appearing in Lagrange's equations are:

$$\frac{\partial \mathcal{L}}{\partial r} = mr\dot\theta^2 + mr \sin^2 \theta\, \dot\varphi^2 - GMmr^{-2},$$

$$\frac{\partial \mathcal{L}}{\partial \theta} = \frac{1}{2} mr^2 \sin 2\theta\, \dot\varphi^2, \qquad\qquad (1.056)$$

$$\frac{\partial \mathcal{L}}{\partial \varphi} = 0,$$

and the Lagrange equations themselves become:

$$\frac{\mathrm{d}}{\mathrm{d}t} (m\dot{r}) - mr\dot\theta^2 - mr \sin^2 \theta\, \dot\varphi^2 + GMmr^{-2} = 0. \qquad (1.057)$$

$$\frac{\mathrm{d}}{\mathrm{d}t} (mr^2\dot\theta) - \frac{1}{2} mr^2 \sin 2\theta\, \dot\varphi^2 = 0. \qquad\qquad (1.058)$$

$$\frac{\mathrm{d}}{\mathrm{d}t} (mr^2 \sin^2 \theta\, \dot\varphi) = 0. \qquad\qquad (1.059)$$

These three equations constitute the equations of motion in the coordinate system in question and, although no attempt will be made to solve them at this time, a comment on equation (1.059) is very appropriate. The latter says that the quantity $mr^2 \sin^2 \theta\, \dot\varphi$, which is simply p_φ, is constant with respect to time. In other words, p_φ in this example is a constant of the motion and the formal expression for it,

$$p_\varphi = mr^2 \sin^2 \theta\, \dot\varphi, \qquad\qquad (1.060)$$

is a first integral of the motion and, of course, a first-order differential equation. (Sometimes the symbol L_z is substituted for p_φ here to emphasize the constancy of this expression.) The three equations of motion may therefore be regarded as (1.057), (1.058), and (1.060), the first two of which are second order and the third, first order; the three equations are, of course, coupled. The identification of p_φ as a constant of the motion is a direct consequence of the fact that $\partial \mathcal{L}/\partial \varphi$ is zero, i.e. that φ is *missing from the Lagrangian*. This simple observation has the force of a general principle:

> *When any coordinate q_j is missing from the Lagrangian, then p_j, the momentum conjugate to q_j, is a constant of the motion.*

In spite of the existence of this rule, the discovery of expressions for the constants of the motion is largely a matter of human ingenuity for even with the Lagrangian method, in which the discovery is automatic, the initial selection of the coordinate system is of crucial importance in determining how many such expressions will be thus discovered. In the Kepler problem, for example, the number of original differential equations (Lagrangian equations) is three. There are therefore three independent first integrals of the motion and six independent constants of the motion in all. If this problem were analyzed in Cartesian coordinates, however, none of the canonical momenta would in general be a constant of the motion and in the spherical polar coordinates used above, only one of the canonical momenta, namely p_φ, is in general such a constant. In either case, another first integral can be made to appear by reorienting the coordinate frame; such a reorientation does not change the *form* of any of the previously derived equations and, with this in mind, (1.058) can be inspected to yield the following trivial solution:

$$\theta = \pi/2; \qquad \dot{\theta} = 0. \tag{1.061}$$

This simply means that the reoriented coordinates have been positioned so that the xy plane, i.e. the plane described by $\theta = \pi/2$, coincides with the orbit plane of the planet. In the reoriented coordinates, p_φ is again a first integral and constant of the motion but the expression to which it is equal is simpler:

$$p_\varphi = mr^2 \dot{\varphi}. \tag{1.062}$$

This expression is actually L, the magnitude of the vector angular momentum. The next first integral is simply:

$$p_\theta = 0 \tag{1.063}$$

and a second integration constant appears when this equation is solved; it is the constant of the motion $\theta = \pi/2$ which has already been mentioned. The third first integral is still not apparent; it is a function called the *Hamiltonian* which, in the case under consideration, is synonymous with the total energy E. The justification for this statement and the rule for recognizing the Hamiltonian as a first integral and constant of the motion are beyond the scope of the present chapter but are amply discussed in the one following.

Since it is well known that a planetary orbit is described in a plane and that the θ motion in the specially oriented coordinates is non-existent, the two trivial constants of the motion, p_θ and θ, are almost never mentioned in textbook analyses. The Kepler problem is usually approached as if it were a problem in two coordinates (r and φ) rather than in three and the non-trivial constants of the motion are therefore four in number; they are embodied in the pairs L and φ_e, E and t_e. The quantities L and E were defined above. Their respective conjugates, φ_e and t_e, are discussed in the subsequent chapter where a complete solution is undertaken. As may be guessed, the latter are the epochal parameters of the planet and refer to the time and position of its perihelion.

It is fortunate that, in the Kepler problem, all of the independent first integrals of the motion are known, i.e. the corresponding constants of the motion can be expressed as functions of q_j, $\dot{q}_j$, and t (although t does not appear in these functions in this particular case). The situation is usually less fortunate when systems of larger numbers of particles are considered. A case in point is the famous problem of three particles moving under their mutual gravitational attraction. In this case, there are nine coordinates, nine equations of motion, and eighteen independent constants of motion in all. Six of these are connected with the motion of the center of mass and are very elementary. Six others are related to the total angular momentum and the total energy and are well known. The remaining six have never been expressed in the general case and the problem is therefore considered "unsolved" from a theoretical point of view.[1] The second-order equations of motion are known, of course, and a computer-implemented solution can always be obtained in any given application of interest.

1.06. Holonomic Constraints

The subject of constraints belongs almost exclusively to the realm of macroscopic physics, i.e. to the physics of gross objects composed of large numbers of particles, and is not absolutely essential as a step toward the understanding of quantum theory. It is so much a part of the tradition of physics, however, and it so clearly brings out the power of the Lagrangian method, that it is most worthy of inclusion. The discussion will be limited to a consideration of holonomic constraints in particular since only these have genuine illustrative value in the present context. The term "holonomic" will be defined shortly.

As indicated earlier, a constraining structure is any entity which limits the diversity of configurations available to a given system. Thus in the naive but often useful model of a rigid body composed of a number of particles maintained at fixed relative distances by massless inextensible bonds, the bonds are the constraining structures. (An elementary example, involving three particles, is illustrated in Fig. 1.08.) Other constraining structures take the form of tracks, guides, bearings, gears, etc., which limit the types of motion available to gross rigid bodies. An example of the latter type is provided by the system illustrated in Fig. 1.09. Here the block B makes a two-line contact with the track PR which, by arrangements that can be readily imagined, holds the center of mass of the block in the YZ plane and prevents the block from rotating about any line except the one marked "axis"

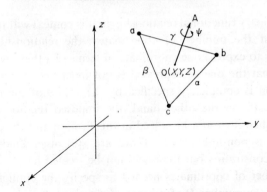

FIG. 1.08. A rigid body composed of three particles.

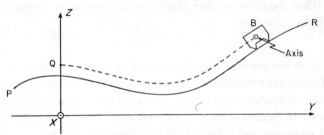

FIG. 1.09. A block free to slide on a track in the YZ plane.

which is parallel to e_X. The latter type of rotation is necessary if the block is to maintain its contact with the track. In this and in other examples to be discussed, the energy dissipated by frictional forces is considered negligible. Although motion involving energy dissipation can sometimes be handled by a modified Lagrangian method, the inclusion of this macroscopic subject would not contribute materially to the final objective of this book which is the treatment of microscopic systems in a quantum context.

If the effect of a restriction imposed upon the system by a constraining structure is expressible as a mathematical relationship among some or all of the coordinates (including, perhaps, the time) *so as to reduce by one the number needed to specify the configuration*, then the restriction in question is called a *holonomic constraint*. It is rather obvious that an ordinary functional relationship in this context will result in a reduction of the number of coordinates; the relationship can be solved so as to express one coordinate in terms of others whereupon it is clear that the one so expressed is not independent and that the configuration is completely specified by the values of the remaining coordinates. If, on the other hand, the imposed relationship is an inequality, then it cannot be solved in this fashion and the constraint in question is non-holonomic.[†] There are still other kinds of non-holonomic constraints but these will not be treated here.

The number of coordinates needed to specify the configuration is often called the number D of *degrees of freedom*. A holonomic constraint reduces D by one; the presence of C such constraints reduces D by C. Thus the total number of degrees of freedom of a system of N particles is given by:

$$D = 3N - C, \qquad (1.064)$$

where C is the number of holonomic constraints. Although C ordinary functional relationships among the coordinates signify the presence of C holonomic constraints, it is well to confirm this in individual cases by independently verifying that $3N - C$ coordinates are indeed necessary and sufficient. This is especially true when the constraints render the original set of coordinates inconvenient and suggest the use of a new set. In illustration of these remarks, consider the bonds connecting the three particles of Fig. 1.08. These enforce the following relationships among the coordinates:

$$\left.\begin{aligned}
(x_a - x_b)^2 + (y_a - y_b)^2 + (z_a - z_b)^2 &= \gamma^2; \\
(x_b - x_c)^2 + (y_b - y_c)^2 + (z_b - z_c)^2 &= \alpha^2; \\
(x_c - x_a)^2 + (y_c - y_a)^2 + (z_c - z_a)^2 &= \beta^2.
\end{aligned}\right\} \qquad (1.065)$$

It is concluded that these are holonomic constraints. Since there were nine coordinates originally, the number of degrees of freedom should

---

[†] A gas molecule confined in a hollow container is an example of a system in which the constraints must be expressed by inequalities and are therefore non-holonomic. In support of this conclusion, one may observe that the number of coordinates needed to specify the configuration of the molecule is not reduced by the presence of the container walls.

be six and six coordinates should suffice to describe the configuration. This is easily seen to be true when the coordinates are chosen as follows:

X, Y, Z, the Cartesian coordinates of O, the center of mass,

θ, φ,   the spherical polar angles (not illustrated) which define the direction of the axis OA,

ψ,   the angle denoting the amount of rotation of the body about OA with respect to some arbitrary reference, e.g. with respect to the plane determined by e_z and OA.

A little reflection shows that the configuration of every isolated rigid body composed of three or more particles held together by inextensible bonds can be described by a similarly chosen set of six coordinates and that every such body must therefore have six degrees of freedom.

One may now analyze the block and track example of Fig. 1.09. Since the block is a rigid body, its internal constraints limit its intrinsic degrees of freedom to six; its external environment supplies five more constraints as follows:

(i) The center of mass of the block must remain in the YZ plane.

(ii) The Z coordinate of the center of mass of the block must be a definite function of the Y coordinate, i.e. the curve QB is described by $Z = Z(Y)$.

(iii) The axis of the block, which points out of the page in Fig. 1.09, must make an angle of 90° with e_Z.

(iv) The plane determined by e_Z and the axis of the block must make an angle of 0° with the plane determined by e_Z and e_X.

(v) The inclination of the block about its own axis is determined by $Z(Y)$, albeit in a rather complicated way. (In first approximation, it is determined by the local value of the slope, dZ/dY.)

The effects of these constraints may be expressed mathematically as follows:

$$X = 0;$$
$$Z = Z(Y);$$
$$\theta = \pi/2;$$
$$\varphi = 0;$$
$$\psi = \psi(Y) \approx \text{arc tan } dZ/dY.$$

(1.066)

The existence of these relationships indicate that the constraints are indeed holonomic and this will be confirmed if it is found that one coordinate is necessary and sufficient to specify the configuration. The confirmation is obvious; the one coordinate might be Y itself or, perhaps, the distance QB measured from an arbitrary reference point to the center of mass along the path of the latter.

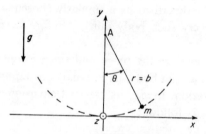

FIG. 1.10. A simple pendulum with coordinates.

The peculiar feature of constrained motion is that the forces exerted by the constraining structures must in general remain unknown until the motion itself is known, i.e. until the problem is solved. As an example, consider the simple pendulum in Fig. 1.10. This pendulum consists of a particle of mass m attached to point A by a massless cord of length b; the particle swings between, say, two glass plates (not illustrated) which further limit its motion to the xy plane. The force exerted by the cord on the particle is unknown until the speed of the latter along its arc has been found. The same is true of the force exerted by the glass plates, although this force can be seen intuitively to be zero if the xyz reference frame to which the plates are attached is inertial.

The first step in solving a problem in constrained motion is, if possible, to choose the generalized coordinates so that the first D of these can vary freely without violating the constraints and the remaining C coordinates cannot vary at all.[†] The mathematical statements, C in number, expressing the constraints will then all be of the form $q_j = $ constant, $j = D+1$, $D+2$, $\ldots$, $3N$. To put the matter very graphically:

$$
\left.
\begin{array}{l}
q_1 \\
q_2 \\
\ldots \\
q_D
\end{array}
\right\} \text{ are the unconstrained generalized coordinates;} \quad (1.067)
$$

$$
\left.
\begin{array}{l}
q_{D+1} = a_1 \\
q_{D+1} = a_2 \\
\quad . \quad . \quad . \\
q_{3N} = a_C
\end{array}
\right\} \begin{array}{l} \text{are the equations which express the} \\ \text{constraints.} \end{array} \quad (1.068)
$$

Thus, for the simple pendulum example, $N = 1$, $C = 2$, and $D = 1$; $q_1 = \theta$ is the one unconstrained coordinate and $q_2 = r = b$ and $q_3 = z = 0$ are the two equations expressing the constraints.

The task at hand is to find the equations of motion governing the D coordinates that are free to vary where D is, of course, the number of degrees of freedom of the system. The final result is quite simple but the reasoning leading thereto is rather subtle. Let the particles of the system be designated a, b, c, $\ldots$, as has been customary, and let the respective vector forces of constraint acting on these particles be F_a^c, F_b^c, F_c^c, $\ldots$. Newton's second law for each of the particles is then expressed by:

$$
\left.
\begin{array}{l}
m_a \ddot{r}_a = -\nabla_a V + F_a^c; \\
m_b \ddot{r}_b = -\nabla_b V + F_b^c; \\
m_c \ddot{r}_c = -\nabla_c V + F_c^c; \\
\quad . \quad . \quad . \quad . \quad . \quad .
\end{array}
\right\} \quad (1.069)
$$

[†] If this step is inconvenient, the problem can also be solved in more conventional coordinates with the aid of Lagrange multipliers. The latter technique will not be discussed here.

This is reminiscent of the system of equations (1.042). Using the same condensation of notation used before, one arrives at the following:

$$m_k \ddot{x}_k = -\frac{\partial V}{\partial x_k} + F_k^c, \tag{1.070}$$

where the F_k^c are the $3N$ Cartesian components of the forces of constraint. If the derivation of Lagrange's equations is now repeated with the additional term in (1.070) carried along, the result is:

$$\frac{d}{dt}\left(\frac{\partial \mathscr{L}}{\partial \dot{q}_j}\right) - \frac{\partial \mathscr{L}}{\partial q_j} = \sum_{k=1}^{3N} F_k^c \frac{\partial x_k}{\partial q_j}. \tag{1.071}$$

It is now helpful to consider an arbitrary imaginary infinitesimal change in the configuration of the system limited only by the stipulations that (i) it does not violate the constraints and (ii) it does not involve the passage of time. Obviously such a change has nothing to do with the normal dynamical metamorphosis of the system; it is, rather, a strictly conceptual device and is called a *virtual displacement*. Such a virtual displacement may be expressed as follows:

$$\left.\begin{array}{l} \delta q_1 \\ \delta q_2 \\ \cdots \\ \delta q_D \end{array}\right\} \text{are arbitrary imaginary infinitesmal changes in the } D \text{ free generalized coordinates;} \tag{1.072}$$

$$\left.\begin{array}{l} \delta q_{D+1} = 0 \\ \delta q_{D+2} = 0 \\ \quad \cdot \quad \cdot \quad \cdot \\ \delta q_{3D} = 0 \end{array}\right\} \text{are changes perforce equal to zero because of the presence of the constraints.} \tag{1.073}$$

$$\delta t = 0. \tag{1.074}$$

Combining the virtual displacement δq_j with equation (1.071), one can say:

$$\sum_{j=1}^{3N} \left[\frac{d}{dt}\left(\frac{\partial \mathscr{L}}{\partial \dot{q}_j}\right) - \frac{\partial \mathscr{L}}{\partial q_j} - \sum_{k=1}^{3N} F_k^c \frac{\partial x_k}{\partial q_j}\right] \delta q = 0, \tag{1.075}$$

since the content of the square brackets is clearly zero. From equations (1.023), it is seen that x_k is in general a function of the q_j and of t, whereupon:

$$\delta x_k = \sum_{j=1}^{3N} \frac{\partial x_k}{\partial q_j} \delta q_j + \frac{\partial x_k}{\partial t} \delta t = \sum_{j=1}^{3N} \frac{\partial x_k}{\partial q_j} \delta q_j. \qquad (1.076)$$

The second statement above is possible when, as here, $\delta t = 0$. Equation (1.076) is a transformation of the description of the virtual displacement from generalized coordinates to Cartesian coordinates. Using it, (1.075) can be written:

$$\sum_{j=1}^{3N} \left[\frac{\mathrm{d}}{\mathrm{d}t} \left(\frac{\partial \mathscr{L}}{\partial \dot{q}_j} \right) - \frac{\partial \mathscr{L}}{\partial q_j} \right] \delta q_j - \sum_{k=1}^{3N} F_k^c \, \delta x_k = 0. \qquad (1.077)$$

The second sum in the left-hand side of this equation can be recognized as a set of vector dot products:

$$\sum_{k=1}^{3N} F_k^c \, \delta x_k = \boldsymbol{F}_a^c \cdot \delta \boldsymbol{r}_a + \boldsymbol{F}_b^c \cdot \delta \boldsymbol{r}_b + \dots, \qquad (1.078)$$

where $\delta \boldsymbol{r}_a$, $\delta \boldsymbol{r}_b$, ... are the virtual displacements of the several particles expressed in vector form. One thing *is* known about the otherwise unknown forces of constraint and that is that they always act at right angles to any conceivable displacement consistent with the constraints under the condition of "stopped time", i.e. to any virtual displacement. Thus it is sometimes said that there is no "virtual work" done by the forces of constraint in a virtual displacement. It follows that both sides of (1.078) vanish and that (1.077) can be rewritten without the second sum. It is also possible, in (1.077), to substitute D for $3N$ as the upper limit in the summation over j since the δq_j are identically zero for $j > D$.

The result is:

$$\sum_{j=1}^{D} \left[\frac{\mathrm{d}}{\mathrm{d}t} \left(\frac{\partial \mathscr{L}}{\partial \dot{q}_j} \right) - \frac{\partial \mathscr{L}}{\partial q_j} \right] \delta q_j = 0. \qquad (1.079)$$

Since the D values of the δq_j are arbitrary, this equation implies that the coefficients in square brackets are themselves identically zero, i.e. *that Lagrange's equations in their customary form are valid for the constrained system.* In a problem involving constraints, therefore, one has only to express the Lagrangian $\mathcal{L} = T - V$ in the D free generalized coordinates, their corresponding generalized velocities and, possibly, the time. The equations of motion governing these coordinates are then obtained by the standard Lagrangian procedure.

The problem of the simple pendulum of Fig. 1.10 will now be dealt with according to the method just outlined. Assuming xyz to be an inertial frame of reference and recognizing that θ is the only free coordinate, one has:

$$T = \tfrac{1}{2}mb^2\dot{\theta}^2. \tag{1.080}$$

$$V = mgb(1-\cos\theta). \tag{1.081}$$

$$\mathcal{L} = \tfrac{1}{2}mb^2\dot{\theta}^2 - mgb(1-\cos\theta). \tag{1.082}$$

The one equation of motion is obviously given by:

$$\frac{\mathrm{d}}{\mathrm{d}t}(mb^2\dot{\theta}) + mgb\sin\theta = 0. \tag{1.083}$$

This is probably more familiar in the form:

$$\ddot{\theta} + \frac{g}{b}\sin\theta = 0. \tag{1.084}$$

1.07. Electromagnetic Applications

The Lagrangian formulation can be applied to systems that are partially or entirely electromagnetic in character provided that *electric energy is regarded as potential energy and magnetic energy as kinetic energy.* As a first and very simple example, consider the *LC* circuit of Fig. 1.11. If the charge q on the upper plate of the capacitor is taken as the one "coordinate" of the system, it is very easy to arrive at the Lagrangian:

$$\mathcal{L} = \underbrace{\tfrac{1}{2}L\dot{q}^2}_{T} - \underbrace{\tfrac{1}{2}C^{-1}q^2}_{V}. \tag{1.085}$$

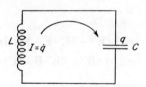

FIG. 1.11. An *LC* resonant circuit viewed as a mechanical system with charge q as its one coordinate. Current I is equal to $\dot{q}$.

From this, the "equation of motion" is readily deduced:

$$L\ddot{q} + C^{-1}q = 0. \tag{1.086}$$

This is easily recognized as the correct equation for the system; similar to equation (1.002) for the harmonic oscillator, it predicts that q will oscillate sinusoidally at a radian frequency $\Omega = (LC)^{-\frac{1}{2}}$.

A more sophisticated example, interesting because it illustrates a number of important physical concepts, is provided by a particle of charge q and mass m moving in a given electromagnetic field which is, in general, time-dependent. As long as electromagnetic radiation from the particle is negligible, as will be assumed, it is legitimate to consider the particle itself as the system and the given field as the source of external force. There are then three degrees of freedom and the Cartesian coordinates of the particle are adequate to describe the configuration.[†] The force exerted on the particle by the given field is found from the Lorentz formula which yields, in MKSA units, the following vector equation of motion:

$$m\ddot{r} = q[\mathbf{E} + (\dot{r} \times \mathbf{B})]. \tag{1.087}$$

Here $\dot{r}$ and $\ddot{r}$ are the velocity and acceleration of the particle with respect to the chosen inertial frame and $\mathbf{E}$ and $\mathbf{B}$ are the given electric and magnetic fields, respectively, in the same frame. If $\mathbf{E}'$ and $\mathbf{B}'$ are the fields generated by the moving particle itself, one may write

[†] If radiation cannot be neglected, the radiated field constitutes an additional physical entity with degrees of freedom of its own and the "configuration" can no longer be adequately described by the three coordinates mentioned above.

for the energies:

$$U_E = \tfrac{1}{2}\varepsilon_0 \int (E'^2 + 2E' \cdot E + E^2)\, d\tau;$$
$$U_B = \tfrac{1}{2}\mu_0^{-1} \int (B'^2 + 2B' \cdot B + B^2)\, d\tau. \tag{1.088}$$

In these and in other expressions to follow, $d\tau$ is a volume element and the integrals are taken over all of space. The third terms in the two integrands are the energies of the given fields and are completely independent of the position and velocity of the particle; these terms are of no further consequence in what follows. The first terms are the self-energies of the fields of the particle and account for part of its mass and part of its mechanical kinetic energy; they also may be ignored because they are automatically incorporated into the Lagrangian through the purely mechanical term in the latter. The second terms are the interaction energies which actually determine the force on the particle. Remembering that magnetic energy is kinetic, the Lagrangian can be written:

$$\mathcal{L} = \underbrace{\tfrac{1}{2}m\dot{r}\cdot\dot{r} + \mu_0^{-1}\int B'\cdot B\, d\tau}_{T} - \underbrace{\varepsilon_0 \int E'\cdot E\, d\tau}_{V}. \tag{1.089}$$

Notice the identification of T and V here. The fields E and B are derivable from a scalar potential φ and a vector potential A as follows:

$$E = -\nabla\varphi - \partial A/\partial t;$$
$$B = \nabla \times A. \tag{1.090}$$

With the aid of some vector manipulations, it is shown in Appendix A that:

$$\varepsilon_0 \int E'\cdot E\, d\tau = q\varphi - \varepsilon_0 \int E'\cdot(\partial A/\partial t)\, d\tau;$$
$$\mu_0^{-1} \int B'\cdot B\, d\tau = q\dot{r}\cdot A + \varepsilon_0 \int A\cdot(\partial E'/\partial t)\, d\tau. \tag{1.091}$$

Here φ and A in the first terms on the right-hand sides are the values of the potentials at the location of the particle. Upon substitution of (1.091) into (1.089), the Lagrangian becomes:

$$\mathcal{L} = \tfrac{1}{2}m\dot{r}\cdot\dot{r} + q\dot{r}\cdot A - q\varphi + \dot{G}(r, t), \tag{1.092}$$

where:

$$G(\mathbf{r}, t) = \varepsilon_0 \int \mathbf{A} \cdot \mathbf{E}' \, d\tau. \tag{1.093}$$

It must now be demonstrated that (1.092) actually leads, through Lagrange's equations, to the vector equation of motion (1.087). The term $\dot{G}(\mathbf{r}, t)$ may be expanded as follows:

$$\dot{G}(\mathbf{r}, t) = \frac{\partial G}{\partial x} \dot{x} + \frac{\partial G}{\partial y} \dot{y} + \frac{\partial G}{\partial z} \dot{z} + \frac{\partial G}{\partial t}. \tag{1.094}$$

One can readily see that $\partial \dot{G}/\partial \dot{x} = \partial G/\partial x$ and that, as in (1.047),

$$\frac{d}{dt} \frac{\partial \dot{G}}{\partial \dot{x}} = \frac{d}{dt} \frac{\partial G}{\partial x} = \frac{\partial \dot{G}}{\partial x}. \tag{1.095}$$

It will be found that $\dot{G}(\mathbf{r}, t)$ is an "empty" term, i.e. a term in the Lagrangian which plays no part in determining the equations of motion.[†] The canonical momentum conjugate to x is:

$$p_x = \frac{\partial \mathscr{L}}{\partial \dot{x}} = m\dot{x} + qA_x + \frac{\partial G}{\partial x}. \tag{1.096}$$

Substitution of this and (1.092) into the appropriate Lagrange equation yields:

$$m\ddot{x} + q\left(\frac{\partial A_x}{\partial x} \dot{x} + \frac{\partial A_x}{\partial y} \dot{y} + \frac{\partial A_x}{\partial z} \dot{z} + \frac{\partial A_x}{\partial t}\right) + \frac{\partial \dot{G}}{\partial x}$$
$$- q\left(\frac{\partial A_x}{\partial x} \dot{x} + \frac{\partial A_y}{\partial x} \dot{y} + \frac{\partial A_z}{\partial x} \dot{z}\right) + q \frac{\partial \varphi}{\partial x} - \frac{\partial \dot{G}}{\partial x} = 0. \tag{1.097}$$

It follows that:

$$m\ddot{x} + q\left[-\dot{y}\left(\frac{\partial A_y}{\partial x} - \frac{\partial A_x}{\partial y}\right) + \dot{z}\left(\frac{\partial A_x}{\partial z} - \frac{\partial A_z}{\partial x}\right) + \frac{\partial \varphi}{\partial x} + \frac{\partial A_x}{\partial t} \right] = 0.$$
$$\tag{1.098}$$

† This term can be removed by shifting to a modified but equally valid set of electrodynamic potentials in what is known as a *gauge transformation*. When, as is often the case, the form (1.092) of the Lagrangian is derived from the vector equation of motion by inspection, $\dot{G}(\mathbf{r}, t)$ is missing.

This becomes:

$$m\ddot{x} = q[\mathbf{E}_x + (\dot{\mathbf{r}} \times \mathbf{B})_x] \tag{1.099}$$

and constitutes the equation of motion for the coordinate x. The corresponding equations for y and z follow similarly and the three together form the vector equation of motion (1.087).

1.08. Hamilton's Principle

Configuration space has been mentioned in Section 1.02. For a system of D degrees of freedom, it is a hypothetical space of D dimensions having the property that, within it, every possible configuration of the system is represented as a single point. (D, of course, becomes equal to $3N$ if there are no constraints.) A symbolic illustration of configuration space for a system of five degrees of freedom is presented in Fig. 1.12.

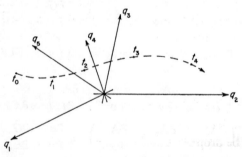

FIG. 1.12. Symbolic illustration of configuration space for a system with 5 degrees of freedom. A path corresponding to $q_j(t)$ is also shown.

The central concept of this section is that of a *path* through configuration space. A path is an imaginary construct created whenever one thinks of the D coordinates as specific functions of time; for present purposes, these functions must be at least once differentiable. By its very definition, a path is scheduled. Thus for every value of t, a configuration $q_j(t)$ and a characteristic point in configuration space are specified. As t increases, the configuration changes and the characteristic point traces out the path; every location upon the latter can therefore be labeled with the associated value of t. The functional forms of the $q_j(t)$ can be chosen arbitrarily within wide limits and a given path need have no relationship to a *dynamic path* which the point in configuration space follows when the system is executing one of its many possible natural motions. As might be expected, however, dynamic paths have a special property not possessed by arbitrary paths and a major objective of this section is to elucidate this property.

Since the D functions $q_j(t)$ are, by hypothesis, at least once differentiable, the D derivatives $\dot{q}_j(t)$ are also defined at every point on a given path. Moreover, since the system Lagrangian is a function of the q_j, the $\dot{q}_j$, and of t, its value is also defined at every point on a path and, under these conditions, one may think of the Lagrangian as a definite function of time. It is clearly possible to integrate the Lagrangian with respect to t along a specified path; the quantity generated thereby is known as the *action* and, if the path is arbitrary, will be denoted by the symbol S_a:

$$S_a = \int_{\substack{\text{arbitrary} \\ \text{path}}} \mathscr{L} \, dt. \tag{1.100}$$

If the path in question is a dynamic path, however, the action has special features which justify a notational change. In this case, the subscript will be dropped and the dynamic path action defined by:

$$S = \int_{\substack{\text{dynamic} \\ \text{path}}} \mathscr{L} \, dt. \tag{1.101}$$

It may be noticed that action has the dimensions of energy times time or, which amounts to the same thing, conventional momentum times distance.

Of great interest in this section is the definite action between an initial configuration C_0 at time t_0 and a final configuration C_f at time t_f, both of which lie on the same dynamic path. This dynamic path connecting C_0 and C_f is called the "true" path and is designated by P

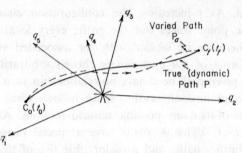

FIG. 1.13. The true (dynamic) path and a varied path in configuration
space.

in Fig. 1.13. Any other path such as P_a which connects the same two
configurations (and same times) is called a "varied" path; varied
paths which are quite close to the true path are most interesting in
this context. (It should be mentioned that a varied path may be geometrically indistinguishable from the true path but yet differ from the
latter because it is traversed on a time schedule which does not,
except at the end points, correspond to the true time schedule.) The
definite action along a varied path, such as P_a, is defined by:

$$\Delta S_a = \int_{\substack{t_0 \\ \text{path } P_a}}^{t_f} \mathcal{L} \, dt. \tag{1.102}$$

If the varied path P_a is made to approach the true path P, then ΔS_a
obviously approaches ΔS, the definite action along the true path.

Let the dynamic path P be described by $q_j(t)$ and the varied path P_a
by $q_{aj}(t)$. One may then say:

$$q_{aj}(t) = q_j(t) + \alpha \eta_j(t). \tag{1.103}$$

Here α is a real coefficient independent of time and the $\eta_j(t)$ form a
set of D real functions of time with the same differentiability as the
$q_j(t)$. The D terms $\alpha \eta_j(t)$ constitute the *variation*, i.e. the difference
between the $q_j(t)$ and the $q_{aj}(t)$. There is an infinitude of possible
varied paths and the $\eta_j(t)$ are correspondingly arbitrary except for the

stipulation that

$$\eta_j(t_0) = \eta_j(t_f) = 0; \tag{1.104}$$

this insures that the varied path begins and ends coincidentally with the true path. The coefficient α controls the "amount" of the variation. If $\alpha = 0$, there is no variation and if $|\alpha|$ increases, the variation increases. The definite action ΔS_a is in general path-dependent and one writes:

$$\Delta S_a = f(\alpha). \tag{1.105}$$

Since the variation disappears with vanishing α, it is clear that $f(0) = \Delta S$. The non-obvious but very interesting fact is that

$$\boxed{f'(0) = 0}, \tag{1.106}$$

which means that the power series development of ΔS_a about $\alpha = 0$ always lacks a linear term. This is one way of stating what is known as HAMILTON'S PRINCIPLE.

Before launching into a proof of Hamilton's principle, it will be desirable to discuss some of its implications. As affirmed above, the power series for ΔS_a begins as follows:

$$\Delta S_a = \Delta S + C_2\alpha^2 + C_3\alpha^3 + \dots \tag{1.107}$$

A plot of ΔS_a with respect to α might therefore appear as the curve AA' in Fig. 1.14 which is drawn so that its first derivative vanishes at the

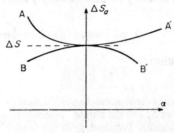

FIG. 1.14. The definite action ΔS_a as a function of α.

origin. It follows that the change in the definite action, $\delta(\Delta S_a)$, in going from the true path to a slightly varied path characterized by an increase in α from zero to $\delta\alpha$ is of order $(\delta\alpha)^2$, i.e. it vanishes to first order in $\delta\alpha$. The definite action is therefore stationary with respect to slight variations away from the true path and, because of the arbitrary character of the $\eta_j(t)$, this is so regardless of the precise functional form of the variation from the point of view of configuration space. For these reasons, Hamilton's principle is often called the PRINCIPLE OF STATIONARY ACTION and is often expressed in notation equivalent to the following:

$$\boxed{\delta(\Delta S_a) = 0}. \tag{1.108}$$

In view of what has been claimed, it is not clear whether ΔS_a will have a minimum at $\alpha = 0$ as in curve AA' of Fig. 1.14, a maximum as in BB', or perhaps even an inflection as in BA' or AB'. Experience shows that the situation depicted by curve AA' almost always obtains. This is so nearly universal that the term "least action" is often substituted for the more correct "stationary action"; Hamilton's principle is then known as the "principle of least action".[†]

As a very simple example of Hamilton's principle, consider a particle of mass m dropped from rest in a uniform gravitational field at $t_0 = 0$. A coordinate system with a downward-pointing z-axis, as in Fig. 1.15, is useful here. The problem may be treated as if there were but one degree of freedom with z as the only coordinate. (The particle could be allowed to fall inside a frictionless tube to provide constraints, although this would actually be unnecessary.) The following are readily found:

$$T = \tfrac{1}{2}m\dot{z}^2; \qquad V = -mgz; \qquad \mathcal{L} = \tfrac{1}{2}m\dot{z}^2 + mgz. \tag{1.109}$$

The equation of motion is:

$$\ddot{z} - g = 0, \tag{1.110}$$

[†] In the earlier history of classical mechanics, "action" was applied to a quantity different from either S or S_a and the phrase "principle of least action" has undergone modification in meaning. Terminology in this book is based upon a usage of the word "action" that is becoming increasingly popular in physics. See [2].

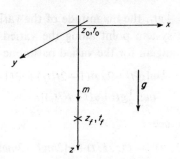

FIG. 1.15. A particle of mass m in free fall.

and the well-known expressions for z and $\dot{z}$, assuming that the motion starts from rest at $z = 0$ and $t = 0$, are:

$$z = \tfrac{1}{2}gt^2; \qquad \dot{z} = gt. \tag{1.111}$$

The Lagrangian, with $z(t)$ and $\dot{z}(t)$ substituted, becomes:

$$\mathcal{L}(z, \dot{z}, t) = \tfrac{1}{2}m(gt)^2 + mg(\tfrac{1}{2}gt^2) = mg^2t^2 \tag{1.112}$$

and the definite action along the true path is given by:

$$\Delta S = \int_0^{t_f} mg^2t^2 \, dt = \tfrac{1}{3}mg^2t_f^3. \tag{1.113}$$

When, as here, configuration space is one-dimensional, a varied path is necessarily one which coincides geometrically with the true path but is traversed on a different time schedule. Let the function $\eta(t)$ be chosen as the simple polynomial:

$$\eta(t) = t - t^2/t_f. \tag{1.114}$$

This has the necessary property of becoming zero at $t = 0$ and at $t = t_f$. A path in configuration space associated with variation $\alpha\eta(t)$ would begin at $t = 0$ and $z_a = 0$ and would end at $t = t_f$ and $z_a = \tfrac{1}{2}gt_f^2$ but, in the interim, would be given by:

$$z_a = \tfrac{1}{2}gt^2 + \alpha(t - t^2/t_f); \tag{1.115}$$

$$\dot{z}_a = gt + \alpha(1 - 2t/t_f). \tag{1.116}$$

Notice that for $\alpha = \frac{1}{2}gt_f$, the magnitude of the variation is such that the progress of the system point along the varied path is uniform. Evaluating the Lagrangian for the varied path, one obtains:

$$\mathscr{L}(z_a, \dot{z}_a, t) = \tfrac{1}{2}m[g^2t^2 + 2\alpha gt(1 - 2t/t_f) + \alpha^2(1 - 2t/t_f)^2]$$
$$+ mg[\tfrac{1}{2}gt^2 + \alpha(t - t^2/t_f)]. \tag{1.117}$$

Rearrangement produces:

$$\mathscr{L}(z_a, \dot{z}_a, t) = \mathscr{L}(z, \dot{z}, t) + \alpha[2mgt - 3mgt^2/t_f]$$
$$+ \tfrac{1}{2}m\alpha^2[1 - (4t/t_f) + (4t^2/t_f^2)]. \tag{1.118}$$

Integration to get the definite action on the varied path yields:

$$\Delta S_a = \Delta S + mg\alpha \int_0^{t_f} (2t - 3t^2/t_f)\,dt$$
$$+ \tfrac{1}{2}m\alpha^2 \int_0^{t_f} [1 - (4t/t_f) + (4t^2/t_f^2)]\,dt. \tag{1.119}$$

As predicted by Hamilton's principle, the coefficient of α vanishes and the expression for the definite action on the varied path becomes:

$$\Delta S_a = \Delta S + \tfrac{1}{6}mt_f\alpha^2. \tag{1.120}$$

This is quadratic in α and is minimum at $\alpha = 0$; the definite action is clearly stationary for the true path and is, in fact, least for this path.

To prove Hamilton's principle generally, one may write for the Lagrangian on the varied path:

$$\mathscr{L}(q_{aj}, \dot{q}_{aj}, t) = \mathscr{L}(q_j + \alpha\eta_j, \dot{q}_j + \alpha\dot{\eta}_j, t). \tag{1.121}$$

This may be expanded in a power series in α, the beginning of which is as follows:

$$\mathscr{L}(q_{aj}, \dot{q}_{aj}, t) = \mathscr{L}(q_j, \dot{q}_j, t) + \sum_{j=1}^{D} \left[\frac{\partial\mathscr{L}}{\partial q_j}\alpha\eta_j + \frac{\partial\mathscr{L}}{\partial \dot{q}_j}\alpha\dot{\eta}_j \right] + \dots. \tag{1.122}$$

Then:

$$\Delta S_a = \Delta S + \alpha \sum_{j=1}^{D} \int_{t_0}^{t_f} \left[\frac{\partial\mathscr{L}}{\partial q_j}\eta_j + \frac{\partial\mathscr{L}}{\partial \dot{q}_j}\dot{\eta}_j \right] dt + \dots. \tag{1.123}$$

The second term is to be integrated by parts. Therefore let:

$$\frac{\partial \mathcal{L}}{\partial \dot{q}_j} = u; \quad du = \frac{d}{dt}\left(\frac{\partial \mathcal{L}}{\partial \dot{q}_j}\right) dt; \quad \dot{\eta}_j\, dt = dv; \quad v = \eta_j. \quad (1.124)$$

One finds that:

$$\int_0^{t_f} \frac{\partial \mathcal{L}}{\partial \dot{q}_j} \dot{\eta}_j\, dt = \left[\frac{\partial \mathcal{L}}{\partial \dot{q}_j} \eta_j\right]_{t_0}^{t_f} - \int_{t_0}^{t_f} \eta_j \frac{d}{dt}\left(\frac{\partial \mathcal{L}}{\partial \dot{q}_j}\right) dt. \quad (1.125)$$

By hypothesis, $\eta_j = 0$ at $t = t_0$ and at $t = t_f$; the term in square brackets therefore vanishes and the following is obtained for the definite action:

$$\Delta S_a = \Delta S - \alpha \sum_{j=1}^{D} \int_{t_0}^{t_f} \eta_j(t) \left[\frac{d}{dt}\left(\frac{\partial \mathcal{L}}{\partial \dot{q}_j}\right) - \frac{\partial \mathcal{L}}{\partial q_j}\right] dt + \ldots. \quad (1.126)$$

By Lagrange's equations, one factor of the integrand is identically zero. It follows that ΔS_a contains no term linear in α; this proves Hamilton's principle.

A rather interesting feature of this proof is the fact that it is valid in reverse. Thus, one may begin by postulating Hamilton's principle, i.e. that (1.126) contains no term linear in α, and from this deduce Lagrange's equations. This is possible because $\eta_j(t)$ is *arbitrary* and if the integral is to vanish regardless of the nature of the variation, the coefficient of $\eta_j(t)$ must vanish. The conclusion to be drawn is that Hamilton's principle rather than Newton's laws could serve as the primary hypothesis of classical mechanics since Lagrange's equations and, with them, the entire theoretical structure of this discipline can be deduced from it. This is an interesting statement in view of the elegance and succinctness of Hamilton's principle.

At present, the chief motivation for studying Hamilton's principle is its association with the concept of action. The latter will be very useful in Chapters 3 and 6.

CHAPTER 2

THE HAMILTONIAN FORMULATION OF MECHANICS

2.01. Hamilton's Equations

Although the configuration of a system is specified by its D generalized coordinates, the complete dynamical state of the system, from which its time development can be deduced, requires for its specification the D generalized coordinates *and* the D generalized velocities. Thus the Lagrangian formulation is based upon the $2D$ quantities, q_j and $\dot{q}_j$, with time entering as a parameter and these quantities are treated as formally independent of one another except when a specific motion of the system is contemplated. In this context, it is appropriate to review the meaning of the following partial derivatives:

$$\left.\begin{array}{l} \dfrac{\partial}{\partial q_j} \text{ implies that all } q_k \text{ except } q_j \text{ are held constant,} \\ \text{all } \dot{q}_k \text{ are held constant, and } t \text{ is held constant.} \\[1em] \dfrac{\partial}{\partial \dot{q}_j} \text{ implies that all } \dot{q}_k \text{ except } \dot{q}_j \text{ are held constant,} \\ \text{all } q_k \text{ are held constant, and } t \text{ is held constant.} \\[1em] \dfrac{\partial}{\partial t} \text{ implies that all } q_k \text{ and all } \dot{q}_k \text{ are held constant.} \end{array}\right\} \quad (2.001)$$

The designation of $\partial \mathcal{L}/\partial \dot{q}_j$ as the generalized momentum p_j is little more than a suitable nomenclature in the Lagrangian formulation even though a particular p_j may assume importance as a constant of

44

the motion. Like the Lagrangian itself, the generalized momenta are functions of q_j, $\dot{q}_j$, and t:

$$
\left.
\begin{aligned}
p_1 &= p_1(q_1, q_2, \ldots q_D; \dot{q}_1, \dot{q}_2, \ldots \dot{q}_D; t). \\
p_2 &= p_2(q_1, q_2, \ldots q_D; \dot{q}_1, \dot{q}_2, \ldots \dot{q}_D; t). \\
&\cdot \quad \cdot \quad \cdot \quad \cdot \quad \cdot \quad \cdot \quad \cdot \quad \cdot \quad \cdot \quad \cdot \quad \cdot \quad \cdot \\
p_D &= p_D(q_1, q_2, \ldots q_D; \dot{q}_1, \dot{q}_2, \ldots \dot{q}_D; t).
\end{aligned}
\right\}
\qquad (2.002)
$$

The Hamiltonian formulation places the generalized momenta in a more fundamental position. It also is based upon $2D$ quantities but these are the D generalized coordinates q_j and the D generalized momenta p_j, i.e. the dynamical variables themselves. As before, time is a parameter. This change in the list of fundamental variables brings with it a change in the very meaning of the derivatives $\partial/\partial q_j$ and $\partial/\partial t$ and introduces a new derivative, namely $\partial/\partial p_j$. In the Hamiltonian formulation:

$$
\left.
\begin{aligned}
&\frac{\partial}{\partial q_j} \text{ implies that all } q_k \text{ except } q_j \text{ are held constant,} \\
&\quad \text{all } p_k \text{ are held constant, and } t \text{ is held constant.} \\
\\
&\frac{\partial}{\partial p_j} \text{ implies that all } p_k \text{ except } p_j \text{ are held constant,} \\
&\quad \text{all } q_k \text{ are held constant, and } t \text{ is held constant.} \\
\\
&\frac{\partial}{\partial t} \text{ implies that all } q_k \text{ and all } p_k \text{ are held constant.}
\end{aligned}
\right\}
\qquad (2.003)
$$

Thus, for all j and k, $\partial \dot{q}_k/\partial q_j$ is identically zero in the Lagrangian formulation but not necessarily so in the Hamiltonian; $\partial p_k/\partial q_j$ is identically zero in the Hamiltonian formulation but not necessarily so in the Lagrangian.

The change from one formulation to the other begins by regarding the $\dot{q}_k$ as unknown in equations (2.002) and solving for them. Since the kinetic energy is quadratic in the $\dot{q}_k$, it follows that the p_j are linear in these quantities and that (2.002) are simultaneous linear equations with coefficients which are in general functions of the q_j and t. They are therefore solvable in a finite number of steps although the algebraic labor involved may be formidable if D is large. The results may be written as follows:

$$\left.\begin{aligned}
\dot{q}_1 &= \dot{q}_1(q_1, q_2, \ldots q_D; p_1, p_2, \ldots p_D; t). \\
\dot{q}_2 &= \dot{q}_2(q_1, q_2, \ldots q_D; p_1, p_2, \ldots p_D; t). \\
&\cdot \cdot \cdot \cdot \cdot \cdot \cdot \cdot \cdot \cdot \cdot \cdot \cdot \cdot \\
\dot{q}_D &= \dot{q}_D(q_1, q_2, \ldots q_D; p_1, p_2, \ldots p_D; t).
\end{aligned}\right\} \tag{2.004}$$

In view of what was said above, these equations must be linear in the p_j with coefficients which are in general functions of the q_j and t.

Central to the Hamiltonian formulation is the function $\mathcal{H}(q_j, p_j, t)$ called the *Hamiltonian*. This function has actually been mentioned earlier in an illustrative example in Section 1.05. The Hamiltonian is defined in terms of functions containing the $\dot{q}_k$, but this is with the implicit understanding that the $\dot{q}_k$ are to be regarded as functions of q_j, p_j, and t according to equations (2.004). Thus the Hamiltonian is ultimately a function of q_j, p_j, and t and is so treated in all operations; every time a partial derivative of $\mathcal{H}$ is taken with respect to one of these quantities, it is taken in the sense of (2.003). The definition of the Hamiltonian is as follows:

$$\boxed{\begin{aligned}
\mathcal{H} &= \left[\sum_{k=1}^{D} p_k \dot{q}_k\right] - \mathcal{L}(q_j, \dot{q}_k, t) \\
&\text{with } \dot{q}_k = \dot{q}_k(q_j, p_j, t) \text{ substituted.}
\end{aligned}} \tag{2.005}$$

Since the $\dot{q}_k$ are linear in the p_j, it follows that the above is quadratic in the p_j, a fact which is corroborated by equations (2.007) below.

The partial derivatives of the Hamiltonian are most interesting. Consider first the partial derivative of $\mathcal{H}$ with respect to a particular one of the generalized momenta, say p_j:

$$\frac{\partial \mathcal{H}}{\partial p_j} = \dot{q}_j + \sum_{k=1}^{D} p_k \frac{\partial \dot{q}_k}{\partial p_j} - \sum_{k=1}^{D} \frac{\partial \mathcal{L}}{\partial \dot{q}_k} \frac{\partial \dot{q}_k}{\partial p_j}. \tag{2.006}$$

Since $\partial\mathscr{L}/\partial\dot{q}_k = p_k$ by definition, the sums cancel one another and one has:

$$\dot{q}_j = \frac{\partial\mathscr{H}}{\partial p_j}\,.$$

(2.007)

These equations are important theoretically in the Hamiltonian formulation even though, in content, they are simply repetitions of equations (2.004). This is seen to be true because the latter are also expressions of the $\dot{q}_j$ in terms of q_j, p_j, and t. On the practical level, agreement between equations (2.007) and (2.004) serves as a check on the derivation of the Hamiltonian for a particular system.

Consider next the partial derivative of $\mathscr{H}$ with respect to a particular one of the generalized coordinates, say q_j:

$$\frac{\partial\mathscr{H}}{\partial q_j} = \sum_{k=1}^{D} p_k \frac{\partial\dot{q}_k}{\partial q_j} - \sum_{k=1}^{D} \frac{\partial\mathscr{L}}{\partial\dot{q}_k}\frac{\partial\dot{q}_k}{\partial q_j} - \frac{\partial\mathscr{L}}{\partial q_j}\,.$$

(2.008)

Notice that $\partial/\partial q_j$ is being taken in the Hamiltonian sense and that the result of this operation on the Lagrangian is represented by everything in the above equation to the right of the first minus sign. The result of the action of $\partial/\partial q_j$ on the Lagrangian in the Lagrangian sense is embodied in the final term alone. For the reason adduced earlier, the two sums cancel and the partial of $\mathscr{H}$ with respect to q_j in the Hamiltonian sense is seen to be equal to minus the partial of $\mathscr{L}$ with respect to q_j in the Lagrangian sense:

$$\frac{\partial\mathscr{H}}{\partial q_j} = -\frac{\partial\mathscr{L}}{\partial q_j}\,.$$

(2.009)

This interesting result shows that if any one of the coordinates is missing from the Lagrangian, it is missing from the Hamiltonian also; the latter situation as well as the former can therefore be used as the criterion which establishes the conjugate momentum as a constant of the motion. From Lagrange's equations, $\partial\mathscr{L}/\partial q_j = \dot{p}_j$, hence:

$$\boxed{\dot{p}_j = -\frac{\partial \mathcal{H}}{\partial q_j}} \ . \tag{2.010}$$

Equations (2.007) and (2.010) are called HAMILTON'S EQUATIONS. The latter of these sets leads to the second-order equations of motion of the system as will be seen in an example and in exercises.

2.02. The Hamiltonian as a Constant of the Motion

Under a certain condition, $\mathcal{H}$ itself is a constant of the motion and, when this is true, it is sometimes called the "Jacobian integral". (Jacobi, in the course of research on the three-body problem, utilized the constancy of this function in an original way.) To investigate the condition in question, consider the partial derivative of $\mathcal{H}$ with respect to t:

$$\frac{\partial \mathcal{H}}{\partial t} = \sum_{k=1}^{D} p_k \frac{\partial \dot{q}_k}{\partial t} - \sum_{k=1}^{D} \frac{\partial \mathcal{L}}{\partial \dot{q}_k} \frac{\partial \dot{q}_k}{\partial t} - \frac{\partial \mathcal{L}}{\partial t} \ . \tag{2.011}$$

Once more, the partial derivative is being taken in the Hamiltonian sense and its effect upon the Lagrangian is represented by everything to the right of the first minus sign. The final term alone is the partial of $\mathcal{L}$ with respect to time in the Lagrangian sense. The end result, after the cancellation of the sums, is like that of (2.009) and with similar stipulations:

$$\frac{\partial \mathcal{H}}{\partial t} = -\frac{\partial \mathcal{L}}{\partial t} \ . \tag{2.012}$$

It appears that time is very similar to a coordinate; if t is missing from the Lagrangian it is also missing from the Hamiltonian as would be true for a coordinate. Now consider the *total* time derivative of $\mathcal{H}$. Remembering that $\mathcal{H} = \mathcal{H}(q_j, p_j, t)$, one has for this derivative:

$$\frac{d\mathcal{H}}{dt} = \dot{\mathcal{H}} = \sum_{k=1}^{D} \left[\frac{\partial \mathcal{H}}{\partial q_k} \dot{q}_k + \frac{\partial \mathcal{H}}{\partial p_k} \dot{p}_k \right] + \frac{\partial \mathcal{H}}{\partial t} \ . \tag{2.013}$$

By Hamilton's equations, the contents of the square brackets becomes $-\dot{p}_k \dot{q}_k + \dot{q}_k \dot{p}_k$ which is obviously zero. The result is:

$$\dot{\mathscr{H}} = \frac{\partial \mathscr{H}}{\partial t}.$$

(2.014)

This says that $\mathscr{H}$ does not vary at all with respect to time unless it contains time explicitly. The criterion is now obvious; if, as is often the case, t is missing from the Hamiltonian (and therefore also from the Lagrangian), then the Hamiltonian itself is a constant of the motion.[†] In the example of the subsequent section, this property is exploited in a way that is characteristic of many problems in mechanics.

The similarity between t and a coordinate has been pointed out. One may also notice the similarity between $\mathscr{H}$ and a momentum since $\mathscr{H}$ assumes the role of a constant of the motion if its associated "coordinate", i.e. time, is absent from $\mathscr{H}$ and $\mathscr{L}$. The association carries a minus sign, however, and $-\mathscr{H}$ is the "momentum" conjugate to t. In this way a strict parallelism between equations (2.010) and (2.014) is preserved.

2.03. Hamiltonian Analysis of the Kepler Problem

The problem of the motion of a planet of relatively negligible mass about a central sun has been discussed in Section 1.05. There it was pointed out that either an arbitrarily oriented or specially oriented coordinate system may be used; the equations of motion have the same form in either system but the specially oriented system has some

---

[†] There are $2D$ independent constants of the motion in all, D of which are independent first integrals of the motion, i.e. first integrals which are not expressible as combinations of other such integrals. If it should happen that all D of the p_j and $\mathscr{H}$ fulfill the criteria for being constants of the motion, then only D of these $D+1$ quantities can be independent. (This situation occurs in the case of an unforced particle.)

very practical advantages. In either of these systems:

$$T = \tfrac{1}{2}m(\dot{r}^2 + r^2\dot{\theta}^2 + r^2 \sin^2 \theta \,\dot{\varphi}^2), \atop V = -GMmr^{-1},\} \qquad (2.015)$$

and the Lagrangian and the generalized momenta are given in equations (1.054) and (1.055), respectively. In this case, it is easy to express the generalized velocities in terms of the generalized momenta, i.e. to arrive at equations (2.004). Thus:

$$\left. \begin{aligned} \dot{r} &= p_r/m; \\ \dot{\theta} &= p_\theta/mr^2; \\ \dot{\varphi} &= p_\varphi/mr^2 \sin^2 \theta. \end{aligned} \right\} \qquad (2.016)$$

From the definition of the Hamiltonian:

$$\mathscr{H} = (p_r\dot{r} + p_\theta\dot{\theta} + p_\varphi\dot{\varphi}) - \tfrac{1}{2}m(\dot{r}^2 + r^2\dot{\theta}^2 + r^2 \sin^2 \theta \,\dot{\varphi}^2) - GMmr^{-1}. \quad (2.017)$$

Substituting from (2.016), one obtains:

$$\mathscr{H} = \left[\frac{p_r^2}{m} + \frac{p_\theta^2}{mr^2} + \frac{p_\varphi^2}{mr^2 \sin^2 \theta} \right]$$
$$- \frac{1}{2} \left[\frac{p_r^2}{m} + \frac{p_\theta^2}{mr^2} + \frac{p_\varphi^2}{mr^2 \sin^2 \theta} \right] - \frac{GMm}{r}. \qquad (2.018)$$

Evidently the second term with square brackets is $-T$ and the first is $2T$. Thus the Hamiltonian is, in this case,

$$\mathscr{H} = 2T - T + V = T + V = E, \qquad (2.019)$$

where E is the total energy. In very "conventional" systems where an inertial coordinate frame is used to describe the configuration, forces of constraint (if present) are supplied by bodies at rest with respect to this frame, and the applied forces are derived from a scalar potential energy function, it can safely be asserted that the Hamiltonian is equal to the total energy. In other cases, such as those involving non-inertial coordinates, moving constraints, magnetic forces, etc., it may not be easy to identify $\mathscr{H}$ with any simple quantitative attribute of the system. Equation (2.018) may be simplified to yield:

$$\mathscr{H} = \frac{1}{2} \left[\frac{p_r^2}{m} + \frac{p_\theta^2}{mr^2} + \frac{p_\varphi^2}{mr^2 \sin^2 \theta} \right] - \frac{GMm}{r} . \qquad (2.020)$$

This result may be checked by an application of (2.007), the first set of Hamilton's equations. Thus for example:

$$\dot{\theta} = \frac{\partial \mathscr{H}}{\partial p_\theta} = \frac{p_\theta}{mr^2} . \qquad (2.021)$$

The equations of motion are obtained from (2.010), the second set of Hamilton's equations:

$$\dot{p}_r = -\frac{\partial \mathscr{H}}{\partial r} ; \qquad \dot{p}_\theta = -\frac{\partial \mathscr{H}}{\partial \theta} ; \qquad \dot{p}_\varphi = -\frac{\partial \mathscr{H}}{\partial \varphi} . \qquad (2.022)$$

Respectively, these yield:

$$\dot{p}_r = \frac{p_\theta^2}{mr^3} + \frac{p_\varphi^2}{mr^3 \sin^2 \theta} - \frac{GMm}{r^2} ; \qquad (2.023)$$

$$\dot{p}_\theta = \frac{p_\varphi^2 \cos \theta}{mr^2 \sin^3 \theta} ; \qquad (2.024)$$

$$\dot{p}_\varphi = 0. \qquad (2.025)$$

In the specially oriented coordinates discussed in Section 1.05, $\theta = \pi/2$ at all times and two independent first integrals of the motion appear immediately, namely $p_\theta = 0$ and $p_\varphi = L$, where L is the magnitude of the vector angular momentum. Only one second-order equation of motion (2.023) remains and, with appropriate substitutions for p_r, p_θ, and p_φ, it becomes:

$$m\ddot{r} = \frac{L^2}{mr^3} - \frac{GMm}{r^2} . \qquad (2.026)$$

Since neither θ nor φ appear in the Hamiltonian (in the specially oriented frame), they cannot appear in equation (2.026) either and the latter can be solved without reference to them. For this reason, coordinates such as θ and φ in the present example are often said to be "ignorable", a term which should not be taken too literally. Direct solution of (2.026) is still not the best approach, however, because it fails to exploit the third independent first integral of the motion, namely $\mathcal{H}$ itself. $\mathcal{H}$, which is equal to E, is known to be a constant of the motion because $\mathcal{H}$ and $\mathcal{L}$ do not contain t. To take advantage of the constancy of $\mathcal{H}$, one should immediately set expression (2.020), again with appropriate substitutions for the constant generalized momenta, equal to E:

$$\underbrace{\frac{1}{2}m\dot{r}^2 + \frac{L^2}{2mr^2}}_{T} \underbrace{- \frac{GMm}{r}}_{V} = E. \tag{2.027}$$

This is a first-order differential equation and a step closer than (2.026) to the final solution. The designation of the energies T and V is for comparison with what follows.

Equation (2.027) contains only r and t and, in this respect, is not different from the equation of motion for a system with but one degree of freedom and with a kinetic energy simply equal to $\frac{1}{2}m\dot{r}^2$. The latter is called the *effective kinetic energy* T' and, on the same viewpoint, the two remaining terms on the left constitute the *effective potential energy* V'. Thus:

$$\underbrace{\frac{1}{2}m\dot{r}^2}_{T'} + \underbrace{\frac{L^2}{2mr^2} - \frac{GMm}{r}}_{V'} = E. \tag{2.028}$$

In the same vein, the term $L^2/2mr^2$ must be regarded as the potential energy of a repelling force and is therefore spoken of as the "centrifugal potential energy". The function $V'(r)$ is illustrated in Fig. 2.01 for three values (including zero) of L. Evidently the problem can be handled from this point by the techniques of Section 1.01. Horizontal lines representing possible values of total energy E may be superposed upon the effective potential energy curve as in Fig. 2.02. The most common type of motion within the solar system is typified by $E = E_e$, producing two turning points r_p and r_a. This corresponds to an *ellip-*

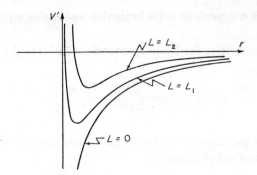

FIG. 2.01. Various effective potential energy curves for the Kepler problem.

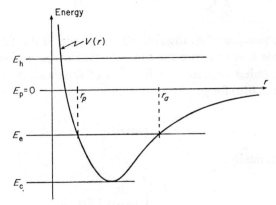

FIG. 2.02. A typical effective potential energy curve for the Kepler problem showing four levels of total energy, each related to a distinct orbit type.

tic orbit with r_p as the perihelion distance and r_a as the aphelion distance. If the energy is lowered to the value E_c, the two turning points coalesce and the planet pursues a *circular* orbit. If the total energy is equal to E_p, i.e. to zero, the situation is that of an object which (theoretically) falls toward the Sun from a state in which $T' = 0$ at infinity and returns to a similar state, all the while following a *parabolic* orbit. If the total energy has a value like E_h, which is positive, the object describes a *hyperbolic* orbit beginning and ending with $T' > 0$ at infinity.

The formal method for handling equation (2.027) begins by solving it for $\dot{r}$:

$$\dot{r} = \left\{ \frac{2}{m} \left[E - \frac{L^2}{2mr^2} + \frac{GMm}{r} \right] \right\}^{\frac{1}{2}}. \tag{2.029}$$

This permits the variables to be separated and time to be expressed as a function of radial distance:

$$t = \int_{r_p}^{r} \left\{ \frac{2}{m} \left[E - \frac{L^2}{2mr^2} + \frac{GMm}{r} \right] \right\}^{-\frac{1}{2}} dr + t_e. \tag{2.030}$$

Notice the position of r_p, the perihelion distance, in this formula and the fact that t_e is therefore the epoch or time of perihelion passage. Implicitly, radial distance is a function of time and, by the use of expression (2.016) for $\dot{\varphi}$:

$$\dot{\varphi} = \frac{L}{m[r(t)]^2}, \tag{2.031}$$

one obtains finally:

$$\varphi = \frac{L}{m} \int_{t_e}^{t} \frac{dt}{[r(t)]^2} + \varphi_e. \tag{2.032}$$

Here φ_e is the value of φ at $t = t_e$, i.e. at the epoch. This completes the formal solution of the problem.

It is interesting to eliminate dt between equations (2.029) and (2.031) to obtain an equation containing only r and φ. The equation so obtained is the differential equation of the orbit:

$$\frac{dr}{d\varphi} = \frac{mr^2}{L} \left\{ \frac{2}{m} \left[E - \frac{L^2}{2mr^2} + \frac{GMm}{r} \right] \right\}^{\frac{1}{2}}. \tag{2.033}$$

The substitution $u = r^{-1}$ is traditional here and greatly simplifies what follows. The equation becomes:

$$\frac{du}{d\varphi} = -\frac{m}{L}\left\{\frac{2}{m}\left[E - \frac{L^2 u^2}{2m} + GMmu\right]\right\}^{\frac{1}{2}}. \tag{2.034}$$

The following is proposed as the form of the solution:

$$u = u_0[1 + \varepsilon \cos(\varphi - \varphi_e)]. \tag{2.035}$$

Substitution shows that this form is indeed correct provided the constants u_0 and ε are given the following values:

$$u_0 = \frac{GMm^2}{L^2}; \tag{2.036}$$

$$\varepsilon = \left(1 + \frac{2EL^2}{G^2 M^2 m^3}\right)^{\frac{1}{2}}. \tag{2.037}$$

The constant ε is called the *eccentricity* of the orbit. An eccentricity greater than unity $(E > 0)$ corresponds to a hyperbolic orbit; an eccentricity equal to unity $(E = 0)$ corresponds to a parabolic orbit; one less than unity but greater than zero, to an elliptic orbit; zero eccentricity, to a circular orbit.

2.04. Phase Space

The concept of a D-dimensional configuration space has already been discussed. In this concept, the configuration of the system is represented by a single point which moves through the space as the configuration evolves. This moving point which represents the instantaneous configuration does not, however, constitute a full dynamical description of the system for if only the configuration is known at a given time t, there is no way of predicting the configuration at time $t + dt$.

A complete dynamical description of a system must include not only the present configuration but also its rate of change. In the Lagrangian formulation, such a description would be simply a list of the q_j and the $\dot{q}_j$. If this list of quantities is known at time t, then the dynamical variables, namely the q_j and the p_j, are also known and vice versa according to equations (2.002) and (2.004). Thus either list could be the complete dynamical description, the latter being more in the style of the Hamiltonian formulation. The adequacy of the Hamiltonian description is confirmed by the observation that if all the values of q_j and p_j are known at time t, then new values of these same variables at time $t + dt$ can easily be found. The reason for this is clear from the following equations; notice that the derivatives $\partial \mathcal{H}/\partial p_j$ and $\partial \mathcal{H}/\partial q_j$ must be known functions if $\mathcal{H}$ itself is known:

$$q_j + dq_j = q_j + \dot{q}_j \, dt = q_j + \frac{\partial \mathcal{H}}{\partial p_j} \cdot dt. \tag{2.038}$$

$$p_j + dp_j = p_j + \dot{p}_j \, dt = p_j - \frac{\partial \mathcal{H}}{\partial q_j} \, dt. \tag{2.039}$$

Obviously this process is self-perpetuating;[†] it follows that the entire future evolution of the system (and the past history also) is contained in the original specification of the $2D$ dynamical variables at time t.

This discussion naturally leads to a new visualization of the dynamical description of the system. In this new visualization, the system is represented as a point in a $2D$-dimensional space known as *phase space*, the "coordinates" of which are the D generalized coordinates *and* the D generalized momenta. Figure 2.03 represents an attempt to illustrate a six-dimensional phase space which would be characteristic of, say, a system consisting of a single physical particle moving without constraints in ordinary three-dimensional space. If the state of the system (physical particle) at a given time is represented by the

---

† This method has been used in conjunction with a computer to achieve a step-by-step solution of a dynamical problem. See ref. 3.

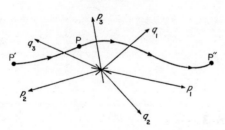

FIG. 2.03. Phase space for a system consisting of one unconstrained
particle.

point P in Fig. 2.03, then, according to the laws of classical mechanics,
the system can have only one course of future development, namely
the path PP″, and only one past history, the path P′P. Thus, at given
time t, there passes one and only one possible dynamical path through
every point of phase space.[†]

The concept of phase space is often mentioned in theoretical discussions. One of its traditional uses is in the formulation of classical
statistical mechanics.

[†] If the Hamiltonian does not contain the time, it is not necessary to include
the phrase, "at given time t".

CHAPTER 3

HAMILTON–JACOBI THEORY

3.01. Canonical Transformations

The relevance of the choice of coordinates to the ease of handling of a mechanical problem has already been emphasized and an example of a coordinate transformation, namely the transformation from the Cartesian to the spherical polar coordinates of the particles of a given system, has been considered. In the present section, an even more general type of transformation is envisioned. This transformation involves a change from a given set of generalized coordinates and generalized momenta to a new set of generalized coordinates and generalized momenta in such a way that all $2D$ of the latter quantities are in general functions of all $2D$ of the former and, possibly, of time. It is, in other words, a change from one set of variables to another in phase space, not merely in configuration space. Suppose that:

$$\left. \begin{array}{l} q_1, q_2, \ldots, q_D \text{ are the old generalized coordinates;} \\ p_1, p_2, \ldots, p_D \text{ are the old generalized momenta;} \end{array} \right\} \quad (3.001)$$

$$\left. \begin{array}{l} Q_1, Q_2, \ldots, Q_D \text{ are the new generalized coordinates;} \\ P_1, P_2, \ldots, P_D \text{ are the new generalized momenta.} \end{array} \right\} \quad (3.002)$$

The equations of transformation may be written as follows:

$$Q_j = Q_j(q_k, p_k, t); \quad (3.003)$$

$$P_j = P_j(q_k, p_k, t). \quad (3.004)$$

58

These relationships, in general, imply their inverses. The latter are:

$$q_k = q_k(Q_j, P_j, t); \qquad (3.005)$$

$$p_k = p_k(Q_j, P_j, t). \qquad (3.006)$$

In these and in other equations to follow, it is understood that all indices range from one to D, inclusive, and that no special index letter is permanently associated with any particular variable or set of variables. The structure of some of the equations to be encountered later precludes any such permanent association.

If the transformation is such that, in the new variables, one can define a new Hamiltonian:

$$\mathcal{K} = \mathcal{K}(Q_j, P_j, t) \qquad (3.007)$$

for which Hamilton's equations are satisfied, i.e. for which:

$$\dot{Q}_j = \frac{\partial \mathcal{K}}{\partial P_j}, \quad \dot{P}_j = -\frac{\partial \mathcal{K}}{\partial Q_j}, \quad \text{and} \quad \dot{\mathcal{K}} = \frac{\partial \mathcal{K}}{\partial t}, \qquad (3.008)$$

then the transformation is said to be *canonical*. Under canonical transformations, the problem can be solved in the new variables just as in the old. One might hope that the problem would be considerably easier in the new variables, but, if it is, the difficulty of finding the transformation is likely to be correspondingly great. Hamilton–Jacobi theory carries this idea to the ultimate. It uses new variables in which the solution is trivial; the entire difficulty of the problem then lies in finding a canonical transformation from the original variables to these new variables.

Conventional coordinate transformations such as the one from Cartesian to spherical polar are special cases in which the new coordinates are functions only of the old coordinates and the time and do not involve the old momenta. The pattern is as follows:

$$Q_j = Q_j(q_k, t); \qquad P_j = P_j(q_k, p_k, t); \qquad (3.009)$$

$$q_k = q_k(Q_j, t); \qquad p_k = p_k(Q_j, P_j, t). \qquad (3.010)$$

Such transformations are called *point transformations* and are always canonical.

It is remarkable that, in general canonical transformations, the new coordinates have no feature which would identify them specifically as coordinates. Neither do the new momenta have any feature which would identify them as momenta. Dimensions are certainly not an adequate guide in this matter. It is possible, for instance, to generate a rather trivial but yet canonical transformation in which the new coordinates become the old momenta and vice versa.

The standard procedure for constructing canonical transformations begins with the assumption that the new Lagrangian may differ from the old by a total time derivative $\dot{G}$ where G is a suitably chosen function of some old and some new variables. The appearance of the empty term in equation (1.092) is a rudimentary manifestation of this concept which, in its full generality, is extraordinarily subtle. One method[4] of proving that transformations so obtained are canonical uses a modified Hamilton's principle on a $2D$-dimensional space; an alternative method, presented in Appendix B, does not invoke any variational principle. In the present section, only the procedural technique for generating canonical transformations — without any serious attempt at justification — is given.

As has been pointed out, a canonical transformation involves the $2D$ old variables and the $2D$ new variables, i.e. a total of $4D$ variables in all. In working with such transformations, one must be willing to consider various ways in which these $4D$ variables can be partitioned into an independent subset and a dependent subset. There are four well-known ways in which this can be done and to each of these there corresponds a suitable function G as shown[†] in Table 3.01. In this book, only the second option is of interest; in other words, the q_j and the P_j will be taken as the independent variables, the p_j and the Q_j as the dependent. The transformation is therefore sought in the following form:

$$p_j = p_j(q_k, P_k, t); \qquad (3.011)$$
$$Q_j = Q_j(q_k, P_k, t). \qquad (3.012)$$

[†] In this chapter, the summation sign without auxiliary symbols means a summation from one to D.

TABLE 3.01. *Data for the Generation of Canonical Transformations*

Independent variables	G-function
q_j, Q_j	$G = G_a(q_j, Q_j, t)$
q_j, P_j	$G = G_b(q_j, P_j, t) - \Sigma Q_j P_j$
p_j, Q_j	$G = G_c(p_j, Q_j, t) + \Sigma q_j p_j$
p_j, P_j	$G = G_d(p_j, P_j, t) - \Sigma Q_j P_j + \Sigma q_j p_j$

Bearing in mind that j can be freely changed to k and vice versa, one may notice that it is theoreticalle possible to solve equations (3.011) simultaneously for the P_k in terms of the q_j and the p_j and thus to arrive at (3.004); similarly, it is possible to solve equations (3.012) simultaneously for the q_k in terms of the Q_j and P_j and thereby arrive at (3.005). Equations (3.004) can then be substituted into (3.012) to yield (3.003) and (3.005) can be substituted into (3.011) to yield (3.006). This is one example of the fact that the transformation, under any of the four choices of independent variables listed in Table 3.01, implies the transformation as originally envisioned. It should be added, however, that the passage from one of these four forms to the original form may be very difficult on the practical level and may well be impossible in terms of recognized functions.

According to the standard procedure mentioned earlier, one writes:

$$\mathcal{L} = \mathcal{L}' + \dot{G}, \tag{3.013}$$

where $\mathcal{L}$ is the Lagrangian in the old variables and $\mathcal{L}'$ is the Lagrangian in the new. By virtue of definition (2.005) of the Hamiltonian:

$$\left. \begin{array}{l} \mathcal{L} = \Sigma p_j \dot{q}_j - \mathcal{H}; \\ \mathcal{L}' = \Sigma P_j \dot{Q}_j - \mathcal{K}. \end{array} \right\} \tag{3.014}$$

Substitution of these together with the second G function of Table 3.01

into (3.013) produces:

$$\Sigma p_j \dot{q}_j - \mathcal{H} = \Sigma P_j \dot{Q}_j - \mathcal{K} + \Sigma \frac{\partial G_b}{\partial q_j} \dot{q}_j + \Sigma \frac{\partial G_b}{\partial P_j} \dot{P}_j + \frac{\partial G_b}{\partial t}$$

$$- \Sigma P_j \dot{Q}_j - \Sigma Q_j \dot{P}_j. \qquad (3.015)$$

It is now required that the coefficients of the $\dot{q}_j$ and of the $\dot{P}_j$ should cancel, leaving a relationship among functions which do not formally contain any of these dotted quantities. The results are:

$$p_j = \frac{\partial G_b}{\partial q_j}; \qquad (3.016)$$

$$Q_j = \frac{\partial G_b}{\partial P_j}; \qquad (3.017)$$

$$\mathcal{K} = \mathcal{H} + \frac{\partial G_b}{\partial t}. \qquad (3.018)$$

Since G_b contains only q_j, P_j, and t, the differentiation indicated in (3.016) produces p_j as a function of these same variables, i.e. it produces (3.011). Similarly, (3.017) produces (3.012) and the transformation is said to have been generated by $G_b(q_j, P_j, t)$. The process of changing equations (3.011) and (3.012) over to the original form of (3.003) and (3.004), or to the inverse thereof, may now be undertaken; as mentioned earlier, this step may be difficult.

3.02. Hamilton's Principal Function and the Hamilton–Jacobi Equation

The introductory discussion of Section 1.01 dealt with the harmonic oscillator, a simple system with one degree of freedom, in some detail. It was emphasized that any particular motion of that system is determined by the values assigned to the two arbitrary and independent constants that are necessarily associated with the solution of the equation of motion. The discussion in question included a method for solving this equation in two stages with one of the constants appearing at each stage. It is now desirable to extend this analysis by examining the problem of an arbitrary system with one degree of freedom and to do so in the context of the Hamiltonian formulation. Evidently only one coordinate q and one momentum p are involved.

It is always possible in principle to find a first integral of the motion, i.e. a function of q, p, and t which remains constant as the motion proceeds. Calling this constant ζ, the first stage in the process of solution is the writing of the expression:

$$\zeta = \zeta(q, p, t). \tag{3.019}$$

Since $p = (\partial \mathcal{L}/\partial \dot{q}) = p(q, \dot{q}, t)$, it follows that the above is a function of q, $\dot{q}$, and t set equal to a constant. It is, therefore, a first-order differential equation. The constant itself, ζ, may be regarded as a constant of integration produced when the second order equation of motion (which exists implicitly whether invoked in this context or not) is replaced by one of first order. Notice that (3.019) may be put in the form $\dot{q} = \dot{q}(\zeta, q, t)$ if this should prove helpful in the operation which follows.

The second stage is the integration of $\dot{q} = \dot{q}(\zeta, q, t)$, i.e. of (3.019), to yield t as a function of q perhaps and, ultimately, q as a function of t. In this step, another constant of integration ξ will appear and q may be written:

$$q = q(\zeta, \xi, t). \tag{3.020}$$

This is the solution sought. It expresses q as a fun ction of time and o the two constants ζ and ξ. These are the constants of the motion; when values are assigned to them, a particular motion is determined.

One may easily differentiate (3.020) to get $\dot{q}$ as a function of ζ, ξ, and t. Again using the fact that $p = p(q, \dot{q}, t)$, it is a matter of simple substitution to obtain the momentum as a function of the same two constants and the time:

$$p = p(\zeta, \xi, t). \tag{3.021}$$

This equation may now be solved for ξ as an unknown to yield $\xi(\zeta, p, t)$. The development is then completed by substituting from expression (3.019) for ζ whereupon ξ appears written in terms of the coordinate, the momentum, and time:

$$\xi = \xi(q, p, t). \tag{3.022}$$

It may be noticed that two of the four important equations appearing above express the canonical variables in terms of the constants and time whereas the other two express the constants in terms of the canonical variables and time. This duality is very important; it will be encountered again in still greater generality and will be exploited in the development of Hamilton–Jacobi theory.

Evidently ζ and ξ are of the same mathematical genus; both are constant functions of the coordinate, the momentum, and the time. This fact is underlined by the realization that if expression (3.022) had been discovered before (3.019), then $\xi(q, p, t)$ would have been considered the first integral of the motion and ζ or some quantity related thereto would have become the second constant. Actually, any functional combination of ζ and ξ could also serve as a first integral since it also would be a constant function of the coordinate, the momentum, and the time. Such a constant, which might be called ζ', would soon be accompanied by a second constant ξ' upon the performance of the necessary analytical steps; the constants of the motion would then be ζ' and ξ', both of which would be functionally related to the old ζ and ξ. It is evident that a constant of the motion is an extraordinarily flexible entity and that, even in a system of one degree of freedom, the two constants of the motion can take on an infinite variety of distinct forms.

It will now be helpful to see how the harmonic oscillator, as a specific example, fits into the pattern of the present discussion. The expression most likely to be discovered as a first integral of the motion is the Hamiltonian itself since the absence of t immediately reveals its constant character. Writing q for x and ζ for $\mathcal{H}$, the first of the four expressions becomes:

$$\zeta = \frac{p^2}{2M} + \frac{Kq^2}{2}. \tag{3.023}$$

One is naturally led to the epoch $t_e = \xi$ as the second constant and, as in Section 1.01, the solution may be written:

$$q = (2\zeta/K)^{\frac{1}{2}} \sin \Omega(t-\xi). \qquad (3.024)$$

Since $p = M\dot{q}$ in this example, one readily finds that:

$$p = (2M\zeta)^{\frac{1}{2}} \cos \Omega(t-\xi). \qquad (3.025)$$

If this is solved for ξ and if ζ is substituted from (3.023), the formula for ξ in terms of q, p, and t is seen to be:

$$\xi = t - \frac{1}{\Omega} \arctan \frac{m\Omega q}{p} . \qquad (3.026)$$

Thus are displayed, for the case in question, the four expressions which were discussed earlier in abstract terms.

As has been repeatedly emphasized, there is nothing unique about the form in which the constants of the motion may appear. Thus, in the harmonic oscillator, an equally valid choice for the first integral of the motion would be the initial momentum:

$$\zeta' = p_0 = p \cos \Omega t + M\Omega q \sin \Omega t. \qquad (3.027)$$

A natural choice for second constant in this case would be the initial displacement:[†]

$$\xi' = q_0 = q \cos \Omega t - \frac{P}{M\Omega} \sin \Omega t. \qquad (3.028)$$

It is left as an exercise to show that ζ' and ξ' can be written as functions of ζ and ξ only, i.e. in the form of expressions which do not contain the canonical variables nor the time. Thus the two pairs of quantities are but two different forms of the same two entities, the constants of the motion.

[†] It is possible for ξ' to be a function of q_0 (or even a function of q_0 and p_0) rather than to be literally equal to q_0. Since the latter alternative is the simplest and most appropriate, however, it should be the one adopted.

The preceding development concerning an arbitrary system of one degree of freedom can be extended to an arbitrary system of D degrees of freedom. In such a case, there are D first integrals:

$$\left.\begin{aligned}
\zeta_1 &= \zeta_1(q_j, p_j, t). \\
\zeta_2 &= \zeta_2(q_j, p_j, t). \\
&\cdot\cdot\cdot\cdot\cdot\cdot \\
\zeta_D &= \zeta_D(q_j, p_j, t).
\end{aligned}\right\} \tag{3.029}$$

These lead, theoretically at least, to the final solution:

$$\left.\begin{aligned}
q_1 &= q_1(\zeta_j, \xi_j, t). \\
q_2 &= q_2(\zeta_j, \xi_j, t). \\
&\cdot\cdot\cdot\cdot\cdot\cdot \\
q_D &= q_D(\zeta_j, \xi_j, t).
\end{aligned}\right\} \tag{3.030}$$

According to (2.002), the momenta are functions of q_j, $\dot{q}_j$, and t and are therefore readily obtainable from the above:

$$\left.\begin{aligned}
p_1 &= p_1(\zeta_j, \xi_j, t). \\
p_2 &= p_2(\zeta_j, \xi_j, t). \\
&\cdot\cdot\cdot\cdot\cdot\cdot \\
p_D &= p_D(\zeta_j, \xi_j, t).
\end{aligned}\right\} \tag{3.031}$$

Finally, (3.031) can in principle be solved for the ξ_j as unknowns to yield $\xi_j = \xi_j(\zeta_j, p_j, t)$ whereupon substitution for ζ_j from (3.029) permits the following to be written:

$$\left.\begin{aligned}
\xi_1 &= \xi_1(q_j, p_j, t). \\
\xi_2 &= \xi_2(q_j, p_j, t). \\
&\cdot\cdot\cdot\cdot\cdot\cdot \\
\xi_D &= \xi_D(q_j, p_j, t).
\end{aligned}\right\} \tag{3.032}$$

These systems of equations are the exact analogs of expressions (3.019) through (3.022). Once more, the $2D$ independent constants of the motion ζ_j and ξ_j can take on an infinite variety of forms. Often some of the ζ_j will be canonical momenta recognized as constants because of the absence of their conjugate coordinates from the Hamiltonian.

It is also conceivable, of course, that the D constants ζ_j could be the initial values p_{0j} of the canonical momenta and that the D constants ξ_j could be the initial values q_{0j} of the coordinates. The constants of the motion could equally well be functional mixtures of these sets of quantities. It must be emphasized that the "method" outlined here is presented to help the reader arrive at a particular theoretical viewpoint; it is not intended as a practical procedure for the solution of any but the most elementary problems, especially if $D > 1$. In the analysis of systems of many degrees of freedom, it is usually not possible to find all D of the expressions for the independent first integrals of the motion (3.029), let alone solve the simultaneous, generally non-linear, first-order differential equations which these expressions represent.

Hamilton–Jacobi theory applies the concepts and methodology of transformation, as described in Section 3.01, to the material contained in the preceding paragraphs. Under this theory, equations (3.029) through (3.032) are regarded as equations of canonical transformation in which the ζ_j play the part of the P_j; the ξ_j, of the Q_j. Obviously, the new "variables" in this arrangement are constants; in spite of this, they can be bona fide solutions of Hamilton's equations provided that the new Hamiltonian does not contain any of them! Thus, if the new Hamiltonian is a function of time only, one has:

$$\xi_j = \frac{\partial \mathcal{K}}{\partial \zeta_j} = 0, \quad \zeta_j = -\frac{\partial \mathcal{K}}{\partial \xi_j} = 0, \quad \dot{\mathcal{K}} = \frac{\partial \mathcal{K}}{\partial t}, \quad (3.033)$$

as required by the Hamiltonian formulation. Evidently, the ζ_j and ξ_j constitute the trivial solution alluded to in Section 3.01.

For the construction of the transformation in question, it is usual to select a generating function of the form $G_b(q_j, \zeta_j, t)$. In the previous section, G_b was regarded as given and the new Hamiltonian was determined by equation (3.018). The situation is now reversed; the new Hamiltonian is prescribed to be a function of time only and (3.018) is invoked as a condition which G_b must satisfy. This condition is:

$$\mathcal{H}(q_j, p_j, t) + \frac{\partial G_b}{\partial t} = \mathcal{K}(t). \quad (3.034)$$

The form of $\mathcal{K}(t)$ is arbitrary since any function of time only would satisfy equations (3.033). It is usual to take fullest advantage of this arbitrariness and simply choose $\mathcal{K}(t) = 0$. The function G_b corresponding to this choice of $\mathcal{K}$ is called *Hamilton's principal function* and is conventionally designated $S(q_j, \zeta_j, t)$. In this context, equations (3.016) and (3.017) become, respectively:

$$p_j = \frac{\partial S}{\partial q_j}, \quad \xi_j = \frac{\partial S}{\partial \zeta_j}, \tag{3.035}$$

and one may substitute $\partial S/\partial q_j$ for p_j in the original Hamiltonian to obtain explicitly the differential equation of which S is a solution. This equation is called the HAMILTON–JACOBI EQUATION:

$$\boxed{\mathcal{H}\left(q_j, \frac{\partial S}{\partial q_j}, t\right) + \frac{\partial S}{\partial t} = 0} \,. \tag{3.036}$$

It may be noticed that the Hamilton–Jacobi equation involves only the symbol S, the coordinates q_j, and the time t; it does not mention the constants ζ_j and gives no clue as to how S might depend upon them. Thus, starting from (3.036) and using no extraneous information, all one can hope to find is a function $S = S(q_j, t)$. This function will indeed involve constant parameters but the identification of these with the ζ_j or with combinations of the ζ_j can be done in infinitely many ways. This variety of possibilities is related to the polymorphism of the constants of the motion and the very process of identifying the ζ_j with the parameters occurring in S involves decisions about the form in which the constants of the motion will eventually appear. To do this in a way that will simplify subsequent steps may require considerable ingenuity. Fortunately, the ease or difficulty of performing this task or even of solving the equations of motion at all is of secondary importance in terms of the ultimate objectives of this book. Of primary importance are the facts that S is a time-dependent scalar field on configuration space, that the components $\partial S/\partial q_j$ of its "gradient" are equal to the canonical momenta of the system, and that its derivative with respect to time is equal to the negative of the Hamiltonian. These are the properties which give significance to S in the transition from classical to quantum mechanics.

Once a function $S(q_j, \zeta_j, t)$ is obtained that (i) satisfies the Hamilton–Jacobi equation and (ii) contains the ζ_j in a convenient way, the canonical transformation can be found by the methods of Section 3.01, at least in principle. Examples of this technique appear in the subsequent section.

3.03. Elementary Properties of Hamilton's Principal Function

The role of Hamilton's principal function as the generator of a transformation which solves the equations of motion has been discussed; its field properties, apparent when it is viewed simply as a function of coordinates and time, have also been mentioned. To appreciate both clearly, it is imperative to consider at first only the simplest

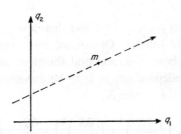

Fig. 3.01. Inertial Cartesian coordinates showing motion of an unforced particle.

of cases and, for this purpose, a system consisting of an unforced particle moving in a plane has been selected. This system can be described in two degrees of freedom and it is convenient to choose the generalized coordinates as synonymous with the inertial Cartesian coordinates pictured in Fig. 3.01. The motion is obviously in a straight line at uniform speed.

It will soon become apparent that infinitely many different functions

$S(q_1, q_2, \zeta_1, \zeta_2, t)$ can be related to the motion of an unforced particle. These will have different dependences upon q_1 and q_2 as well as upon ζ_1 and ζ_2. Every distinct dependence upon q_1 and q_2 corresponds to a distinct *family of motions* for each of which the lines of constant S spread in wave-like fashion, always remaining perpendicular to the possible trajectories of the particle. These trajectories (which are straight lines because of the unforced nature of the motion) will be mutually parallel in one type of family, divergent from a common point in another type, and will possess neither of these attributes in a third. Similar features are, of course, present in the S functions of forced particles but are difficult to elucidate because of the mathematical complexities associated with even simple examples.

The canonical momenta for the system in question are the ordinary Cartesian components, $p_1 = m\dot{q}_1$ and $p_2 = m\dot{q}_2$. Setting V equal to zero, the Hamiltonian becomes:

$$\mathscr{H} = \frac{1}{2m}(p_1^2 + p_2^2). \qquad (3.037)$$

This contains neither q_1, q_2, nor t and therefore p_1, p_2, and $\mathscr{H}$ itself are constants of the motion. Of course, these three quantities are connected by the above equation and therefore any two (and only two) of them are independent, as is to be expected. The Hamilton–Jacobi equation for the system is:

$$\frac{1}{2m}\left[\left(\frac{\partial S}{\partial q_1}\right)^2 + \left(\frac{\partial S}{\partial q_2}\right)^2\right] + \frac{\partial S}{\partial t} = 0. \qquad [(3.038)$$

Since $\mathscr{H}$ is a constant of the motion equal, say, to $\mathscr{H}_0$, it follows quite obviously from (3.036) that:

$$\frac{\partial S}{\partial t} = -\mathscr{H}_0 \qquad (3.039)$$

This implies that S must be of the form:

$$S(q_j, \zeta_j, t) = W(q_j, \zeta_j) - \mathscr{H}_0 t, \qquad (3.040)$$

where $W(q_j, \zeta_j)$ contains only coordinates and constants. $W(q_j, \zeta_j)$ is called *Hamilton's characteristic function*; it appears whenever, as here, the Hamiltonian does not contain the time. Substitution of (3.040) into (3.038) shows that W obeys the following equation:

$$\frac{1}{2m} \left[\left(\frac{\partial W}{\partial q_1} \right)^2 + \left(\frac{\partial W}{\partial q_2} \right)^2 \right] = \mathcal{H}_0. \tag{3.041}$$

This equation is a condition governing the magnitude of the gradient of W and, in the present case, it requires this magnitude to be constant, i.e. $|\nabla W|(=2m\mathcal{H}_0)^{\frac{1}{2}}$. Equation (3.041) admits an infinite number of different solutions, each yielding a different S and each relating to a different family of motions. Of these, the simplest and most basic is the following:

$$W = Aq_1 + Bq_2, \tag{3.042}$$

where A and B are constants. It is natural to identify these constants as ζ_1 and ζ_2, whereupon:

$$W = \zeta_1 q_1 + \zeta_2 q_2. \tag{3.043}$$

Substitution into (3.041) reveals that $\mathcal{H}_0 = (\zeta_1^2 + \zeta_2^2)/2m$; this permits $\mathcal{H}_0$ to be eliminated from S. The latter then becomes a function of $q_1, q_2, \zeta_1, \zeta_2$, and t, as required by theory and the result is:

$$S = \zeta_1 q_1 + \zeta_2 q_2 - \frac{\zeta_1^2 + \zeta_2^2}{2m} t. \tag{3.044}$$

The function S is now complete and one may apply the standard procedures of Section 3.01 to find the canonical transformation which it generates. Because of the simplicity of the example, some of the steps are trivial. First, an application of $p_j = \partial S/\partial q_j$ gives (3.047) below; after this, $\xi_j = \partial S/\partial \zeta_j$ yields $\xi_1 = q_1 - \zeta_1 t/m$ and $\xi_2 = q_2 - \zeta_2 t/m$. Rearrangements and substitutions eventually produce the fully formed transformation. For emphasis, the latter is displayed so as to show all four of the functional dependences even though some

6*

of these are redundant:

$$\left.\begin{array}{l} \zeta_1 = p_1; \\ \zeta_2 = p_2. \end{array}\right\} \qquad (3.045)$$

$$\left.\begin{array}{l} q_1 = \xi_1 + \dfrac{\zeta_1}{m}\, t; \\[2ex] q_2 = \xi_2 + \dfrac{\zeta_2}{m}\, t. \end{array}\right\} \qquad (3.046)$$

$$\left.\begin{array}{l} p_1 = \zeta_1; \\ p_2 = \zeta_2. \end{array}\right\} \qquad (3.047)$$

$$\left.\begin{array}{l} \xi_1 = q_1 - \dfrac{p_1}{m}\, t; \\[2ex] \xi_2 = q_2 - \dfrac{p_2}{m}\, t. \end{array}\right\} \qquad (3.048)$$

Notice that (3.046) constitutes the solution of the equations of motion. The constants ζ_j are the conserved Cartesian components of momentum and the constants ξ_j are the initial values of the coordinates.

It will be helpful to discuss S as a function of coordinates and time, i.e. as a *field*, and not merely as a tool for finding the solutions of the equations of motion. It is expedient to do this by carefully examining the lines of constant S, first at a given instant and later as a function of time. Using the vectors $\boldsymbol{\zeta} = e_1\zeta_1 + e_2\zeta_2 = e\zeta$ and $\boldsymbol{q} = e_1 q_1 + + e_2 q_2$, one may write:

$$S = \boldsymbol{\zeta}\cdot\boldsymbol{q} - \frac{\zeta^2}{2m}\, t. \qquad (3.049)$$

Notice that $\boldsymbol{\zeta}$ is equal to the constant vector momentum of the particle and that $\boldsymbol{q}$ is its instantaneous vector displacement. Dividing by the magnitude of $\boldsymbol{\zeta}$ and rearranging:

$$e\cdot\boldsymbol{q} = \frac{S}{\zeta} + \frac{\zeta}{2m}\, t. \qquad (3.050)$$

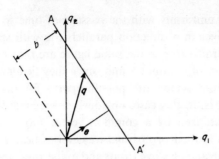

FIG. 3.02. Equation of line AA' is $e \cdot q = b$.

This equation may be compared with the equation $e \cdot q = b$ of a straight line perpendicular to e and lying at a distance b from the origin. Such a line is illustrated by AA' in Fig. 3.02. From these considerations, it is evident that at any particular instant of time the lines of constant S are parallel straight lines having the momentum unit vector e as their common perpendicular. The distance from the origin to each line is given by:

$$b = \frac{S}{\zeta} + \frac{\zeta}{2m} t. \tag{3.051}$$

A set of lines of constant S is illustrated, at some given time t, by the solid lines in Fig. 3.03. Notice that lines representing larger values of S are situated at greater distances from the origin and that these distan-

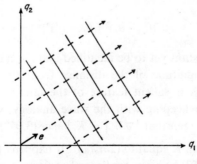

FIG. 3.03. Lines of constant S (solid) and possible particle trajectories (dashed) for the S of equation (3.044).

ces all increase uniformly with the passage of time so that the entire set of lines moves in a direction parallel to e with speed $\zeta/2m$. The dashed lines parallel to e in the same figure are the *orthogonal trajectories* of the lines of constant S and, since they lie along the direction of the momentum vector, are possible paths for the particle itself. The function S is, in this case, obviously associated with a family of motions characterized by a common vector momentum. The lines of constant S and their orthogonal trajectories bear a strong resemblance to the advancing wave fronts and to the rays, respectively, of a two-dimensional *plane wave*, e.g. a suitably excited wave on the surface of a body of water or on a stretched membrane.

The careful reader will have noticed that, whereas the speed of the particle is ζ/m, the speed of advance of the lines of constant S is one-half this value or $\zeta/2m$. This disparity is not a defect in the theory but rather a very important paradox, also related to the wave analogy, which will be resolved in Chapter 6; evidently the relationship between the movement of the lines of constant S and the motion of the particle is more subtle than would appear at first glance.

It is interesting to go back now to equation (3.041) and seek another solution $W(q_1, q_2, \zeta_1, \zeta_2)$ with a different dependence upon q_1 and q_2. This will lead to an associated wave-like behavior different from that illustrated in Fig. 3.03. It is not difficult to find another function of the two coordinates which has a gradient of constant magnitude. One such function is:

$$W = \pm K[(q_1-q_1')^2+(q_2-q_2')^2]^{\frac{1}{2}} = \pm Kr, \qquad (3.052)$$

where K is a constant yet to be identified and r is a convenient shorthand for the magnitude of the distance from the fixed point (q_1', q_2'). This point is a salient feature of the family of motions to be discussed and, in keeping with the wave analogy, will be called the *focus*. It is easily seen that $|\nabla W| = K$, hence (3.052) is indeed a solution of (3.041) with K equal to $(2m\mathcal{H}_0)^{\frac{1}{2}}$. It happens that the undetermined sign in (3.052) is minus for lines of constant S which are collapsing toward the focus and plus for similar lines which are expanding away from the focus; to avoid undue prolixity, only the plus sign will be considered in what follows.

The problem of identifying either ζ_1, ζ_2, or some combination of both with K is now at hand. Once more the path of simplicity will be taken and K will be set equal to ζ_1; this implies that ζ_2 is missing from W. Evidently $\mathcal{H}_0 = \zeta_1^2/2m$ and Hamilton's principal function becomes:

$$S = \zeta_1[(q_1 - q_1')^2 + (q_2 - q_2')^2]^{\frac{1}{2}} - \frac{\zeta_1^2}{2m} t. \qquad (3.053)$$

The quantities q_1' and q_2', which can be eliminated by shifting the origin of coordinates, are not canonical constants. This S function is clearly different from (3.044), yet it is equally applicable to the motion of an unforced particle in a plane. The process of finding the transformation begins by invoking $p_j = \partial S/\partial q_j$. This yields:

$$\left.\begin{aligned} p_1 &= \zeta_1 \frac{q_1 - q_1'}{r} \; ; \\[2mm] p_2 &= \zeta_1 \frac{q_2 - q_2'}{r} \, . \end{aligned}\right\} \qquad (3.054)$$

It continues by applying $\xi_j = \partial S/\partial \zeta_j$, which gives:

$$\left.\begin{aligned} \xi_1 &= r - \frac{\zeta_1}{m} t; \\[2mm] \xi_2 &= 0. \end{aligned}\right\} \qquad (3.055)$$

Equations (3.054) may be solved for ζ_1 by squaring and adding; the result confirms that ζ_1, given by (3.058) below, is indeed the momentum magnitude. Since ζ_2 does not appear at all in the above, it may be defined at the discretion of the analyst as any function of p_1, p_2, q_1, q_2, and t which is independent of ζ_1. The obvious choice is to define ζ_2 as the angle between the momentum vector and e_1, i.e. as arc tan (p_2/p_1). From this and (3.054), it follows that:

$$\left.\begin{aligned} \frac{q_1 - q_1'}{r} &= \frac{p_1}{\zeta_1} = \cos \zeta_2; \\[2mm] \frac{q_2 - q_2'}{r} &= \frac{p_2}{\zeta_1} = \sin \zeta_2. \end{aligned}\right\} \qquad (3.056)$$

Solving the first of equations (3.055) for r, one finds that:

$$r = \xi_1 + \frac{\zeta_1}{m} t. \tag{3.057}$$

With the aid of a few rearrangements and substitutions, the complete transformation can now be assembled as follows:

$$\left. \begin{aligned} \zeta_1 &= (p_1^2 + p_2^2)^{\frac{1}{2}}; \\ \zeta_2 &= \arctan (p_2/p_1). \end{aligned} \right\} \tag{3.058}$$

$$\left. \begin{aligned} q_1 &= q_1' + \left(\xi_1 + \frac{\zeta_1}{m} t\right) \cos \zeta_2; \\ q_2 &= q_2' + \left(\xi_1 + \frac{\zeta_1}{m} t\right) \sin \zeta_2. \end{aligned} \right\} \tag{3.059}$$

$$\left. \begin{aligned} p_1 &= \zeta_1 \cos \zeta_2; \\ p_2 &= \zeta_1 \sin \zeta_2. \end{aligned} \right\} \tag{3.060}$$

$$\left. \begin{aligned} \xi_1 &= \left[(q_1 - q_1')^2 + (q_2 - q_2')^2\right]^{\frac{1}{2}} - \frac{(p_1^2 + p_2^2)^{\frac{1}{2}} t}{m}; \\ \xi_2 &= 0. \end{aligned} \right\} \tag{3.061}$$

Once more, the second set of equations in the transformation, namely (3.059), constitutes the solution of the equations of motion. The particle momentum is characterized in magnitude and angle by the constants ζ_1 and ζ_2, respectively, but the function S mentions only the former of these. Thus, in the associated family of motions, all members have a trajectory which passes through the focal point (q_1', q_2') and all have momentum of magnitude ζ_1 but each member has an arbitrary value of ζ_2. The constant ξ_1 is obviously the initial value of r; specification of ζ_2 and ξ_1 therefore serves to select a particular motion out of the family. The trivial constant ξ_2 plays no part.

The equation for the lines of constant S is found more readily in this case than in the former one; it is obtained by solving (3.053) for r:

$$r = \frac{S}{\zeta_1} + \frac{\zeta_1}{2m} t. \tag{3.062}$$

This shows that the lines in question are circles of increasing radii, portions of which are illustrated in Fig. 3.04 along with their orthogonal trajectories. Once more the speed paradox is evident since the particle travels at speed ζ_1/m but the circle radii change at the rate $\zeta_1/2m$. The wave analogy for this family of motions is obviously a wave with circular fronts which expand away from the focal point.

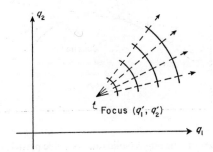

FIG. 3.04. Lines of constant S (solid) and possible particle trajectories (dashed) for the S of equation (3.053).

The two examples discussed here in detail do not exhaust the variety of S-functions and related families of motions which can be associated with an unforced particle. Another possibility, for example, is one in which the particle trajectories are neither parallel nor confocal but rather are distributed as in Fig. 3.05. It is possible to characterize this family by the *envelope curve* to which each of the trajectories is tangent and the latter may, as in optics, be called a *caustic*.

This section has shown, by specific examples, how partial differentiations performed upon Hamilton's principal function S provide the material from which a canonical transformation to the constants ζ_j and ξ_j can be constructed and how this transformation contains solutions to the equations of motion. It has also revealed some of the field properties of S by showing how the lines of constant S continually move along their orthogonal trajectories, the latter being possible trajectories of the particle itself. Finally, the enormous flexibility of S has been emphasized. There is no such thing as *the S-func-*

tion for a given system; any one of an infinite number of such functions can be employed and each represents a different family of motions related through some common set of properties.

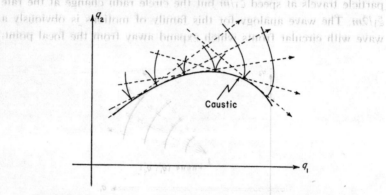

FIG. 3.05. Lines of constant S (solid) and possible particle trajectories (dashed) for an S function more general than either (3.044) or (3.053). Constant S lines have cusps at the caustic.

3.04. Field Properties of Hamilton's Principal Function in the Context of Forced Motion

The field properties of Hamilton's principal function, which were treated in a limited context in the previous section, will be developed more fully in this one. The system will once more consist of a single particle but the motion will be in ordinary three-dimensional space; forces will be present and inertial Cartesian coordinates, called by their usual names of x, y, and z, will be employed. Since the manner in which the constants of the motion ζ_j enter into S is not of interest here, the latter will simply be regarded as a function of coordinates and time, i.e. as $S(r, t)$.

The Hamiltonian for the system is given by:

$$\mathcal{H} = \frac{1}{2m}\,(p_x^2 + p_y^2 + p_z^2) + V(r, t).\qquad(3.063)$$

This is not a constant of the motion unless t is missing from V. The equations $p_j = \partial S / \partial q_j$ become:

$$\mathbf{p} = \nabla S,$$ (3.064)

and to this should be added:

$$\mathcal{H} = -\frac{\partial S}{\partial t}.$$ (3.065)

Equation (3.064) shows that the particle momentum is always perpendicular to the surfaces of constant S and it follows that the possible particle paths must, in all cases, be the orthogonal trajectories to these surfaces. The Hamilton–Jacobi equation is clearly given by:

$$\frac{1}{2m} |\nabla S|^2 + V(\mathbf{r}, t) + \frac{\partial S}{\partial t} = 0.$$ (3.066)

In view of (3.064) and (3.065), this equation simply reiterates $T + V - \mathcal{H} = 0$; when solved for $|\nabla S|$, it yields:

$$|\nabla S| = [2m(\mathcal{H} - V)]^{\frac{1}{2}}.$$ (3.067)

This becomes very easy to interpret if $\mathcal{H}$ is a constant of the motion for then $|\nabla S|$ is a function of position only and is large where V is small and small where V is large. The situation is illustrated qualitatively in Fig. 3.06 where the region of large potential energy is at the top of the drawing, that of small potential energy at the bottom. The force on the particle obviously acts from top to bottom. An S-function and its associated family of motions has been selected so that the particle trajectories are generally from upper left to lower right. The surfaces of constant S must be crowded more closely together where $|\nabla S|$ is relatively large, i.e. at the bottom of the illustration. Moreover, when the orthogonal trajectories are curved as they are here, they must necessarily be concave toward the region where the surfaces of constant S are closer together; this is precisely the region of lower potential energy and the region toward which the force acts. The geometry of these trajectories is therefore in harmony with the direction of the force.

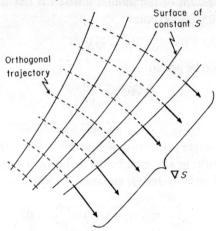

Large Potential Energy

Surface of
constant S

Orthogonal
trajectory

∇S

Small Potential Energy

FIG. 3.06. A section through several surfaces of constant S (solid lines) applicable to a particle moving in three-dimensional space; surface of greatest S value is at right. Orthogonal trajectories (possible particle paths) are shown as dashed lines. Arrows indicate ∇S, which is equal to particle momentum, at five typical points on right-hand surface. Force acts generally downward.

The movement of the surfaces of constant S may be analyzed by imagining a hypothetical observer who travels through configuration space and, while so doing, remains upon a given such surface so that for him $dS/dt = 0$. (The motion of such an observer is not dynamic, i.e. he cannot "let himself go" and expect to travel in this manner.) Letting v represent the observer's velocity, dS/dt is given by the usual formula for the substantial derivative and is then deliberately set equal to zero:

$$\frac{dS}{dt} = v \cdot \nabla S + \frac{\partial S}{\partial t} = 0. \qquad (3.068)$$

This becomes:

$$v \cdot p = \mathscr{H}. \qquad (3.069)$$

At every point in the space and at every time, p and $\mathscr{H}$ are determinate quantities; the magnitude of v is therefore minimum if v is parallel to p. This corresponds to the situation in which the hypothetical observer moves along an orthogonal trajectory for if he were to travel in any other direction he would have to go faster to keep up with the surface of constant S. This v is therefore the local velocity with which the surface itself is deemed to move and is the quantity sought. It follows that $(v_{\text{surface}})p = \mathcal{H}$ and that the vector form of this velocity may be expressed by:

$$v_{\text{surface}} = e\,\frac{\mathcal{H}}{p} = e\,\frac{\mathcal{H}}{mv_{\text{particle}}}, \qquad (3.070)$$

where e is at once the unit normal to the surface of constant S and the unit tangent to the orthogonal trajectory.

The speed paradox is now exhibited in full generality; the surfaces of constant S not only move at a speed which is in general different from that of the particle, they move at a speed which is inversely proportional to the latter! One may now recall the examples of Section 3.03 in which $V = 0$ and $\mathcal{H} = p^2/2m$. If equation (3.070) is applied to these examples, it asserts that $v_{\text{surface}} = e\frac{1}{2}v_{\text{particle}}$, which was indeed found to be the case. As intimated earlier, the speed paradox is important in the derivation of wave mechanics and its resolution will be accomplished in the course of that derivation.

3.05. Hamilton's Principal Function and the Concept of Action

In the previous section it was mentioned that a hypothetical observer who travels so as to make dS/dt vanish cannot be traveling dynamically. Suppose, however, that another observer does travel dynamically, i.e. that he travels along an orthogonal trajectory to the surfaces of constant S and does so on the time schedule called for by the equations of motion. In short, he travels along a dynamic path as that term was used in Section 1.08. This observer will either overtake or be overtaken by the surfaces of constant S and it is interesting to ask what, for him, will be the value of dS/dt.

To answer this question with maximum generality, let the space in question be the configuration space of the generalized coordinates upon which $S(q_j, t)$ is a time-dependent scalar field. (As in the previous section, the constants of the motion ζ_j are not of interest here.) The substantial derivative in this context is expressed by:

$$\frac{dS}{dt} = \sum \frac{\partial S}{\partial q_j}\, \dot{q}_j + \frac{\partial S}{\partial t}. \tag{3.071}$$

Using relations which, by now, should be very familiar, the above can be rewritten as:

$$\frac{dS}{dt} = \sum p_j \dot{q}_j - \mathcal{H} = \mathcal{L}. \tag{3.072}$$

This asserts an important and interesting fact, namely that dS/dt as experienced by a dynamically moving observer is equal to the *value of the Lagrangian* at the point and time in question. Integrating with respect to time, one finds that:

$$S = \int_{\substack{\text{dynamic} \\ \text{path}}} \mathcal{L}\, dt, \tag{3.073}$$

which is the familiar definition of action. It follows that, to within a constant which is inconsequential, Hamilton's principal function is equal to the *action evaluated upon a dynamic path*. It is no accident, then, that the same symbol, S, has been used for both quantities. Notice that the action evaluated upon an arbitrary path, called S_a in Section 1.08, does not have the properties of Hamilton's principal function and cannot be identified therewith.

The reader is cautioned against assuming that one may begin with a valid solution $S(q_j, t)$ of the Hamilton–Jacobi equation, differentiate this with respect to time to obtain $\dot{S}(q_j, \dot{q}_j, t)$ and expect thereby to find the Lagrangian for the system. Equation (3.072) asserts an equality of values only, not an identity of functional forms. Like the function $\dot{G}$ of equation (1.092), $\dot{S}$ is the total time derivative of a function of coordinates and time; when substituted into Lagrange's equations, it yields only identities and fails to produce the equations of motion. Thus $\dot{S}$ is like an empty term in the context of Lagrange's equations and obviously cannot be a Lagrangian.

CHAPTER 4

WAVES

4.01. Waves on a String under Tension

Much can be learned about waves from the study of a system consisting of an infinitely flexible[†] string subjected only to a longitudinal tension. A portion of such a string is illustrated in Fig. 4.01. It will be assumed that the motion is predominantly transverse and occurs in a single plane, namely the ux plane shown. The quiescent state of

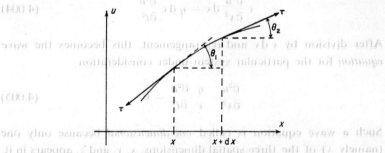

FIG. 4.01. Portion of string under tension with free body diagram of element between x and $x+dx$.

the string is described by $u = 0$; the disturbed state, by $u = u(x, t)$. It will be further assumed that the local angle of inclination, θ, is at most a very small fraction of a radian. The linear density of the string

[†] Adjacent elements of the string exert no torques on one another.

is denoted by η and the tension by τ. The mass of the element of string between x and $x+\mathrm{d}x$ is $\eta\,\mathrm{d}x$ and, from Fig. 4.01, the force on this element in the direction of increasing u is:

$$\mathrm{d}F = \tau(\sin\theta_2 - \sin\theta_1). \qquad (4.001)$$

Because of the smallness of θ, this is approximately:

$$\mathrm{d}F = \tau(\tan\theta_2 - \tan\theta_1)$$

$$= \tau\left[\left(\frac{\partial u}{\partial x}\right)_2 - \left(\frac{\partial u}{\partial x}\right)_1\right]$$

$$= \tau\frac{\partial^2 u}{\partial x^2}\,\mathrm{d}x. \qquad (4.002)$$

By Newton's law of motion:

$$\mathrm{d}F = \eta\,\mathrm{d}x\,\frac{\partial^2 u}{\partial t^2} \qquad (4.003)$$

and, combining equations (4.002) and (4.003), one obtains:

$$\tau\frac{\partial^2 u}{\partial x^2}\,\mathrm{d}x = \eta\,\mathrm{d}x\,\frac{\partial^2 u}{\partial t^2}. \qquad (4.004)$$

After division by $\tau\,\mathrm{d}x$ and rearrangement, this becomes the *wave equation* for the particular system under consideration:

$$\frac{\partial^2 u}{\partial x^2} - \frac{\eta}{\tau}\,\frac{\partial^2 u}{\partial t^2} = 0. \qquad (4.005)$$

Such a wave equation is called *one-dimensional* because only one (namely x) of the three spatial dimensions, x, y, and z, appears in it.

Solutions of (4.005) can take on an infinite variety of different forms. Of all these forms, one in particular can claim in some sense to be fundamental because of the ease with which it can be used as a "building block" for the construction of other forms. The solution in question is the *complex exponential*:

$$u(x, t) = A_0 \exp i(kx - \omega t + \varphi). \qquad (4.006)$$

Here A_0, φ, k, and ω are real constants. A_0 is called the *amplitude*; φ, the *phase*; k, *the wave number*;[†] and ω, the *radian frequency* or (for brevity) simply the *frequency*. From (4.006):

$$\frac{\partial^2 u}{\partial x^2} = -k^2 u, \quad \frac{\partial^2 u}{\partial t^2} = -\omega^2 u, \tag{4.007}$$

and the wave equation is satisfied provided that

$$-k^2 u + \frac{\eta}{\tau}\,\omega^2 u = 0. \tag{4.008}$$

On dividing by u and solving for ω, this becomes:

$$\omega = \pm (\tau/\eta)^{\frac{1}{2}} k. \tag{4.009}$$

It may be observed that A_0 and φ are independently arbitrary but that k and ω are related by equation (4.009). The latter is a very important equation and will be discussed at greater length later.

Since the form of the string can never be complex, (4.006) alone can never be a physically realizable solution. A simple superposition of two solutions like (4.006), however, one with negative values of k, ω, and φ, does indeed yield a real solution. Thus:

$$\begin{aligned}
u(x, t) &= \tfrac{1}{2} A_0 \exp i(kx - \omega t + \varphi) + \tfrac{1}{2} A_0 \exp i(-kx + \omega t - \varphi) \\
&= A_0 \cos(kx - \omega t + \varphi) \\
&= \mathrm{Re}\, A_0 \exp i(kx - \omega t + \varphi).
\end{aligned} \tag{4.010}$$

This function is perfectly acceptable and can represent the form of a sinusoidally driven string which is either infinite in length or which has a non-reflective terminating device at its remote end.

Returning to (4.006), one may plot the real and imaginary parts as in Fig. 4.02. Let the crests of the real part, $x_{-1}, x_0, x_1, x_2, \ldots$, be selected as features to be observed. These crests are simply the values of x

---

[†] The application of the term "wave number" to the quantity k in (4.006) is conventional in this area of physics even though the same term has been used extensively in spectroscopy with a slightly different meaning.

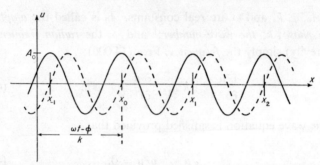

FIG. 4.02. Real part (solid line) and imaginary part (dashed line) of complex exponential wave shown at a particular instant of time. Drawing is based upon the assumption that k is positive.

for which $\cos(kx - \omega t + \varphi) = 1$ and are determined by the relation:

$$kx_n - \omega t + \varphi = 2\pi n. \tag{4.011}$$

This yields:

$$x_n = (\omega t - \varphi + 2\pi n)/k. \tag{4.012}$$

Notice that the crest that has been (arbitrarily) chosen as the zeroth is located at abscissa $(\omega t - \varphi)/k$. If ω and k are of the same sign, this crest — and with it the entire waveform — moves to the right as time passes; if ω and k are of opposite sign, the entire waveform moves to the left.

Two aspects of (4.006) are particularly interesting: (i) the distance between adjacent crests and (ii) the velocity with which the waveform moves. The former quantity is called the *wavelength* λ and is given by:

$$\lambda = x_{n+1} - x_n = 2\pi/k. \tag{4.013}$$

The latter is, by long-standing convention, called the *phase velocity* v_p and is obtained by differentiating (4.012) with respect to t:

$$v_p = \frac{dx_n}{dt} = \frac{\omega}{k}. \tag{4.014}$$

It is sometimes convenient to define an "actual" frequency v, measured in cycles per second, which is smaller than ω by a factor of 2π, i.e. $v = \omega/2\pi$. Using this, the phase velocity may be written as $v_p = \lambda v$ and $|v|$ is seen to be the number of real crests which pass a stationary observer per second. The fact that, on these definitions, both k and ω (and therefore both λ and v) can take either sign should not be allowed to become an obstacle to understanding. Mathematically speaking, both positive and negative values of both k and ω are required if solutions of maximum generality are to be built out of complex exponential waves like (4.006). This was apparent in the construction of the real function (4.010).

The physical characteristics of the wave-supporting structure determine the wave equation and, through it, the relationship between radian frequency and wave number. This, obviously, is also a relationship between actual frequency and wavelength and, to adopt a term from the optical tradition, is called a *dispersion relation*. Thus equation (4.009) is the dispersion relation for the simple string under tension. The two branches of this relation are plotted with respect to k in Fig. 4.03. Dispersion relation (4.009) is unique in that each one

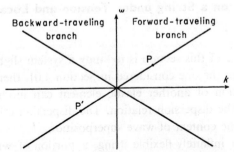

Fig. 4.03. Dispersion relation for the string under tension. The two points P and P′ represent two complex exponential waves which could be superposed to form a real forward-traveling sinusoidal wave.

of its two branches determines a phase velocity that is independent of the value of k. Thus, on the forward-traveling branch $\omega = (\tau/\eta)^{\frac{1}{2}}k$, on the backward-traveling branch $\omega = -(\tau/\eta)^{\frac{1}{2}}k$, and for the two branches:

$$v_p = \pm (\tau/\eta)^{\frac{1}{2}}, \tag{4.015}$$

respectively. The content of this formula is physically reasonable; a large tension τ is associated with a large value of $|v_p|$ whereas a large density η, which would tend to make the system sluggish, is associated with a small value.

As a final comment, it should be emphasized that the term "phase velocity" is appropriately applied only to examples which, like the complex exponential wave of (4.006), have a well-defined crest and trough structure discernible over an interval at least several wavelengths long. This implies that, if the wave is a superposition, its components have individual phase velocities that are very nearly the same. If such is not the case, then the wave disturbance in question does not have a readily definable phase velocity and the very concept ceases to be useful.

4.02. Waves on a String under Tension and Local Restoring Force

The objective of this section is to study a system slightly more sophisticated than the one considered in Section 4.01, thereby showing how the addition of another elastic element can modify the wave equation and the dispersion relation. The dispersion relation is very important in the context of wave superposition.

Consider an infinitely flexible string, a portion of which is illustrated in Fig. 4.04a, under a longitudinal tension τ as before but also connected to a rigid beam B by an elastic web W. This web, which is capable of both extension and compression, is unstressed when the string is in its quiescent state described by $u = 0$. It may be helpful to think of the web as the limiting case of a set of springs, illustrated in Fig. 4.04b, as the elastic constant of each individual spring tends to zero and the number of springs per unit length tends to infinity. The effect of the web is described by a parameter γ denoting force per

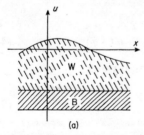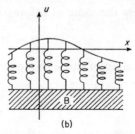

(a)   (b)

FIG. 4.04. String under tension with (a) elastic web which supplies local restoring force and (b) analog of web consisting of a large number of springs.

unit length per unit transverse displacement. As in other cases (Section 1.01, for example), this elastic component will be assumed massless; all the mass will be regarded as concentrated in the string itself which, as before, has linear density η. Using the notation of Fig. 4.01, the upward force on an element of string lying between x and $x+\mathrm{d}x$ in the present example is:

$$\mathrm{d}F = \tau(\sin\theta_2 - \sin\theta_1) - \gamma u\,\mathrm{d}x. \tag{4.016}$$

This becomes:

$$\mathrm{d}F = \left[\tau\frac{\partial^2 u}{\partial x^2} - \gamma u\right]\mathrm{d}x. \tag{4.017}$$

Applying Newton's law of motion as before, one obtains:

$$\left[\tau\frac{\partial^2 u}{\partial x^2} - \gamma u\right]\mathrm{d}x = \eta\,\mathrm{d}x\,\frac{\partial^2 u}{\partial t^2}. \tag{4.018}$$

Once more the wave equation is obtained by dividing by $\tau\,\mathrm{d}x$ and rearranging; it is clearly different from equation (4.005):

$$\frac{\partial^2 u}{\partial x^2} - \frac{\eta}{\tau}\frac{\partial^2 u}{\partial t^2} - \frac{\gamma}{\tau}u = 0. \tag{4.019}$$

As in the case of the string under tension alone, complex exponential solutions are of fundamental importance. The function itself is precisely the same as before:

$$u(x, t) = A_0 \exp i(kx - \omega t + \varphi); \qquad (4.020)$$

substitution of this into the wave equation yields:

$$-k^2 u + \frac{\eta}{\tau}\, \omega^2 u - \frac{\gamma}{\tau}\, u = 0. \qquad (4.021)$$

By cancelling u and solving for ω, one obtains the dispersion relation:

$$\omega = \pm \left(\frac{\tau}{\eta}\, k^2 + \frac{\gamma}{\eta} \right)^{\frac{1}{2}}. \qquad (4.022)$$

The two branches of this function are plotted in Fig. 4.05. The curve is an hyperbola asymptotic to the simpler dispersion relation of Fig. 4.03 in the limit of large $|k|$, i.e. at short wavelength where tension is

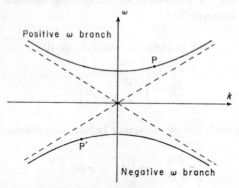

FIG. 4.05. Dispersion relation for string under both tension and local restoring force. The two points P and P' represent two complex exponential waves which could be superposed to form a real forward-traveling sinusoidal wave.

much more important than the local restoring force supplied by the web. It is interesting that the two branches of the dispersion relation are classified by a different criterion than formerly. Here, one has the positive ω branch and the negative ω branch; in each branch both

forward-traveling and backward-traveling waves occur. The phase
velocity, $v_p = \omega/k$, undergoes enormous variation with respect to k
as shown in Fig. 4.06. A real sinusoidal wave on the structure in question will necessarily involve both a positive and a negative value of ω
and both a positive and a negative value of k, just as before.

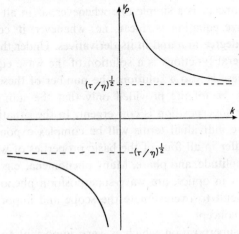

FIG. 4.06. Phase velocity plotted as a function of k for waves on a
string under both tension and local restoring force. Plot is for
positive ω branch only.

The wave solutions at $k = 0$ are worthy of comment. At this value
of k, $\omega = \pm (\gamma/\eta)^{\frac{1}{2}}$ and a real superposition of solutions from the
two branches takes the following form:

$$u(t) = \tfrac{1}{2}A_0 \exp i[-(\gamma/\eta)^{\frac{1}{2}}t + \varphi] + \tfrac{1}{2}A_0 \exp i[(\gamma/\eta)^{\frac{1}{2}}t - \varphi]$$

$$= A_0 \cos [(\gamma - \eta)^{\frac{1}{2}}t - \varphi]. \tag{4.023}$$

This is a sinusoidal function of time only; it indicates that all parts
of the string execute simple harmonic motion in unison. Notice that
the frequency is determined only by the web constant and the density
and that the tension is of no consequence. This state of *synchronized
vibration* is interpreted as a wave traveling with infinite phase velocity.

4.03. The Superposition of Waves

The word superposition has been used in a few instances already and its meaning is undoubtedly clear; there is said to be a superposition whenever two or more waves exist simultaneously on the same system. The concept is a simple one whenever, as in all cases in this book, the wave equation is linear, i.e. whenever it contains only terms of first degree in u and in its derivatives. Under this condition, if each of several functions is a solution of the wave equation, then the sum of these is also a solution; the number of these component functions may be infinite provided only that the sum (or integral) expressing the superposition is convergent. In the situations encountered here, the individual terms will be complex exponential waves which may differ in all four of the basic properties of wave number, frequency, amplitude, and phase. Many phenomena, e.g. interference and diffraction in optics, are wave-superposition phenomena and it would be difficult to overestimate the scope and importance of the superposition concept.

A type of superposition which is very important to the present development is one in which several (usually infinitely many) individual waves differing only slightly in wave number and frequency are present. Such an assembly of waves is characterized by an average wave number k_0 about which the individual wave numbers are clustered. The whole set is called a *wave packet*. Generally speaking, the region of substantial constructive interference among the component waves is limited in spatial extent and is flanked by infinite regions of destructive interference in which activity is negligible. A typical wave packet as it would appear at some particular time t is illustrated in Fig. 4.07.

A wave packet involving an infinite number of component waves is described by means of an *amplitude density function* $A_{0k}(k)$ and a *phase distribution function* $\varphi(k)$. Here "density" means density with respect to k considered as a continuous variable; the effective amplitude between wave numbers k and $k+dk$ is $A_{0k}\,dk$. For interesting

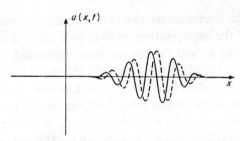

FIG. 4.07. Real part (solid line) and imaginary part (dashed line) of a typical wave packet composed of an infinite number of complex exponential waves. Appearance of packet at a particular instant of time is shown. Drawing is based upon the assumption that k_0 is positive.

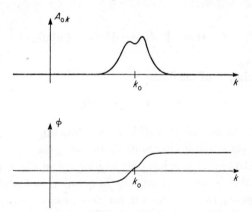

FIG. 4.08. A possible amplitude density and phase distribution for a wave packet.

packets, the amplitude density is relatively small except in a limited region surrounding the average[†] wave number k_0 and might appear as in Fig. 4.08. A typical phase distribution is also shown in this figure. The actual superposition takes the form of an integral over k:

$$u(x, t) = \int_{-\infty}^{\infty} A_{0k} \exp i[kx - \omega(k)t + \varphi(k)] \, dk. \qquad (4.024)$$

[†] A definition of the average wave number will be supplied when the present theory is applied to quantum phenomena.

Notice that the dispersion relation $\omega = \omega(k)$ appears explicitly in this expression. If the superposition involves only one branch of the dispersion relation, as will be true in most applications in this book, then expression (4.024) is adequate as it stands. If the superposition involves both branches, however, a second integral must be added to (4.024) to express the contribution from the second branch.

In dealing with complex exponential waves, it is often desirable to define a complex "amplitude" $A = A_0 \exp i\varphi$. This quantity combines both amplitude and phase into a single symbol and facilitates the writing of such waves. Applying the same idea to amplitude densities, one defines $A_k = A_{0k} \exp i\varphi(k)$ whereupon (4.024) may be expressed more simply as follows:

$$u(x, t) = \int_{-\infty}^{\infty} A_k \exp i[kx - \omega(k)t] \, dk. \qquad (4.025)$$

A wave disturbance of the form:

$$u(x, t) = f(x - vt), \qquad (4.026)$$

where f is any single-valued well-behaved[†] function of the argument indicated, moves without change of shape along the x-axis at velocity v. To demonstrate this, one has only to imagine an observer who also travels along the x-axis at the velocity in question and whose position is therefore given by $x = x_0 + vt$; for this observer $u(x, t)$ has the fixed value $f(x_0)$. The complex exponential wave (4.006), which can be written as:

$$u(x, t) = A_0 \exp ik(x - v_p t + \varphi/k), \qquad (4.027)$$

is clearly a function of $(x - v_p t)$ and is therefore an example of (4.026) with $v = v_p$. For a system like the string under tension alone (where v_p is indepedent of k) a packet composed of any number of, say, forward-traveling complex exponential waves will likewise be a function of $(x - v_p t)$ since v_p is the same for every component in the packet. In this case, the entire packet is an example of (4.026); it will travel

---

[†] Differentiable to the extent required.

at velocity v_p and retain its original shape indefinitely. This elementary state of affairs obtains only for systems with dispersion relations like (4.009) for only then is (4.026) in its general sense a solution of the wave equation.

If the dispersion relation for a wave-supporting system is more sophisticated than (4.009), the various components of a wave packet travel at different velocities and a very important question arises, namely, at what velocity does the outline or *envelope* of the packet (which demarks the region of constructive interference) travel? To investigate this question, let the appropriate branch of the dispersion relation be expanded in a Taylor series about the average wave number k_0:

$$\omega(k) = \omega(k_0) + (k - k_0)\left(\frac{d\omega}{dk}\right)_{k_0} + \frac{1}{2}(k - k_0)^2\left(\frac{d^2\omega}{dk^2}\right)_{k_0} + \ldots. \quad (4.028)$$

This may be rewritten with the simpler symbols ω_0, ω_0', ω_0'', ... to denote the various derivatives of ω evaluated at k_0. Thus:

$$\omega(k) = \omega_0 + (k - k_0)\omega_0' + \tfrac{1}{2}(k - k_0)^2\omega_0'' + \ldots \quad (4.029)$$

If the amplitude density function is relatively compact in k space, as is assumed here, there is very little contribution from wave components with large values of $|k - k_0|$ and one may use only a few terms in (4.029). For present purposes, actually, only the first two terms will be used. Denoting $k - k_0$ by k', the superposition of (4.025) can be written as an integral over k' as follows:

$$u(x, t) = \int_{-\infty}^{\infty} A_k \exp i[(k' + k_0)x - (\omega_0 + k'\omega_0')t] \, dk'. \quad (4.030)$$

The quantity $\exp i(k_0 x - \omega_0 t)$ is not a function of k' and can be factored out of the integral:

$$u(x, t) = \underbrace{\exp i(k_0 x - \omega_0 t)}_{\text{undulatory factor}} \underbrace{\int_{-\infty}^{\infty} A_k \exp ik'(x - \omega_0' t) \, dk'}_{\text{envelope}}. \quad (4.031)$$

This result is seen to be a complex exponential wave based upon k_0 and ω_0 times a function expressed by an integral. The latter, under the assumption of a relatively compact amplitude density function, is slowly varying compared to the former and is to be identified as the envelope of the packet. The complex exponential wave has been designated as the *undulatory factor* because it gives the packet its wave-like appearance. It is well to refer again to Fig. 4.07 at this point. The undulatory factor obviously travels at its usual phase velocity of ω_0/k_0 but the envelope, which is a function of $(x-\omega_0't)$, moves at a velocity equal to ω_0'. The latter is conventionally called the *group velocity* v_g. Thus:

$$v_g = \left(\frac{d\omega}{dk}\right)_{k_0}. \tag{4.032}$$

To an observer traveling at the group velocity, the packet envelope will be at rest but the wave which forms its undulatory factor will move forward through the packet if the phase velocity is greater than the group velocity or backward if it is less. In the special case in which the dispersion relation is like (4.009), $d\omega/dk$ is equal to ω/k and phase and group velocities are equal; the packet then moves like a rigid object as was mentioned in the discussion following equation (4.026).

The foregoing treatment of group velocity was predicated on the assumption that terms of second and higher order in (4.029) can be neglected. Although this is usually a workable assumption, there are cases in which it is not applicable and the packet envelope can then no longer be written precisely as a function of $(x-\omega_0't)$. In these cases, the envelope undergoes distortion, usually consisting of reduction of amplitude and broadening, as it moves.

With the aid of formula (4.032) and dispersion relation (4.022), it is possible to calculate the group velocity for a wave disturbance on a string under both tension and local restoring force as discussed in Section 4.02. The result is:

$$v_g = \pm \frac{\tau}{\eta} k \left[\frac{\tau}{\eta} k^2 + \frac{\gamma}{\eta}\right]^{-\frac{1}{2}}. \tag{4.033}$$

This group velocity is plotted as a function of k in Fig. 4.09 for the positive ω-branch and, as such, should be compared with the phase velocity for the same system plotted in Fig. 4.06. This system, or any system with a dispersion relation like (4.022), has the interesting property that (on a given branch) the product of phase and group velocities is constant for all values of k.

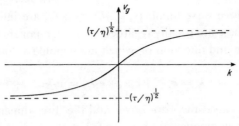

Fig. 4.09. Group velocity plotted as a function of k for waves on a string under both tension and local restoring force. Plot is for positive ω branch only.

4.04. Extension to Three Dimensions; Plane Waves

Perhaps the simplest type of classical wave in three dimensions is the acoustic or sound wave in a gas. For the undulatory quantity in this case it is convenient to choose the local departure of the pressure from its average value. Thus:

$$u(\mathbf{r}, t) = P(\mathbf{r}, t) - P_0. \tag{4.034}$$

This quantity is a scalar and the associated wave is said to be a *scalar wave*. The applicable wave equation[†] contains the average density ϱ_0 and the bulk modulus B:

$$\nabla^2 u - \frac{\varrho_0}{B} \frac{\partial^2 u}{\partial t^2} = 0. \tag{4.035}$$

[†] Since acoustic concepts are introduced here merely to provide a model for the study of three-dimensional waves, the use of space to derive (4.035) does not seem justified. For such a derivation, the reader is referred to ref. 5.

It is reasonable to seek a solution to this equation which is as nearly as possible analogous to the complex exponential solution (4.006) for one-dimensional wave motion. The solution in question is easy to construct; it is called a *complex exponential plane wave* or, for brevity, a *plane wave* and is given by the following:

$$u(r, t) = A_0 \exp i(k_x x + k_y y + k_z z - \omega t + \varphi). \tag{4.036}$$

Notice that three wave numbers, k_x, k_y, and k_z, are involved in this expression. These are usually regarded as the components of a vector, the magnitude and direction of which are denoted k and e_k, respectively:

$$k = e_x k_x + e_y k_y + e_z k_z = e_k k. \tag{4.037}$$

This vector is called the *wave vector* and the three-dimensional space formed by its components when treated as "coordinates" is called k-space or *wave vector space*. With the aid of the wave vector, the expression for a plane wave can be written in the compact form:

$$u(r, t) = A_0 \exp i(k \cdot r - \omega t + \varphi). \tag{4.038}$$

An even more compact form is achieved by substituting $A = A_0 \exp i\varphi$, whereupon:

$$u(r, t) = A \exp i(k \cdot r - \omega t). \tag{4.039}$$

It is a simple matter to verify that (4.036) or either of the expressions equivalent to it isa souution of wave equation (4.035). Taking the second partial derivativ with respect to each of the spatial coordinates in turn, one has:

$$\frac{\partial^2 u}{\partial x^2} = -k_x^2 u, \text{ etc.} \tag{4.040}$$

Therefore:

$$\nabla^2 u = -k^2 u, \quad \frac{\partial^2 u}{\partial t^2} = -\omega^2 u, \tag{4.041}$$

and the wave equation is satisfied provided that

$$-k^2 u + \frac{\varrho_0}{B} \omega^2 u = 0. \tag{4.042}$$

This is the dispersion relation; when solved for ω, it becomes:

$$\omega = \pm \left[\frac{B}{\varrho_0}\right]^{\frac{1}{2}} k = \pm \left[\frac{B}{\varrho_0}(k_x^2 + k_y^2 + k_z^2)\right]^{\frac{1}{2}}. \qquad (4.043)$$

As in an earlier example, the dispersion relation has a positive ω-branch and a negative ω-branch; on the former the wave travels in the direction of k, and on the latter in the direction of $-k$. For economy of presentation, from now on all discussions will apply specifically to the positive ω-branch unless a statement to the contrary is made. Notice that, in (4.043), ω is cast in the role of a scalar function on k space, i.e. a function of k_x, k_y, and k_z.

For the one-dimensional waves of Sections 4.01 and 4.02, the crests of the real part were the points for which $\cos(kx - \omega t + \varphi) = 1$ and nth crest, denoted x_n, was obtained by setting the argument of the cosine equal to $2\pi n$. If the same technique is applied to the three-dimensional wave as expressed in (4.038), the following condition characterizes the nth crest of the real part:

$$k \cdot r_n - \omega t + \varphi = 2\pi n. \qquad (4.044)$$

Division by k results in the following form in which the right-hand side, which has the dimensions of length, is called b_n:

$$e_k \cdot r_n = (\omega t - \varphi + 2\pi n)/k = b_n. \qquad (4.045)$$

From the background acquired in Section 3.03, it is easy to see that this is the equation of a plane which is perpendicular to e_k and which lies at a distance b_n (measured along e_k) from the origin. The quantity b_n increases with time. It follows that the nth crest, which has the form of an infinite plane and is called a *wave front*,[†] moves in the direction e_k. The same appellation applies to all the crests, which form a family of parallel planes and move in similar fashion. The unit vector e_k, denoting the direction of movement, is called the *wave normal*. The orthogonal trajectories of the wave fronts, i.e. lines perpendicular to these fronts, are known as *rays*[†] and are conceptually useful.

[†] The terms *wave front* and *ray* were introduced in Section 3.03; in the present section they can be more specifically related to the physical structure of the wave.

In the three-dimensional plane wave case, the wavelength is defined as the perpendicular distance between adjacent wave fronts. Thus:

$$\lambda = b_{n+1} - b_n = 2\pi/k. \qquad (4.046)$$

The speed with which the wave fronts move along the wave normal is readily found by differentiating b_n with respect to time and, as before, is called the phase velocity:

$$v_p = \frac{\mathrm{d}b_n}{\mathrm{d}t} = \frac{\omega}{k}. \qquad (4.047)$$

This quantity may be endowed with vector character by annexing e_k:

$$v_p = \frac{e_k \omega}{k} = \frac{k\omega}{k^2}. \qquad (4.048)$$

In three dimensions as in one, the superposition of waves is a very important concept. A wave packet in three dimensions is a multifarious construct since it contains, in general, an infinite number of component plane waves with wave vectors which differ not only in magnitude but also in direction. Such a packet is expressed by the following superposition formula:

$$u(r, t) = \int_{-\infty}^{\infty} \int_{-\infty}^{\infty} \int_{-\infty}^{\infty} A_k \exp i[k \cdot r - \omega(k)t] \, \mathrm{d}k_x \, \mathrm{d}k_y \, \mathrm{d}k_z. \qquad (4.049)$$

This involves a complex distribution function A_k which has appreciable magnitude in only a restricted region of k space roughly centered about the average wave vector k_0. The above formula can be written more simply if $\mathrm{d}\tau_k$, understood as a "volume" element in k space, is substituted for $\mathrm{d}k_x \, \mathrm{d}k_y \, \mathrm{d}k_z$:

$$u(r, t) = \int_{\substack{\text{all } k \\ \text{space}}} A_k \exp i[k \cdot r - \omega(k)\, t] \, \mathrm{d}\tau_k. \qquad (4.050)$$

The dispersion relation, here written $\omega(k)$, appears in these formulas.

The question of the velocity of travel of the packet envelope is, again, a very important one and can be answered in terms analogous to those employed in the one-dimensional case. Let $\omega(\boldsymbol{k})$ be expanded in a three-dimensional Taylor series about the average wave vector $\boldsymbol{k}_0$:

$$\omega(\boldsymbol{k}) = \omega(\boldsymbol{k}_0) + (k_x - k_{0x}) \left(\frac{\partial \omega}{\partial k_x} \right)_{k_0} + (k_y - k_{0y}) \left(\frac{\partial \omega}{\partial k_y} \right)_{k_0}$$

$$+ (k_z - k_{0z}) \left(\frac{\partial \omega}{\partial k_z} \right)_{k_0} + \frac{1}{2} (k_x - k_{0x})^2 \left(\frac{\partial^2 \omega}{\partial k_x^2} \right)_{k_0}$$

$$+ (k_x - k_{0x})(k_y - k_{0y}) \left(\frac{\partial^2 \omega}{\partial k_x \partial k_y} \right)_{k_0} + \dots . \quad (4.051)$$

No attempt is made to write out all the second-order terms since these are to be neglected anyway. To first order, the above may be written more compactly as follows:

$$\omega(\boldsymbol{k}) = \omega(\boldsymbol{k}_0) + (\boldsymbol{k} - \boldsymbol{k}_0) \cdot (\nabla_k \omega)_{k_0}$$

$$= \omega_0 + (\boldsymbol{k} - \boldsymbol{k}_0) \cdot \boldsymbol{\omega}_0', \quad (4.052)$$

where ω_0 means $\omega(\boldsymbol{k}_0)$ and the vector $\boldsymbol{\omega}_0'$ means the gradient of ω in $\boldsymbol{k}$ space evaluated at $\boldsymbol{k}_0$. With the aid of a new variable $\boldsymbol{k}' = \boldsymbol{k} - \boldsymbol{k}_0$ and the symbols defined above, the expression for the packet becomes:

$$u(\boldsymbol{r}, t) = \int\limits_{\substack{\text{all } \boldsymbol{k} \\ \text{space}}} A_k \exp i[(\boldsymbol{k}' + \boldsymbol{k}_0) \cdot \boldsymbol{r} - (\omega_0 + \boldsymbol{k}' \cdot \boldsymbol{\omega}_0')t] \, d\tau_k. \quad (4.053)$$

The quantity $\exp i(\boldsymbol{k}_0 \cdot \boldsymbol{r} - \omega_0 t)$ is not a function of $\boldsymbol{k}'$ and can be factored out. The result is:

$$u(\boldsymbol{r}, t) = \underbrace{\exp i(\boldsymbol{k}_0 \cdot \boldsymbol{r} - \omega_0 t)}_{\substack{\text{undulatory} \\ \text{factor}}} \underbrace{\int\limits_{\substack{\text{all } \boldsymbol{k} \\ \text{space}}} A_k \exp i\boldsymbol{k}' \cdot [\boldsymbol{r} - \boldsymbol{\omega}_0' t] \, d\tau_k}_{\text{envelope}}. \quad (4.054)$$

The undulatory factor and the envelope can be easily recognized here. The former is a complex exponential plane wave based upon $\boldsymbol{k}_0$ and ω_0; it travels at its usual phase velocity $k_0 \omega_0 / k^2{}_0$. The envelope, expressed by the integral, is a function of the vector argument $(\boldsymbol{r} - \boldsymbol{\omega}_0' t)$ and moves with vector velocity $\boldsymbol{\omega}_0'$ as can be seen by reasoning analogous to that employed in Section 4.03. The velocity $\boldsymbol{\omega}_0'$ is therefore the group velocity of the packet and, as indicated earlier, is equal to the gradient of ω in k-space evaluated at k_0:

$$v_g = (\nabla_k \omega)_{k_0}. \qquad (4.055)$$

In anisotropic media, even the direction of travel of the undulatory factor usually does not coincide with the direction of travel of the envelope, i.e. phase and group velocities are, in general, neither equal in magnitude nor parallel in direction.

Unfortunately, the rather featureless dispersion relation of the acoustic wave, namely equation (4.043), leads to a group velocity that fails to exhibit the interesting properties which this quantity can possess in more general cases. It would be instructive, perhaps, to study more sophisticated examples of classical wave motion but it is questionable if such a digression would be worthwhile especially since the background already presented should be adequate for making the transition to quantum theory. The latter discipline contains its own store of examples.

4.05.   Quasi-Plane Waves; the Short Wavelength Limit

The acoustic wave equation (4.035), which is typical of many wave equations in physics, admits an infinite variety of undulatory solutions in unbounded space, i.e. solutions in which the pattern of crests and troughs, of wave fronts and rays, is evident over (theoretically) infinite regions of space and time. The plane wave is, of course, the prime example of this type of solution but other examples are not difficult to find. One which is well known is the spherical wave:

$$u(r, t) = \frac{C_0}{r} \exp i(\alpha r - \beta t + \varphi), \qquad (4.056)$$

where C_0, α, β, and φ are real constants and $\beta = \pm (B/\varrho_0)^{\frac{1}{2}}\alpha$. In still other examples, the wave fronts have cylindrical or spheroidal shapes or, perhaps, have forms not describable by any of the standard mathematical surfaces. A pattern resembling Fig. 3.05, for instance, is a possibility. In regions of space remote (in terms of wavelengths) from singularities, such waves are *quasi-plane*, i.e. they resemble a plane wave arbitrarily exactly in neighborhoods that are sufficiently small.

Any wave (in fact, any function of x, y, z, and t) can always be written in complex exponential form with the aid of two real functions $A_0(r, t)$ and $\sigma(r, t)$ as follows:

$$u(r, t) = A_0(r, t) \exp i\sigma(r, t). \tag{4.057}$$

This is, after all, just a resolution of the generally complex $u(r, t)$ into magnitude and angle. If $u(r, t)$ is a truly plane wave, then A_0 is absolutely constant and σ, equal to $k \cdot r - \omega t + \varphi$, is absolutely linear in coordinates and time. If the wave is quasi-plane, on the other hand, these statements become only approximate. In any event, one may define a *local wave vector* k and a *local frequency* ω so that an infinitesimal increase in σ is expressed in the same terms as if the wave were plane. Such an infinitesimal increase is in general given by:

$$d\sigma = (\nabla\sigma) \cdot dr + \frac{\partial\sigma}{\partial t} \, dt. \tag{4.058}$$

If the wave were plane, the corresponding expression would be:

$$d\sigma = k \cdot dr - \omega dt. \tag{4.059}$$

Equating of the coefficients in these two expressions yields $k_x = \partial\sigma/\partial x$, etc. The conclusions may be summarized as follows:

$$\left.\begin{aligned} k &= \nabla\sigma; \\ \omega &= -\frac{\partial\sigma}{\partial t}. \end{aligned}\right\} \tag{4.060}$$

The function $\sigma(r, t)$, upon which these results depend, is called the *eikonal*.[†]

[†] From the Greek word for "image", so named because of applications in geometrical optics. Sometimes $\sigma(r, t)$ in (4.057) is replaced at the outset by $f(r) - \omega t$ where ω is a constant; $f(r)$ is then called the eikonal. The resulting treatment is less general than the one given here.

It is interesting to apply equations (4.060) to the spherical wave; when this is done, the results $k = e_r\alpha$ and $\omega = \beta$ are readily obtained. The amplitude A_0 is equal to C_0/r and may be described as a gently varying function at large r where the wave is indeed quasi-plane.

A wave front for a quasi-plane wave can be defined analogously to a wave front for a plane wave, i.e. as a surface upon which $\cos \sigma = 1$. Thus the nth such front is characterized by:

$$\sigma = \sigma_n = 2\pi n, \tag{4.061}$$

where n is an integer. A few such surfaces and some of their orthogonal trajectories or rays are illustrated for a portion of a quasi-plane wave field in Fig. 4.10. Notice that the local wave normal e_k is at once the direction of the gradient of σ and the unit tangent to the ray

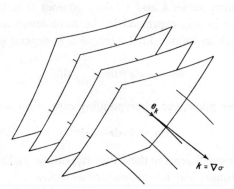

FIG. 4.10. Wave fronts (surfaces of constant σ) and some of their orthogonal trajectories (rays) for a quasi-plane wave.

through the point in question. A local wavelength $\lambda = 2\pi/k$ may also be defined and, to what is usually a very excellent approximation, it is the distance from one wave front to the next adjacent wave front measured along a ray. The wave fronts move with the local phase velocity $k\omega/k^2$.

Additional insight into the nature of quasi-plane waves is gained by taking derivatives of (4.057) with the intention of substituting these into the wave equation (4.035); this task begins as follows:

$$\frac{\partial u}{\partial x} = \left[\frac{\partial A_0}{\partial x} + iA_0 \frac{\partial \sigma}{\partial x}\right] \exp i\sigma. \tag{4.062}$$

$$\frac{\partial^2 u}{\partial x^2} = \left[\frac{\partial^2 A_0}{\partial x^2} - A_0 \left(\frac{\partial \sigma}{\partial x}\right)^2 + 2i \frac{\partial A_0}{\partial x} \frac{\partial \sigma}{\partial x} + iA_0 \frac{\partial^2 \sigma}{\partial x^2}\right] \exp i\sigma. \tag{4.063}$$

$$\nabla^2 u = [\nabla^2 A_0 - A_0 |\nabla\sigma|^2 + 2i\nabla A_0 \cdot \nabla\sigma + iA_0 \nabla^2\sigma] \exp i\sigma. \tag{4.064}$$

Similarly:

$$\frac{\partial^2 u}{\partial t^2} = \left[\frac{\partial^2 A_0}{\partial t^2} - A_0 \left(\frac{\partial \sigma}{\partial t}\right)^2 + 2i \frac{\partial A_0}{\partial t} \frac{\partial \sigma}{\partial t} + iA_0 \frac{\partial^2 \sigma}{\partial t^2}\right] \exp i\sigma. \tag{4.065}$$

Factorization of A_0 from the square brackets permits the two previous equations to be written as follows:

$$\left.\begin{array}{l} \nabla^2 u = \left[\dfrac{\nabla^2 A_0}{A_0} - |\nabla\sigma|^2 + \dfrac{2i}{A_0} \nabla A_0 \cdot \nabla\sigma + i\nabla^2\sigma\right] u; \\[3mm] \dfrac{\partial^2 u}{\partial t^2} = \left[\dfrac{1}{A_0} \dfrac{\partial^2 A_0}{\partial t^2} - \left(\dfrac{\partial \sigma}{\partial t}\right)^2 + \dfrac{2i}{A_0} \dfrac{\partial A_0}{\partial t} \dfrac{\partial \sigma}{\partial t} + i \dfrac{\partial^2 \sigma}{\partial t^2}\right] u. \end{array}\right\} \tag{4.066}$$

In a quasi-plane wave, the eikonal varies rapidly but nearly linearly with respect to coordinates and time; at a given time, its rate of change with respect to distance along a ray is 2π per wavelength and, at a given point in space, its rate of change with respect to time is -2π per period. In sharp contrast to this, the amplitude A_0 changes relatively slightly in a distance equal to a wavelength or in a time interval equal to a period. Thus in (4.066), the normalized derivatives $\nabla^2 A_0/A_0$ and $\nabla A_0/A_0$ are very small and the second derivatives of σ are similarly small. Only the terms consisting of the squared first derivatives of σ are large and, if these alone are retained, equations (4.066) may be written:

$$\left.\begin{array}{l} \nabla^2 u = -|\nabla\sigma|^2 u = -k^2 u; \\[3mm] \dfrac{\partial^2 u}{\partial t^2} = -\left(\dfrac{\partial \sigma}{\partial t}\right)^2 u = -\omega^2 u. \end{array}\right\} \tag{4.067}$$

These expressions are rigorously true for a genuine plane wave as can be seen by comparison with (4.041). For a quasi-plane wave they are useful approximations which improve if, other things being equal, the wavelength is made shorter and the frequency higher. Thus they also become rigorous for a quasi-plane wave as k and ω approach infinity. This limit is, appropriately, called the *short wavelength limit*.

Equations (4.067) may now be substituted into the wave equation. If the expressions involving k and ω are used, the dispersion relation is retrieved and it may be inferred that the latter holds for quasi-plane waves as well as for plane waves to a very good approximation. If, on the other hand, the expressions involving the derivatives of σ are employed, an equation governing σ is obtained; this is called the *eikonal equation*. Thus, as an example, the eikonal equation for the acoustic wave equation (4.035) may be written:

$$|\nabla\sigma|^2 - \frac{\varrho_0}{B}\left(\frac{\partial\sigma}{\partial t}\right)^2 = 0. \tag{4.068}$$

The eikonal equation is useful because, by solving it, an eikonal function can be found and the corresponding wave front and ray structure determined without actually solving the wave equation itself. This technique is particularly valuable if the medium is inhomogeneous, i.e. if, in the above example, ϱ_0/B varies from place to place and possibly even from time to time. Truly plane waves are, of course, impossible in an inhomogeneous medium but the quasi-plane waves which can exist are locally indistinguishable from plane waves in small neighborhoods and exhibit almost all of the important properties of such waves. Thus quasi-plane waves can be superposed to form a packet which, at a given time and place, travels at the local group velocity. The latter will vary as the medium varies and the packet will, in general, follow a curved path at non-uniform speed.

The purpose of this chapter, particularly the present section, has been to present wave theory in such fashion as to anticipate its application to particle motion. Some notion of this application may already be taking shape in the mind of the reader; its full development is reserved for Chapter 6.

CHAPTER 5

HISTORICAL BACKGROUND OF THE QUANTUM THEORY

5.01. Isothermal Cavity Radiation

Up to this point, the chief endeavor has been to present logical accounts both of classical analytical mechanics and of wave theory with the implication that an amalgamation of the two is in prospect. The amalgamation in question, which was discovered by Schrödinger, constitutes an approach to quantum theory known specifically as *wave mechanics*, an approach to which this book is largely devoted. Schrödinger's work, which made use of Planck's experimentally determined fundamental constant, must be viewed in the context of important contributions by both predecessors and contemporaries; like most great advances, it represented the culmination of many years of scientific effort inspired ultimately by the challenge of unexplained or only partially explained empirical data. It would therefore be artificial (and perhaps unnecessarily difficult) to plunge immediately into a study of wave mechanics without reviewing several of the details of this challenge. In the course of this review, to which the present chapter is devoted, it is hoped that the reader will acquire an appreciation of the real crises which began to develop in physics in the latter part of the nineteenth and earlier part of the twentieth centuries. These crises, which were in many respects distinct from those ushered in by the theory of relativity, could be met only by going beyond the established disciplines of classical mechanics and classical electro-dynamics. For economy and continuity, in brief accounts of this type, it is usual to excerpt only the more relevant portions of the original works and to paraphrase these considerably. The present narrative is no exception to this rule and anyone wishing to investigate the contributions of this period in historical depth is encouraged to consult the original papers.[†]

The first of the aforementioned crises to be clearly recognized as such occurred in conjunction with the theory of *black body radiation*. This phenomenon is intimately associated with *isothermal cavity radiation*, the latter being somewhat easier to understand. Consider an evacuated hollow cavity with walls maintained at absolute temperature θ as in Fig. 5.01. Objects such as A, B, ... suspended in the cavity by thermally insulating threads and having initial temperatures different from θ are observed, after the passage of time, to attain the temperature θ; in other words, they eventually come into thermal equilibrium with the walls. The agency responsible for energy transfer

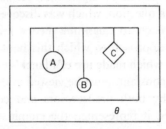

FIG. 5.01. Cavity with walls maintained at temperature θ. Suspended bodies, which eventually attain thermal equilibrium with the walls, are also shown.

across the vacuum between the walls and the suspended bodies is thermal (i.e. electromagnetic) radiation and the unoccupied spaces within the cavity support a continual traffic of electromagnetic waves.

Once thermal equilibrium has been established, many of the phe-

---

[†] For a very helpful commentary on this era with a collection of original papers (in English translation, where necessary), see ref. 6.

nomena within the cavity become unusually simple. The electromagnetic radiation becomes homogeneous in energy density and isotropic with regard to the direction of travel of the waves in any given neighborhood. Every small element of area, whether located on one of the walls or on one of the suspended bodies, receives an isotropic flux of radiant energy and sends an equal and equally isotropic flux back into the cavity. This restitution of radiant energy by the boundary elements is, in general, partly by reflection and partly by absorption and re-emission. With respect to isotropic radiation incident on such an element, one may define a reflection cofficient R and an absorption coefficient A such that if a total energy per unit time P is incident, then PR is reflected and PA is absorbed and re-emitted.[†] Evidently $R + A = 1$. The existence of this energy flux balance is corroborated by observation of a furnace heated to incandescence. In such an environment, all surface elements acquire the same uniformly bright appearance regardless of position or angle of view and it is impossible to tell whether the radiation received at a given observation point from a particular element has been absorbed and re-emitted or merely reflected. Another way to express this matter is to say that, with respect to the ingress and egress of radiant energy, a boundary element is not different from one side of any element of area on, say, an imaginary surface cutting directly through the cavity space. It follows that such an imaginary surface could be replaced by a material wall with no detectable change in the radiation field filling the now smaller cavity and that *the properties of the radiation field are therefore independent of the size and shape of the cavity as well as of the material and surface quality of its walls.*

An even stronger statement about the energy flux balance in the equilibrium state can be made, namely that every boundary element is returning to the radiation field as much energy per unit time as it is receiving *in any given frequency interval,*[‡] e.g. in the interval between

[†] Neither the reflected flux nor the re-emitted flux is necessarily isotropic by itself but the sum of the two is isotropic.

[‡] For traditional reasons and for ease of comparison with other treatments, the actual frequency $\nu = \omega/2\pi$ will be used often in this and subsequent sections.

v and $v+dv$. This fact was appreciated by Kirchhoff as early as 1859 although it was stated in terms of the equality of absorption and emission rather than in terms of total flux received and returned. If this balance did not obtain for each spectral component, the density of radiant energy within the cavity space would become abnormally higher at some frequencies and abnormally lower at others, whereupon, by clever use of frequency sensitive filters, energy could be continuously transferred to an adjacent cavity having a slightly higher temperature and no anomalies in its frequency distribution. This would violate the second law of thermodynamics and will not happen. It follows that, at any given temperature θ, there can be one and only one distribution of radiant energy with respect to frequency, i.e. there can be but one possible function $U_{rv}(v, \theta)$ representing energy per unit volume per unit frequency interval.

The radiation field is a thermodynamic system in its own right. Besides having an energy density, it exerts a pressure and has a temperature, the latter being equal to the temperature of the walls with which it is in equilibrium. The use of the term *isothermal cavity radiation* to describe this system whose characteristics are determined primarily by the electromagnetic properties of the vacuum and not by the specific properties of any particular material substance, is seen to be most appropriate.

Since surfaces differ in their ability to absorb radiation (and to reflect it) it is interesting to consider an idealized surface which absorbs all the radiation incident upon it and, moreover, does so at all

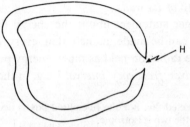

FIG. 5.02. Cavity with small hole H. All but a negligible fraction of the external radiation incident upon this hole is absorbed.

frequencies. For such a surface, $R = 0$ and $A = 1$; a body whose surface has these properties is conventionally called a *black body*, although the term *black surface* is really more accurate and will be used here. No truly black surface exists in nature; however, a small hole in a relatively large and preferably irregular cavity as illustrated in Fig. 5.02 forms, when viewed from the outside, an arbitrarily good approximation to an element of black surface. This is because only a negligible fraction of the radiation falling upon such a hole is reflected back out through the hole.[†] The latter statement is true even though the walls themselves may have a reflection coefficient close to unity. If the same cavity is maintained at temperature θ, the small hole will, when viewed from the outside, have the emissive properties of a black surface element at temperature θ. The spectral distribution of

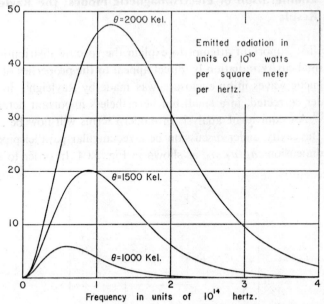

FIG. 5.03. Spectral density of black surface radiation.

[†] The hole diameter must be considerably larger than the wavelength at the lowest frequency of importance. This criterion is easily satisfied for thermal experiments likely to be of interest in this context.

the radiation thus emitted will be the same as that of the isothermal radiation within the cavity; the spectrum of the latter may therefore be inferred from an investigation of the former. In 1899 Lummer and Pringsheim made reliable determinations of this spectrum (illustrated for a few representative temperatures in Fig. 5.03) and it became an important challenge to explain this spectrum or the corresponding energy density distribution $U_{Tv}(v, \theta)$ on theoretical grounds. The failure of classical attempts at an explanation and the success of Planck's quantum hypothesis marked the first confrontation between the germinal ideas of the quantum theory and the largely unreceptive scientific community.

5.02. Enumeration of Electromagnetic Modes; the Rayleigh–Jeans Result

The first important attempt to explain the spectral distribution of isothermal cavity radiation by direct appeal to the properties of electromagnetic waves in an enclosure was made by Rayleigh[7] in 1900 and later corrected, in a small but nevertheless important detail, by Jeans.[8] An equivalent analysis in modern terms will now be given.

Let the cavity under discussion be a rectangular parallelepiped of inner dimensions a_1, a_2, a_3, as shown in Fig. 5.04. In order to make

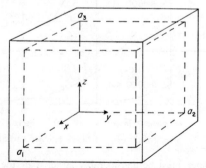

F<small>IG</small>. 5.04. Cavity in form of rectangular parallelepiped with perfectly conducting walls and inner dimensions a_1, a_2, a_3.

the analysis possible, perfectly conducting walls will be assumed at first; later a minor modification which overcomes the effects of this unrealistic assumption will be introduced and discussed. The radiation field inside this cavity is assumed to consist of an infinite number of electromagnetic standing waves or *modes of oscillation* similar to those found in microwave resonant cavities. The electric field **E** and the magnetic field **B** in the cavity interior obey the following five equations, namely Maxwell's equations for free space and the electromagnetic wave equation which is derivable therefrom:

$$\nabla \cdot \mathbf{E} = 0. \tag{5.001}$$

$$\nabla \cdot \mathbf{B} = 0. \tag{5.002}$$

$$\nabla \times \mathbf{E} = -\frac{\partial \mathbf{B}}{\partial t}. \tag{5.003}$$

$$\nabla \times \mathbf{B} = \frac{1}{c^2} \frac{\partial \mathbf{E}}{\partial t}. \tag{5.004}$$

$$\nabla^2 \mathbf{E} - \frac{1}{c^2} \frac{\partial^2 \mathbf{E}}{\partial t^2} = 0. \tag{5.005}$$

Here c is the speed of light in vacuum.

The applicable boundary conditions include, first of all, the well-known statement that the tangential components of E vanish on a perfectly conducting, i.e. perfectly reflecting, surface. On such a surface, then:

$$\mathbf{E} = e_n \mathbf{E}_n, \tag{5.006}$$

where e_n is the unit normal and $\mathbf{E}_n$ is the normal component of **E**. The second boundary condition is readily deduced from this and from equation (5.001); if n is perpendicular distance from the surface, it follows that:

$$\frac{\partial \mathbf{E}_n}{\partial n} = 0. \tag{5.007}$$

To summarize, the electric field at a perfectly conducting surface has only a normal component and the normal derivative of this component vanishes.

It is useful to assume a product solution of (5.005):

$$\mathbf{E}(\mathbf{r}, t) = \mathbf{R}(\mathbf{r})f(t); \quad f = f_0 \sin(\omega t + \varphi). \qquad (5.008)$$

The quantity $f(t)$, to which $\mathbf{E}$ is everywhere proportional, serves as the "coordinate" of the mode in much the same way that the quantity q was the "coordinate" for the LC circuit in Section 1.07. The following is the solution of (5.005) which has the form of (5.008) and at the same time satisfies the boundary conditions:

$$\left.\begin{array}{l} E_x = g_1 \cos(m_1\pi x/a_1) \sin(m_2\pi y/a_2) \sin(m_3\pi z/a_3)f; \\ E_y = g_2 \sin(m_1\pi x/a_1) \cos(m_2\pi y/a_2) \sin(m_3\pi z/a_3)f; \\ E_z = g_3 \sin(m_1\pi x/a_1) \sin(m_2\pi y/a_2) \cos(m_3\pi z/a_3)f. \end{array}\right\} \qquad (5.009)$$

Here g_1, g_2, and g_3 are the rectangular components of a unit vector $\mathbf{g}$, the direction of which is arbitrary within limits discussed below. The quantities m_1, m_2, and m_3 are non-negative integers, called mode numbers, which denote the number of half-cycles of field variation in each of the three coordinate directions. Substitution of (5.009) into (5.005) reveals the following condition which must be satisfied by ω and the three m_j:

$$(m_1\pi/a_1)^2 + (m_2\pi/a_2)^2 + (m_3\pi/a_3)^2 = \omega^2/c^2. \qquad (5.010)$$

From this it follows that the mode numbers and the dimensions of the cavity determine the frequency ν according to the relation:

$$\nu = \tfrac{1}{2}c[(m_1/a_1)^2 + (m_2/a_2)^2 + (m_3/a_3)^2]^{\tfrac{1}{2}}, \qquad (5.011)$$

as they do in any situation involving standing waves.

It is essential that the electric field satisfy (5.001) as well as (5.005). Substitution of the solution (5.009) into the former of these yields the following condition which must therefore be obeyed:

$$-[(m_1\pi g_1/a_1) + (m_2\pi g_2/a_2) + (m_3\pi g_3/a_3)]$$
$$\times \sin(m_1\pi x/a_1) \sin(m_2\pi y/a_2) \sin(m_3\pi z/a_3)f = 0. \qquad (5.012)$$

This will indeed be satisfied if at least one of the following is true:

$$\text{(i) One of the } m_j \text{ is equal to zero;} \qquad (5.013)$$

$$\text{(ii) } \frac{m_1 g_1}{a_1} + \frac{m_2 g_2}{a_2} + \frac{m_3 g_3}{a_3} = 0. \qquad (5.014)$$

By inspection of (5.009), it is obvious that no mode can exist if more than one of the m_j are equal to zero. However, if precisely one of these numbers is zero, as suggested by (i), it follows that only one of the three Cartesian components of $\mathbf{E}$ exists and that the mode is especially simple. Suppose, for example, that $m_3 = 0$; then:

$$\left.\begin{array}{l} E_x = 0; \qquad E_y = 0; \\[4pt] E_z = g_3 \sin(m_1 \pi x/a_1) \sin(m_2 \pi y/a_2) f; \\[4pt] \nu = \tfrac{1}{2}c[(m_1/a_1)^2 + (m_2/a_2)^2]^{\frac{1}{2}}. \end{array}\right\} \qquad (5.015)$$

The component g_3 is superfluous here; it may be set equal to unity which is tantamount to making g parallel to e_z. The field configuration at the time of maximum f is illustrated for two of these simple modes in Fig. 5.05.

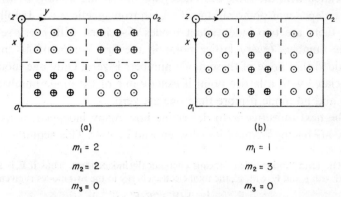

(a)

$m_1 = 2$
$m_2 = 2$
$m_3 = 0$

(b)

$m_1 = 1$
$m_2 = 3$
$m_3 = 0$

Fig. 5.05. Field lines at any observation plane parallel to the xy plane at time of maximum f for two simple cavity modes. Small circles with dots indicate electric lines pointed toward reader; with crosses, away from reader. Dashed lines indicate planes of zero E (nodal planes).

In the general case, none of the m_j is equal to zero and condition (ii) comes into play. To discuss this case efficiently, let a vector M be defined as follows:

$$M = e_x \frac{m_1}{a_1} + e_y \frac{m_2}{a_2} + e_z \frac{m_3}{a_3}.$$ (5.016)

This vector might be called the *mode vector* since it characterizes the mode by virtue of the m_j which appear in its components. With the aid of M and g, equations (5.011) and (5.014) may be expressed very succinctly:

$$\nu = \tfrac{1}{2}cM.$$ (5.017)

$$M \cdot g = 0.$$ (5.018)

The mode vector M is a lattice vector (a vector from the origin to a lattice point) in a space coordinatized by the components of M as illustrated in Fig. 5.06a. According to (5.018), the unit vector g must lie in a plane Q perpendicular to M as shown in Fig. 5.06b. A second unit vector g', also illustrated in Fig. 5.06b, is perpendicular to both M and g and represents a second independent[†] mode associated with the same M. These two modes are analogous to the two independent polarizations of an electromagnetic wave. It follows that there are two independent modes associated with every set of mode numbers (every lattice point in the first octant of M space) provided that none of these mode numbers is zero. Only one mode is associated with a lattice point if exactly one of the mode numbers is zero and no mode if more than one are zero.

The next objective is to determine how many independent modes there are having frequencies between ν and $\nu + \Delta\nu$. This acquires more

---

† The term "independent" means energetically independent. Thus if E is associated with g and E' with g', the total electric energy in the two modes is given by:

$$V_{\text{tot}} = \tfrac{1}{2}\varepsilon_0 \int (E^2 + 2E \cdot E' + E'^2) \, d\tau.$$

The term involving $E \cdot E'$ integrates to zero provided $g \cdot g' = 0$ whereupon the total energy becomes the sum of the individual energies. Similar results characterize the magnetic energy. Thus each of the two modes possesses its energy independently of the other and, in fact, independently of all other modes in the cavity.

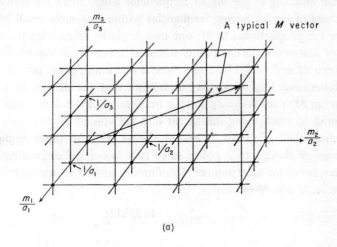

(a)

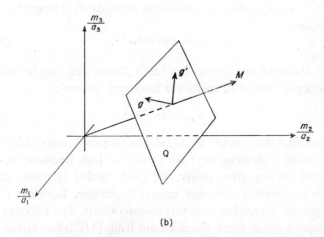

(b)

FIG. 5.06. The space of the mode vector **M** showing (a) lattice points determined by integral values of the mode numbers and (b) the plane Q perpendicular to **M** with the two mutually perpendicular unit vectors **g** and **g′**.

precise meaning at the higher frequencies where there are enormous numbers of modes having frequencies within extremely small bands. Since v is proportional to M, one may begin by calculating the number of first-octant lattice points whose distance from the origin lies between M and $M + \Delta M$. Since there is one lattice point per unit cell of dimensions $1/a_1$, $1/a_2$, $1/a_3$, it follows that the density of lattice points in M space is $a_1 a_2 a_3$ and the number of such points in question is found by multiplying this factor into the volume of an octant of a spherical shell of radius M and thickness ΔM. All but a negligible fraction of these lattice points have two associated electromagnetic modes, hence the total number ΔN of modes with M vector magnitudes between M and $M + \Delta M$ is:

$$\Delta N = 2a_1 a_2 a_3 \frac{4\pi M^2 \Delta M}{8}. \tag{5.019}$$

This becomes:

$$\Delta N = \pi \tau M^2 \Delta M, \tag{5.020}$$

where $\tau = a_1 a_2 a_3$ is the physical volume of the cavity. Using $M = 2v/c$, one obtains:

$$\Delta N = 8\pi \tau v^2 \Delta v c^{-3}. \tag{5.021}$$

Finally, the result can be expressed as a density denoting the number of modes per volume of cavity per frequency interval:

$$N_{\tau v} = 8\pi v^2 c^{-3}. \tag{5.022}$$

It is seen that the number of modes per frequency interval increases quadratically with frequency; the density at high frequency is enormous and, in any given cavity, the total number of modes having frequencies between zero and infinity is infinite. Equation (5.022) is a rigorous expression with permanent validity. The difficulty with the Rayleigh–Jeans result did not stem from (5.022) but rather from the fact that the then accepted method for predicting the distribution of energy among the various modes was incorrect.

In a cavity with perfectly conducting walls, the various modes are completely independent of one another. In such idealized circumstances, whatever modes are originally excited will persist indefinitely and whatever modes are originally unexcited will remain unexcited indefinitely. On the other hand, a very small bit of ordinary matter such as a grain of dust introduced into the cavity will couple the modes together however slightly and, given sufficient time, will erase all memory of the original pattern of excitation. Thus the coupling mechanism will absorb energy from some modes and deliver energy to others producing, finally, an equilibrium situation in which a certain fraction of the total energy is permanently associated with the modes in each frequency interval. The dust grain and the assembly of chaotically excited electromagnetic modes is a valid example of isothermal cavity radiation; since the dust grain is present, the perfect reflectivity of the remaining wall surface is of no consequence. The problem of finding the equilibrium distribution of energy among the various modes will now be considered.

The magnetic field of an electromagnetic cavity mode may be found from the known electric field by invoking the Maxwell equation (5.003). Thus one first finds $-\nabla \times \mathbf{E}$ from (5.009) and then integrates this with respect to time. Since $\int f \, dt = -\omega^{-2} \dot{f}$, the result is proportional to $\dot{f}$:

$$
\left.
\begin{aligned}
\mathbf{B}_x &= \pi(\mathbf{M}\times\mathbf{g})_x \omega^{-2} \sin{(m_1\pi x/a_1)} \cos{(m_2\pi y/a_2)} \cos{(m_3\pi z/a_3)} \dot{f}; \\
\mathbf{B}_y &= \pi(\mathbf{M}\times\mathbf{g})_y \omega^{-2} \cos{(m_1\pi x/a_1)} \sin{(m_2\pi y/a_2)} \cos{(m_3\pi z/a_3)} \dot{f}; \\
\mathbf{B}_z &= \pi(\mathbf{M}\times\mathbf{g})_z \omega^{-2} \cos{(m_1\pi x/a_1)} \cos{(m_2\pi y/a_2)} \sin{(m_2\pi z/a_3)} \dot{f}.
\end{aligned}
\right\} \quad (5.023)
$$

With reference to Section 1.07, the potential and kinetic energies of the fields in the cavity can be respectively identified as follows:

$$
\left.
\begin{aligned}
V &= \tfrac{1}{2}\varepsilon_0 \int_0^{a_3}\int_0^{a_2}\int_0^{a_1} \mathbf{E}^2 \, dx \, dx \, dz; \\
T &= \tfrac{1}{2}\mu_0^{-1} \int_0^{a_3}\int_0^{a_2}\int_0^{a_1} \mathbf{B}^2 \, dy \, dy \, dz.
\end{aligned}
\right\} \quad (5.024)
$$

Here ε_0 and μ_0 are the usual MKSA parameters representing the permittivity and permeability of free space, respectively. Since $|\mathbf{M}\times\mathbf{g}| =$

9*

$= M = \omega/\pi c$ and $\varepsilon_0 = 1/\mu_0 c^2$, it is a simple matter to obtain the following results for the case[†] at hand:

$$V = \tfrac{1}{16}\varepsilon_0\tau f^2 = \tfrac{1}{16}\varepsilon_0\tau f_0^2 \sin^2(\omega t+\varphi); \left.\vphantom{\begin{matrix}a\\b\end{matrix}}\right\}$$
$$T = \tfrac{1}{16}\varepsilon_0\tau\omega^{-2}\dot{f}^2 = \tfrac{1}{16}\varepsilon_0\tau f_0^2 \cos^2(\omega t+\varphi).$$
$$(5.025)$$

As might be expected, the total energy of the mode has the constant value $\tfrac{1}{16}\varepsilon_0\tau f_0^2$. The potential and kinetic energies have exactly the same relationship to f and $\dot{f}$ as they have to x and to $\dot{x}$, respectively, in a harmonic oscillator and the electromagnetic mode is, literally, such an oscillator. Thus the Lagrangian for the mode may be written:

$$\mathscr{L} = \tfrac{1}{16}\varepsilon_0\tau(\omega^{-2}\dot{f}^2 - f^2). \tag{5.026}$$

Continuing, one may define the momentum p_f conjugate to f as follows:

$$p_f = \frac{\partial\mathscr{L}}{\partial\dot{f}} = \frac{1}{8}\varepsilon_0\tau\omega^{-2}\dot{f}. \tag{5.027}$$

The Hamiltonian, finally, is given by:

$$\mathscr{H} = 4\omega^2\varepsilon_0^{-1}\tau^{-1}p_f^2 + \tfrac{1}{16}\varepsilon_0\tau f^2. \tag{5.028}$$

The Hamiltonian derived in the preceding paragraph is the key to Rayleigh's and Jeans' result. It is a well-known prediction of classical statistical mechanics — the only statistical mechanics in existence at the time — that if a number of systems are in thermal equilibrium at temperature θ, the average energy of a given system is $\tfrac{1}{2}k\theta$ times the number of quadratic terms in its Hamiltonian,[9] k being Boltzmann's constant. This is the celebrated principle of *equipartition of energy* which had proved to be valid in many instances. According to this principle, every electromagnetic mode in the cavity should assume an average energy equal to $k\theta$. Rayleigh and Jeans therefore arrived at the following expression for the energy density in the

---

[†] This result is valid when none of the mode numbers is zero. If one of the mode numbers is zero, then $\tfrac{1}{16}$ must be replaced by $\tfrac{1}{8}$ in (5.025). This is a very minor point and has no effect upon the conclusions of this section.

cavity per frequency interval:

$$U_{\tau\nu} = 8\pi\nu^2 c^{-3} k\theta. \tag{5.029}$$

As the parabolic appearance of the curves of Fig. 5.03 at small ν would indicate, this expression does agree with experiment at the low frequencies but it yields values for the energy density that are much too high at the higher frequencies. Upon integration over all frequencies in $0 \leqslant \nu < \infty$ it predicts an infinite energy density for all $\Theta > 0$; if valid, any finite portion of the vacuum would be an infinitely efficacious heat sink capable of absorbing arbitrarily large quantities of energy without undergoing a temperature rise. It is evident that the equipartition principle is somehow at fault. The correct expression for the average energy per mode must eventually become a very rapidly decreasing function of frequency in order that $U_{\tau\nu}$ can have a dependence upon ν like that illustrated in Fig. 5.03.

5.03. Planck's Quantum Hypothesis

Planck, also in the year 1900, presented a law for the spectrum of black surface radiation which did agree with experiment in all measurable details.[10] Although originally derived on the basis of a model of an isothermal cavity with walls composed of harmonically vibrating "atoms" coupled to one another via electromagnetic radiation in the hollow space, his result can also be obtained from the model used in the previous section, i.e. from an assembly of weakly coupled electromagnetic modes in a cavity with otherwise perfectly reflecting walls. The latter model, which is more familiar and convenient, will be employed in the present discussion.

Consider a small frequency interval between ν and $\nu + \Delta\nu$. The number of modes with frequencies in this interval is an integer ΔN which, at large ν, is well represented by the expression $8\pi\tau\nu^2\Delta\nu c^{-3}$. Suppose that, at any given time, these modes collectively possess a total energy ΔU and, in order to permit a combinatorial analysis of the situation, let ΔU be equal to $n\varepsilon$ where n is a non-negative integer and ε is a small

energy unit whose size is fixed for the frequency interval under consideration. The number of meaningfully different ways w in which these
n energy units can be distributed among the ΔN modes must now be
determined. The problem is similar to that of determining the number
of meaningfully different ways in which n dollars can be distributed
among ΔN bank accounts since a dollar, like an energy unit, is in
principle indistinguishable from other dollars when several are put
together to form a sum. The value of w can be derived by considering
an assembly of $n+\Delta N-1$ objects, of which n are energy units and
$\Delta N-1$ are imaginary separators dividing the set of energy units into
ΔN disjoint subsets. The situation is pictured schematically in Fig.
5.07 in which the open circles represent energy units and the vertical

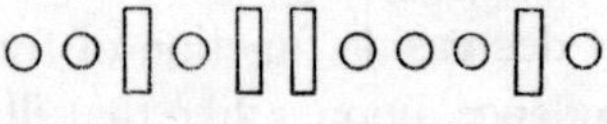

Fig. 5.07. Schematic model for calculating the number of ways of
distributing n energy units among ΔN modes. In this illustration,
$n = 7$ and $\Delta N = 5$. The seven open circles symbolize energy units
and the four vertical bars are separators which divide the set of
circles into five disjoint subsets. Here the populations of the subsets
are (from left to right) 2, 1, 0, 3, and 1, respectively.

bars represent separators. The total number of permutations of all
the objects is $(n+\Delta N-1)!$ but, of these, both the $n!$ permutations of
the energy units and the $(\Delta N-1)!$ permutations of the separators are
without physical significance. It follows that w is given by:

$$w = \frac{(n+\Delta N-1)!}{n!\,(\Delta N-1)!},$$ (5.030)

and if one considers all the frequency intervals Δv_1, Δv_2, Δv_3, ... in
$0 \le v < \infty$, the total number of ways of making all the various distributions of energy units is:

$$W(n_i) = \prod_i \frac{(n_i+\Delta N_i-1)!}{n_i!\,(\Delta N_i-1)!}.$$ (5.031)

As already mentioned, the ΔN_i in the above expression have a fixed
relationship to the corresponding Δv_i; they are therefore to be regard-

ed as constants. The n_i, on the other hand, should be regarded as variables and therefore W may be thought of as $W(n_i)$. The coupling mechanism permits the exchange of energy between modes in one frequency interval and those in another, i.e. it permits changes in the various values of the n_i subject only to the conservation of total energy:

$$U(n_i) = \sum_i n_i \varepsilon_i = U_0. \qquad (5.032)$$

The set of equilibrium values of the n_i is the set of most probable values, namely that set for which W is maximum. The procedure for finding this maximizing set is given in Appendix C and the result, denoted n_i', is expressed by the following formula:

$$n_i' = \frac{\Delta N_i}{(\exp \beta \varepsilon_i) - 1}. \qquad (5.033)$$

According to Appendix C, β can theoretically be determined from the amount of total energy U_0; as will be seen, however, it is much easier to determine this quantity through its relationship to the temperature, a relationship which becomes apparent in the limit of small ε_i. From (5.033), one may write for the average energy per mode in the ith frequency interval:

$$\bar{u}_i = \frac{n_i' \varepsilon_i}{\Delta N_i} = \frac{\varepsilon_i}{(\exp \beta \varepsilon_i) - 1}. \qquad (5.034)$$

As mentioned earlier, the motivation for introducing energy units in the first place was "to permit a combinatorial analysis". Now that an expression for the average energy has been derived, one might expect to let $\varepsilon_i \to 0$ since there is no apparent reason for retaining the rather artificial energy units. When this step is taken, however, the result is disappointing since:

$$\lim_{\varepsilon_i \to 0} \bar{u}_i = \lim_{\varepsilon_i \to 0} \frac{\varepsilon_i}{(\exp \beta \varepsilon_i) - 1} = \frac{1}{\beta}, \qquad (5.035)$$

which indicates that the average energy per mode is a constant and that energy is partitioned equally among the modes as the Rayleigh-

Jeans result had asserted. Planck reluctantly adopted the point of view that the size of the energy units should not be zero but rather should be small at the low frequencies where equipartition of energy does in fact occur and large at the high frequencies where it does not. Accordingly, he decided to let the size of the energy units be proportional to frequency and wrote:

$$\boxed{\varepsilon_i = h\nu_i} \, , \tag{5.036}$$

where h is a constant of proportionality with the dimensions of energy times time, i.e. of action. On this viewpoint, the electromagnetic modes in the cavity do not gain or lose energy continuously as would be expected but do so rather in discrete units or *quanta*, the size of which depends upon the frequency. In this way, the quantum hypothesis was first introduced into physics as an *ad hoc* principle without which it was impossible to arrive at a successful theory of isothermal cavity radiation.

To continue with the derivation, one may observe that in the limit of low frequency (5.036) asserts that $\varepsilon_i \rightarrow 0$. Then there is indeed equipartition of energy and $\bar{u}_i$ must equal $k\theta$. It follows from (5.035) that:

$$\frac{1}{\beta} = k\theta \tag{5.037}$$

and the identification of β is thus accomplished. When this and (5.036) are substituted into (5.035), a more informative expression for the average energy per mode is obtained. This expression is:

$$\bar{u}_i = \frac{h\nu_i}{(\exp h\nu_i/k\theta) - 1} \, . \tag{5.038}$$

The subscript i is really not needed here; the average energy per mode is simply a function of frequency and temperature as indicated. Upon multiplication by the number of modes per frequency interval per volume of cavity, given by (5.022), the famous Planck law for the energy density of isothermal cavity radiation is obtained:

$$U_{\tau\nu} = \frac{8\pi\nu^2}{c^3} \frac{h\nu}{(\exp h\nu/k\theta) - 1}.$$
(5.039)

By adjusting the value of h, now known as *Planck's constant*, this formula can be fitted to an experimentally determined black surface spectrum. Thus did Planck conclude that $h = 6.55 \times 10^{-34}$ joule second, a result remarkably close to the modern value of $6.626196 \times \times 10^{-34}$ joule second.[11]

5.04. The Photoelectric Effect

The photoelectric effect is another phenomenon which was carefully investigated during the era treated in this chapter and which also proved to be unexplainable without the use of quantum concepts. The effect may be defined as the ejection of electrons from a metallic sample in response to irradiation by light; its precise observation

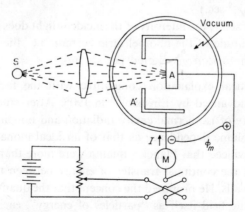

FIG. 5.08. Diagram of photoelectric apparatus. Cup-shaped electrode A′ collects electrons ejected from A. Symbol with plus and minus signs defines φ_m as the potential of the conductor leading to A minus the potential of the conductor leading to A′.

requires a clean surface in a vacuum environment and a source of monochromatic light, e.g. the surface A and the source S illustrated in Fig. 5.08. Electrons ejected with sufficient kinetic energy will overcome the electrostatic field maintained between A and the cup-shaped electrode A′ and will register as a current on the sensitive meter M. By gradually making A′ more negative with respect to A and measuring φ_m, the potential difference at which the current stops, it is possible to infer T_{max}, the maximum kinetic energy of the ejected electrons. By 1903 the photoelectric effect had been investigated by a few experimenters and the following facts were known:

(i) There is a threshold frequency ν_0, dependent only upon the material of the irradiated electrode, which must be exceeded by ν, the frequency of the incident light, if any photoelectrons are to be ejected.

(ii) T_{max} is related only to the frequency difference $\nu - \nu_0$ and not to the intensity of the incident light.

(iii) The time delay between the application of light and the ejection of electrons is extremely short. (In 1903 this time delay was known to be less that 10^{-3} sec; it is now known to be less than 3×10^{-9} sec.)

(iv) For fixed ν, the intensity of the incident light does determine the magnitude of the photoelectric current, i.e. the total number of photoelectrons ejected per second.

A theoretical explanation compatible with the facts mentioned above was advanced by Einstein[12] in 1905. After studying the entropy density of isothermal cavity radiation and noticing that it had the same volume dependence as that of an ideal monatomic gas, he became convinced that Planck's quanta were more than just units of exchange in the continual transfer of energy between radiation field and cavity walls. He proposed the concept that the quanta are actually present in the field itself as "particles of energy", each with energy content $h\nu$, which are emitted, transmitted, and absorbed as indivisible entities. This viewpoint did not repudiate the highly sucessful Maxwellian theory of electromagnetic fields but suggested that field phenomena are in some sense an averaged description of the behavior of the more fundamental energy particles or *photons* as they came to be called at a much later date. The concept is therefore dualistic in that it attributes both wave and particle aspects to light; bizarre though it is, it has proved fruitful in hundreds of situations and the term "photon" has become an indispensable part of the vocabulary of physics. Einstein described the photoelectric effect as an interaction in which a photon from the radiation field is absorbed by one of the electrons in the metal; the electron in question then experiences an energy increase of $h\nu$. Recognizing that even the most energetic of the internal electrons must overcome a potential barrier as it comes out of the surface, he wrote his celebrated PHOTOELECTRIC EQUATION which, in present notation, is:

$$T_{max} = h\nu - e\varphi_w \qquad (5.040)$$

Here e is the magnitude of the charge of the electron and φ_w, which has the dimensions of volts, is called the *work function* of the material in question;[†] $e\varphi_w$ is the height in energy units of the barrier mentioned above. At the threshold frequency ν_0, the photon energy is just equal to this barrier height and at this frequency $T_{max} = 0$. Evidently ν_0 is equal to $e\varphi_w/h$ and the photoelectric equation can be rewritten in the form:

$$T_{max} = h(\nu - \nu_0), \qquad (5.041)$$

which agrees with (ii). The relationship between T_{max} and the measured potential difference φ_m at which the current just stops must be examined carefully if A and A′ are made of dissimilar metals, a situation which is recommended for experimental reasons.[13] One has:

$$T_{max} = e \int_{A}^{A'} \mathbf{E} \cdot d\mathbf{l} = e[\varphi_m + \varphi'_w - \varphi_w]. \qquad (5.042)$$

The integral in this equation involves the electrostatic field alluded to

[†] For most metals, φ_w is of the order of 1 to 6 volts.

earlier and is taken from a point just outside the surface of A to a similar point just outside the surface of A'. It is equal to the conventionally measured voltage[†] φ_m plus the work function of A' minus the work function of A and is often referred to as the "true stopping potential". The values of the work functions are not determinable with high accuracy but this is unimportant in the present context. What is important is that the right-hand sides of (5.041) and (5.042), when equated, yield a linear relationship between φ_m and ν. Thus:

$$e \, \Delta\varphi_m = h \, \Delta\nu, \qquad (5.043)$$

and the slope $\Delta\varphi_m/\Delta\nu$, which is accurately determinable, gives the quotient h/e.

The photoelectric effect was subjected, over a period of years, to an exhaustive experimental investigation by Millikan[14] using ingenious techniques and invoking his own previously determined value of the electronic charge. In spite of Millikan's personal disinclination to believe in the photon concept, the outcome was a complete vindication of Einstein's equation. As a consequence, the fundamental character of Planck's constant was more firmly established and the credibility of a radiation field composed of energy quanta was enhanced.

It should be pointed out that the crystal lattice of the photoemitter is an indispensable participant in the photoelectric process. If the lattice were not involved, it would be impossible to explain the delivery of the total photon energy to the ejected electron or to explain the fact that the momentum of this electron can be opposite in direction to that of the incident photon. (An interaction between a photon and an essentially isolated electron is called a Compton event and has a very different outcome, as will be seen in Section 5.06.) An analysis of the actual mechanism of the photoelectric interaction is an advanced

[†] The voltage φ_m is the potential of the *Fermi level* of A minus the potential of the *Fermi level* of A'; it may well be a negative quantity. (For present purposes, the Fermi level can be defined as the highest energy level within a metal, at low or moderate temperatures, at which a copious supply of electrons exists.)

problem and the photoemissive properties of materials have been studied extensively in order to gain greater understanding of the solid state.[15]

5.05. Bohr's Explanation of the Hydrogen Spectrum

In the latter part of the nineteenth century, one of the most important challenges in physics was that of understanding the origin of the discrete frequencies (lines) in the spectrum of light emitted by atoms under suitable conditions of excitation. The enormous jumble of lines in most spectra made the problem formidable indeed and early efforts were aimed at cataloging these lines in the hope of discovering some regularities among them; since wavelength was directly measurable and frequency was not, it was inevitable that experimental results and theoretical speculation in this area would, initially at least, be expressed in terms of wavelength.

In 1885 Balmer discovered a numerical law relating the wavelengths of the four most easily visible lines in the hydrogen spectrum and predicted the existence of other lines in the same family or *series*; this prediction was subsequently verified. The law in question, written in modern terminology, is as follows:

$$\frac{1}{\lambda_{n2}} = R\left(\frac{1}{2^2} - \frac{1}{n^2}\right); \qquad n = 3, 4, 5, \ldots, \infty. \qquad (5.044)$$

The significance of the subscript $n2$ will become evident as the discussion proceeds. The constant R was named after Rydberg who also worked extensively in this area and who found similar formulas for other atomic species; the value of this constant which best fitted the experimental data was approximately 1.097×10^7 meter^{-1}. It may be noticed that (5.044) predicts an infinite number of lines forming what is now called the "Balmer series". With increasing n, the lines of the series become ever more densely spaced as they approach a limiting wavelength $\lambda_{\infty 2} = 3646 \times 10^{-10}$ meter or, as it is often written, 3646 Å.

In 1906 Lyman discovered a series in the ultraviolet region of the hydrogen spectrum with wavelengths given by:

$$\frac{1}{\lambda_{n1}} = R\left(\frac{1}{1^2} - \frac{1}{n^2}\right); \qquad n = 2, 3, 4, \ldots, \infty \qquad (5.045)$$

and, in 1908, Paschen discovered still another series, this time in the infrared:

$$\frac{1}{\lambda_{n3}} = R\left(\frac{1}{3^2} - \frac{1}{n^2}\right); \qquad n = 4, 5, 6, \ldots, \infty. \qquad (5.046)$$

Other series in the spectrum of hydrogen have since been found with wavelengths which, following the same scheme of designation, would be written λ_{n4}, λ_{n5}, etc. Thus it is apparent that the wavelengths of all the lines in this spectrum can be written in terms of two integers, n and n', as follows:

$$\frac{1}{\lambda_{nn'}} = R\left(\frac{1}{n'^2} - \frac{1}{n^2}\right); \qquad n = n'+1, \quad n'+2, \quad n'+3, \ldots \infty;$$

$$n' = 1, 2, 3, \ldots, \infty. \qquad (5.047)$$

At about this time Ritz enunciated the proposition, now called the *Ritz combination principle*, that the reciprocal wavelengths in the spectrum of any given element can be expressed as differences of quantities called *spectral terms*. Since there are many fewer terms than lines, a large amount of information about a spectrum can be recorded very economically by simply tabulating the values of its terms. It is convenient to regard spectral terms as negative quantities for a reason that will soon become apparent; on this basis, the hydrogenic terms (which are particularly simple functions of an index number) may be written:

$$\tau_n = -R/n^2. \qquad (5.048)$$

The reciprocal wavelengths may therefore be expressed as:

$$\frac{1}{\lambda_{nn'}} = \tau_n - \tau_{n'}, \qquad (5.049)$$

which agrees with (5.047). Notice that the first subscript refers to the algebraically larger (less negative) term and the second to the algebraically smaller (more negative) term. Since it is always reciprocal wavelengths which appear in these formulas, it is apparent that the physical quantity really involved is the frequency. The latter is exhibited if (5.049) is multiplied on each side by c:

$$\nu_{nn'} = c\tau_n - c\tau_{n'}. \tag{5.050}$$

Several of the terms and related lines of the hydrogen spectrum are shown in Fig. 5.09. In this figure, the terms are represented by horizontal marks and the spectral lines are symbolized by arrows originating on the larger term and terminating on the smaller term.

Not long after these discoveries, Rutherford[16] investigated the scattering of α-particles by atoms in a thin foil. The occasional occurrence of large scattering angles led to the inevitable conclusion that almost the entire mass of the atom is concentrated in a central body of very small dimensions called the nucleus. The electrons, which are relatively very light, were presumed to move around the nucleus with a sort of planetary motion. In spite of the success of this atomic model in explaining the scattering of highly energetic charged particles, it did not account, in any obvious way, for the spectrum produced by even the simplest atom, namely the hydrogen atom with its one planetary electron. Classically, one would expect that the electromagnetic radiation from an electric charge in cyclic motion would consist of a fundamental frequency and, possibly, a set of higher frequencies (harmonics) which are multiples thereof. In the spectra of hydrogen and of other elements, however, the distribution of frequencies does not even approximately fit this description. Classical physics would also predict the eventual collapse of the atom as the electron radiates away its energy and falls into the positive nucleus but the high final frequency and large total energy output necessarily associated with such a collapse are never observed and atoms are found to be stable for arbitrarily long periods of time either singly or as constituents of molecules or solids.

It was Bohr[17] who first produced an even partially satisfactory

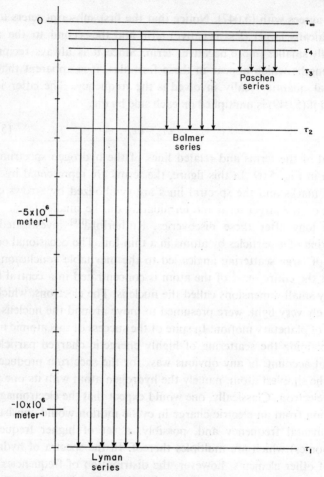

FIG. 5.09. Diagram showing basic positions of several of the spectral terms (τ_n) for hydrogen. Spectral lines in three well-known series are indicated by arrows. Frequency of line is proportional to length of arrow.

explanation for the hydrogen spectrum; his explanation, although not the final answer, permitted an actual calculation of the hydrogenic spectral terms and was a very important step forward for its time. Taking inspiration from Planck and Einstein, Bohr reasoned that the radiation from an atom might also consist of energy quanta (photons) and that in the production of a spectral line, say that of frequency $v_{nn'}$, large numbers of atoms in the luminous sample emit photons of energy $hv_{nn'}$. On multiplication of (5.050) by h, it is seen that this photon energy is the difference of two energies which can be associated with the two spectral terms:

$$hv_{nn'} = hc\tau_n - hc\tau_{n'} = E_n - E_{n'}. \tag{5.051}$$

Bohr identified these energies, E_n and $E_{n'}$, with two different levels of the total energy of electron motion in the atom. Thus he was led to postulate that a hydrogenic term of index n corresponds to a definite state of the atom in which the electron has total orbital energy E_n given by:

$$E_n = hc\tau_n = -Rch/n^2. \tag{5.052}$$

A state of this type is called a *stationary state*. Bohr further postulated that, in a stationary state, the electron pursues a classical orbit without the emission of electromagnetic radiation and that the atom can therefore remain in such a state for an indefinite period of time. The actual process of radiation, i.e. the emission of energy quantum $hv_{nn'}$, was considered by Bohr to be a *transition process not describable by classical physics* in which the atom passes from its initial state of energy E_n to a final state of lower energy $E_{n'}$. The state of lowest energy, namely E_1, is the *ground state*. An atom with energy E_1 can make no further transitions and, if left alone, will remain in the ground state perpetually; it cannot again emit electromagnetic radiation unless it is once more raised to a higher state (an excited state) by some external agent, e.g. by collision with another atom.

The most ingenious feature of the Bohr theory was not the enunciation of these postulates but rather the calculation of the values of the energy levels of hydrogen or, if one prefers, the theoretical determination of the constant R. The following is a modern adaptation of Bohr's analysis which was remarkable in that so much was obtained from limited initial information.

Let the electron orbit in the nth stationary state be a circular orbit of radius r_n. Equating electric force to centripetal force, one obtains:

$$\frac{e^2}{4\pi\varepsilon_0 r_n^2} = mr_n\omega_n^2. \qquad (5.053)$$

Here m is the mass of the electron and ω_n is its orbital angular velocity.[†] The kinetic energy is $\frac{1}{2}mr_n^2\omega_n^2$ which, with the aid of (5.053), may be written as $e^2/8\pi\varepsilon_0 r_n$. The potential energy is $-e^2/4\pi\varepsilon_0 r_n$ and therefore the total energy, which is negative, is given by:

$$E_n = T_n + V_n = \frac{e^2}{8\pi\varepsilon_0 r_n} - \frac{e^2}{4\pi\varepsilon_0 r_n} = -\frac{e^2}{8\pi\varepsilon_0 r_n}. \qquad (5.054)$$

One may now eliminate r_n between (5.053) and (5.054) to obtain the following formula for ω_n directly in terms of E_n or of R and n:

$$\omega_n = \frac{4\pi\varepsilon_0(-2E_n)^{\frac{3}{2}}}{m^{\frac{1}{2}}e^2} = \frac{4\pi\varepsilon_0(2Rch)^{\frac{3}{2}}}{m^{\frac{1}{2}}e^2n^3}. \qquad (5.055)$$

Consider now a stationary state of very large n. By (5.052), the corresponding E_n must be very close to zero (in other words, very high), and, by (5.054), the corresponding r_n must be quite large. An electron in a large circular orbit is comparable to an electron moving on any gently curving path through, say, some macroscopic piece of vacuum apparatus and one intuitively feels that all aspects of such a situation should be in accord with the predictions of classical physics, i.e. the electron should very gradually spiral inward toward the nucleus all the while emitting rather weak radiation of ever increasing frequency and intensity. At any time, *the instantaneous frequency of this radiation*

[†] Like the planetary analyses of Sections 1.05 and 2.03, this is based upon the assumption that the central body (nucleus) has a relatively infinite mass and is therefore motionless. To make a more accurate analysis, the method of Section 12.01 must be used; if this method is employed, the only change is the replacement of the electron mass m by the "reduced mass" μ in all equations. If M stands for the nucleus, $\mu = Mm/(M+m)$, and, for hydrogen, is smaller than m by about six parts in 10,000.

should be equal to the instantaneous orbital frequency of the electron.
According to the quantization concept, however, the electron is really
making one small transition after another, passing successively from
the nth state to the $n-1$st state to the $n-2$nd state, etc. At large n,
the states are very closely spaced and this sequence of small transitions
creates the illusion that the phenomenon is as described classically.
The thinking employed here is an application of a very powerful
principle now known as *Bohr's correspondence principle* which states
that in circumstances where a given system may be regarded as classical, the behavior predicted by quantum theory must be the same as
that expected under the laws of classical physics.

To exploit the ideas of the preceding paragraph quantitatively, one
may notice that the frequency of the observed radiation in a transition
from state n to state $n-1$ is given by:

$$v_{n,\,n-1} = \frac{E_n - E_{n-1}}{h} = Rc\left[\frac{1}{(n-1)^2} - \frac{1}{n^2}\right] = \frac{Rc(2n-1)}{n^2(n-1)^2}. \quad (5.056)$$

In the limit of large n, this becomes:

$$\lim_{n\to\infty} v_{n,\,n-1} = \frac{2Rc}{n^3}. \quad (5.057)$$

This frequency is supposed to be equal to the average orbital frequency
of the electron, i.e. approximately equal to the orbital frequency in
either the initial or the final state since there is very little difference
between the two. One may therefore set $2Rc/n^3$ equal to $\omega_n/2\pi$; both
expressions are proportional to n^{-3}:

$$\frac{2Rc}{n^3} = \frac{2\varepsilon_0(2Rch)^{\frac{3}{2}}}{m^{\frac{1}{2}}e^2 n^3}. \quad (5.058)$$

After cancelling n^{-3} this equation can be solved for R in terms of
fundamental constants. Thus is achieved a major objective of the
analysis:

$$R = \frac{me^4}{8\varepsilon_0^2 ch^3}. \quad (5.059)$$

Utilizing this result,[†] E_n and r_n may be similarly expressed with the aid of the index number (quantum number) n:

$$E_n = -\frac{me^4}{8\varepsilon_0^2 h^2 n^2} . \tag{5.060}$$

$$r_n = \frac{\varepsilon_0 h^2 n^2}{\pi m e^2} . \tag{5.061}$$

Substitution of the value for F_n into (5.055) yields ω_n in the same terms:

$$\omega_n = \frac{\pi m e^4}{2\varepsilon_0^2 h^3 n^3} . \tag{5.062}$$

Since the angular momentum of the electron in the nth orbit is $mr_n^2\omega_n^2$, one finds for this significant quantity a very simple expression:

$$L_n = n\frac{h}{2\pi} . \tag{5.063}$$

Bohr recognized the importance of this result and elevated it to the status of a universal postulate in the latter part of his first 1913 paper. Contrary to what is often believed, however, he did not assume it at the outset but arrived at it by the arguments which, in substance, have been reproduced here.

The central concept of Bohr's work is the stationary state which may be defined as a hypothetical privileged state of motion, exempt from the requirement that accelerating charged particles must radiate but otherwise strictly obedient to the well-known laws of classical mechanics. In view of the initial success of this concept, it was inevitable that research efforts in the years immediately following 1913 should attempt to delineate it more precisely, i.e. to express with the utmost possible generality the "quantum conditions" which characterize a stationary state and distinguish it from all other types of motion. These conditions were necessarily given in the form of postu-

---

[†] This quantity is properly designated R_∞ since it is based on the assumption of an infinite nuclear mass; its value is 10973731.2 meter^{-1}. The Rydberg constant for hydrogen, $R_H = \mu e^4/8\varepsilon_0^2 ch^3$, is equal to 10967758 meter^{-1}; see ref. 11.

lates to be verified by subsequent comparisons between theoretical deductions and spectroscopic observations. They were independently enunciated by Ishiwara,[18] Wilson,[19] and Sommerfeld;[20] Sommerfeld, however, worked extensively on their theoretical implications and is given major credit for their development.

A Bohr–Sommerfeld stationary state is a bound state in which each of the D free coordinates is involved in oscillatory or rotatory motion. These motions in general have different periods but, in particular cases, some or all may have the same period. The universal quantum conditions assert, for each degree of freedom individually, that:

$$\boxed{\oint p_j \, \mathrm{d}q_j = n_j h} \,, \tag{5.064}$$

where n_j is a quantum number that can assume only integer values. The quantity $\oint p_j \, \mathrm{d}q_j$ is taken over a complete period of q_j and is called a *phase integral*. The D-phase integrals are constants of the motion and are equal to the canonical momenta P_j in a regime[†] in which the transformed Hamiltonian contains, in general, all of these quantities but none of the Q_j. The latter are then the linear functions of time, $Q_j = (t/\varkappa_j) + \xi_j$, where the $\varkappa_j$ are the periods alluded to above. Bohr's angular momentum rule (5.063) is easily recognized as a special case of (5.064) in which q is the angular coordinate φ and p is its conjugate momentum L. The latter, by virtue of the quantum condition, must take one of the discrete values L_n.

The discovery and exploitation of the universal quantum conditions brought to a culmination what is now known as the "old quantum theory". This theory, although rich in conceptual content[‡] and impressive in some of its accomplishments, was destined to be only a temporary expedient. In the hands of Sommerfeld, who analyzed the hydrogenic states relativistically, it yielded a numerically correct explanation for the splitting of certain of the spectral terms into a number of different but closely spaced levels, a phenomenon known

---

† In this regime, the P_j and the Q_j are collectively called *action and angle variables*.

‡ See ref. 6, especially chapter 5.

as the *fine structure*. This explanation, however, was later shown to be fortuitous. The reasons for the inadequacy of the Bohr–Sommerfeld theory will become clear when the Schrödinger theory of the hydrogenic atom is studied in Chapter 12. It will then be seen that Bohr's quasi-classical stationary states are but obscure foreshadowings of the actual stationary states of wave mechanics.

In spite of its ephemeral status, the attitude of physicists toward the work of Bohr and Sommerfeld has been most appreciative. Judged in reference to its time, when quantum research was largely a groping for *ad hoc* hypotheses with which to explain the origin and details of spectra, it was a great achievement. Furthermore, it did much to pave the way for future developments.

5.06. The Compton Effect

Further evidence for the existence of energy quanta in electromagnetic radiation was provided in a very graphic way by the scattering of X-rays from relatively weakly bound electrons. Under such conditions, the scattered X-ray undergoes a frequency decrease (wavelength increase) which depends upon the angle of scattering and suggests that a transfer of energy to the electron has taken place. In 1923 Compton[21] observed this phenomenon and was able to interpret it with the aid of the photon concept.

The Compton effect cannot be understood without reference to the special theory of relativity. According to that theory, if a particle has rest mass m_0, moving mass m, velocity v, momentum p, and total (rest mass plus kinetic) energy[†] U, the following relations obtain:

$$m = m_0(1 - v^2/c^2)^{-\frac{1}{2}};\qquad (5.065)$$

$$p = mv;\qquad (5.066)$$

$$U = mc^2.\qquad (5.067)$$

[†] The sum of the rest mass and kinetic energies is usually denoted "E". To avoid confusion with $T + V$, which would be especially troublesome in Chapter 6, the symbol "U" has been adopted instead.

These formulas are well known and may be found in most elementary physics texts. If (5.065) is multiplied on both sides by $c^2(1 - v^2/c^2)^{\frac{1}{2}}$ and the resulting equation is then squared, one finds that:

$$m^2c^4 - m^2v^2c^2 = m_0^2c^4. \tag{5.068}$$

This is easily seen to be:

$$U^2 - (pc)^2 = U_0^2, \tag{5.069}$$

where $U_0 = m_0c^2$ is that part of the total energy of the particle which is associated with its rest mass. According to equation (5.069), the quantities U_0 and pc can be thought of as the sides of a right triangle having U as its hypotenuse. An electron has rest mass and a typical triangle diagram for it, in which all quantities have been given the subscript e, is illustrated in Fig. 5.10a. (The line segment representing

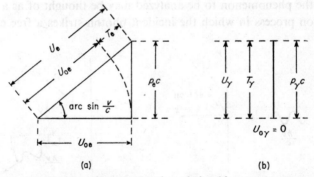

FIG. 5.10. Triangle diagrams showing relationships among rest mass energy U_0, total energy U, kinetic energy T, and momentum p. Diagram (a) is for an electron, (b) for a photon.

the kinetic energy T_e of the elecron should be noticed.) A photon, on the other hand, should have a speed equal to the speed of light and its rest mass should therefore be zero according to (5.068). Using the subscript γ for such a particle, it is seen that $U_\gamma = p_\gamma c$ and that the triangle degenerates to a single vertical line as in Fig. 5.10b. The energy U is entirely kinetic since there is no rest mass energy; according to Sections 5.03 and 5.04, it is given by:

$$U_\gamma = h\nu. \qquad (5.070)$$

It follows that the momentum of the photon is, in magnitude,

$$p_\gamma = h\nu/c. \qquad (5.071)$$

Knowledge of these simple mechanical attributes of electron and photon is essential in the analysis which follows.

In Compton scattering, the energy $h\nu$ of the incident photon is often of the order of 10,000 or more electron volts; this is much larger than the energy required to separate completely a loosely bound electron, e. g. a valence electron, from an atom or a solid. Such an electron is, relatively speaking, almost free and the very simple case in which it is assumed to be entirely free may be studied with profit. Thus the phenomenon to be analyzed may be thought of as a simple collision process in which the incident photon strikes a free electron

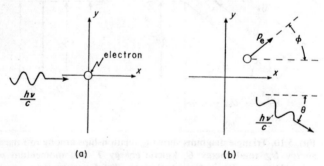

(a)   (b)

FIG. 5.11. Collision between a photon and an electron. Pre-collision state is shown in (a), post-collision state in (b).

at rest in the laboratory frame of reference as in Fig. 5.11a. The electron acquires some momentum p_e and another photon of lower energy and therefore of lower frequency ν' is emitted from the site of the collision as in Fig. 5.11b. The momentum of the incident photon, the momentum of the scattered photon, and the momentum acquired by the electron form a vector triangle as in Fig. 5.12; by the law of cosines:

$$p_e^2 = \left(\frac{h\nu}{c}\right)^2 + \left(\frac{h\nu'}{c}\right)^2 - 2\frac{h\nu h\nu'}{c^2}\cos\theta. \qquad (5.072)$$

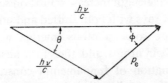

FIG. 5.12. Vector relationship involving the momentum of the incident photon ($h\nu/c$), the momentum of the scattered photon ($h\nu'/c$), and the momentum acquired by the electron (p_e).

This equation expresses the conservation of momentum. The conservation of energy is even simpler:

$$h\nu + U_{0e} = h\nu' + U_e, \qquad (5.073)$$

where U_{0e} is the initial energy of the electron (consisting of only rest mass energy) and U_e is its final energy. It is convenient to multiply equation (5.072) by c^2 thereby converting the dimensions of every term to energy squared. Once this is done, the quantity $U_e^2 - U_{0e}^2$ may be substituted for $(p_e c)^2$ and the result written:

$$U_e^2 - U_{0e}^2 = (h\nu)^2 + (h\nu')^2 - 2h\nu h\nu' \cos\theta. \qquad (5.074)$$

Let (5.073) now be solved for U_e and the result substituted into (5.074); this yields:

$$(h\nu)^2 + (h\nu')^2 + U_{0e}^2 + 2h\nu U_{0e} - 2h\nu h\nu' - 2h\nu' U_{0e} - U_{0e}^2$$
$$= (h\nu)^2 + (h\nu')^2 - 2h\nu h\nu' \cos\theta. \qquad (5.075)$$

Cancellation and rearrangement reduce this to:

$$(h\nu - h\nu')U_{0e} = h\nu h\nu'(1 - \cos\theta). \qquad (5.076)$$

If this equation is now divided by $h\nu\nu' m_0 c$, the result is:

$$\frac{c}{\nu'} - \frac{c}{\nu} = \frac{h}{m_0 c}(1 - \cos\theta), \qquad (5.077)$$

which may be rewritten very neatly as follows:

$$\lambda' - \lambda = \varDelta\lambda = \frac{h}{m_0 c} (1 - \cos \theta).$$

(5.078)

This expresses the wavelength increase (shift) on scattering from a free electron. Scattered X rays with both shifted and unshifted wavelengths are observed because some photons interact with electrons which remain bound to their atoms and therefore absorb only negligible energy. The combination of fundamental constants $h/m_0 c$ has the dimensions of length and is called the *Compton wavelength* of the electron. Its value is 0.02426 Å.

It should be mentioned that the stratagem of picturing photon and electron as one would picture the participants in a collision of macroscopic objects is unrealistic and that Fig. 5.11 is to be regarded as strictly symbolic. If the initial momenta of photon and electron are defined precisely, then, according to the principles of quantum mechanics, the positions of the two particles at any given time are indefinite[†] and it is impossible to calculate the outcome, e.g. the angle θ, for any particular event. It is possible, however, to calculate the probability distribution for θ. The analysis given here legitimately assumes that the initial and final energies and momenta of the participants are definite and are known; it then applies the usual conservation laws to these quantities and obtains results in agreement with experimental observation. Thus one is able to verify the mechanical properties of the photon through its participation in a collision with another particle.

[†] The reason for this statement will be appreciated only after considerably more material, particularly that of Chapters 10 and 11, has been studied.

5.07. The de Broglie Relations and the Davisson–Germer Experiment

In 1924 de Broglie[22] proposed that a wave which he called the "phase wave" can be associated wih a material particle in presumably much the same way that a particle, i.e. a photon, can be associated with an electromagnetic wave. By a relativistic analysis of the characteristics of this phase wave, de Broglie was able to deduce some very general conclusions.

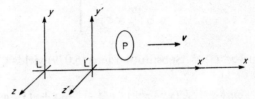

FIG. 5.13. Particle P at rest in proper frame L' and moving with respect to laboratory frame L with velocity v.

In Fig. 5.13, let P represent un unaccelerated material "particle" which, as suggested by the illustration, may somehow involve spatial extension in spite of its name. This particle has velocity v with respect to the laboratory and, for simplicity, the laboratory reference frame (Lorentz frame L) has been oriented so that v has only an x component. Lorentz frame L' moves to the right also with velocity $v = e_x v$ and therefore constitutes the *proper frame* for the particle, i.e. the frame in which the latter is at rest. It is assumed that the phase wave is a plane wave traveling in the x direction with respect to L; its wave vector may therefore be denoted $e_x k$ and its eikonal accordingly becomes $\sigma = kx - \omega t + \varphi$.

Events are absolutes; if an observer belonging to frame L arrives at the knowledge that the real part of the phase wave has a crest at space-time point A, then any other observer (regardless of the reference frame to which he belongs) must agree that this is what took place at A although the values of x and of t which he assigns to A will differ from those assigned by the L observer. Thus the *value* taken by σ at a particular space-time point will be the same whether σ is written as a function of the space-time coordinates of frame L or of frame L'. (In the conventional jargon of relativity theory, σ is a *scalar invariant*.) Quantitatively, if k' is the wave number and ω' is the frequency in the L' frame, then:

$$kx - \omega t + \varphi = k'x' - \omega't' + \varphi'. \tag{5.079}$$

By the Lorentz transformation:

$$\left.\begin{aligned} x' &= \gamma x - \gamma vt; \\ t' &= -\frac{\gamma v}{c^2}x + \gamma t, \end{aligned}\right\} \tag{5.080}$$

where $\gamma = (1 - v^2/c^2)^{-\frac{1}{2}}$. Substitution into (5.079) yields:

$$kx - \omega t + \varphi = k'(\gamma x - \gamma vt) - \omega'\left(-\frac{\gamma v}{c^2}x + \gamma t\right) + \varphi'. \tag{5.081}$$

Since this must hold for arbitrary values of x and t, coefficients and constants may be equated on each side with the following results:

$$\left.\begin{aligned} \varphi &= \varphi'; \\ k &= \gamma k' + \frac{\gamma v}{c^2}\omega'; \\ \omega &= \gamma vk' + \gamma\omega'. \end{aligned}\right\} \tag{5.082}$$

Thus is established the relationship between the wave number and frequency of the phase wave in one frame of reference and the corresponding quantities in another frame of reference.

De Broglie was greatly influenced by (i) the conclusion from earlier quantum-theoretic results that with energy U there is to be associated a (radian) frequency[†] $\omega = U/\hbar$ and (ii) the conclusion from special relativity that with mass m there is to be associated an energy $U = mc^2$. He postulated that in the L' or proper frame, where the particle is at rest, the phase wave is a *non-traveling disturbance with a null*

[†] The symbol $\hbar = h/2\pi$.

wave number (similar to the wave solution studied in Section 4.02 in which all parts of the system vibrate in unison) and *with a frequency related to the rest mass energy* of the particle. Thus:

$$k' = 0; \qquad \omega' = m_0 c^2 / \hbar. \tag{5.083}$$

Substitution of these values into equations (5.082) shows that, in the L frame, k is proportional to the *momentum* of the particle and ω is proportional to its *total energy as reckoned in L:*

$$k = \frac{m_0 \gamma v}{\hbar} = \frac{mv}{\hbar} = \frac{p}{\hbar} \quad ;$$
$$\omega = \frac{m_0 \gamma c^2}{\hbar} = \frac{mc^2}{\hbar} = \frac{U}{\hbar} . \tag{5.084}$$

These are the celebrated DE BROGLIE RELATIONS. By employing a more general form of the Lorentz transformation, it can readily be shown that the former of these has vector character and can therefore be written:

$$\mathbf{k} = \frac{\mathbf{p}}{\hbar} . \tag{5.085}$$

Also of interest is the corollary relationship associating the wavelength of the phase wave with the momentum magnitude; since $k = 2\pi/\lambda$, it follows that:

$$\lambda = \frac{h}{p} . \tag{5.086}$$

The wavelength exhibited here is called the *de Broglie wavelength.*

A startling conclusion to be drawn from equations (5.084) is that the phase velocity of the phase wave is greater than the speed of light:

$$v_p = \frac{\omega}{k} = \frac{c^2}{v} . \tag{5.087}$$

This result was not regarded as disastrous, especially since de Broglie went on to show that the group velocity of a packet of phase waves is in fact equal to the velocity of the particle and is therefore less than the speed of light. According to (5.069):

$$U = [p^2c^2 + (m_0c^2)^2]^{\frac{1}{2}}. \tag{5.088}$$

Upon division by $\hbar$, this becomes:

$$\omega = [k^2c^2 + (m_0c^2/\hbar)^2]^{\frac{1}{2}}. \tag{5.089}$$

The vector group velocity is simply the gradient of this expression in k space. Since $\nabla_k k^2 = 2k$, one has:

$$v_g = \nabla_k \omega = \frac{kc^2}{\omega} = \frac{pc^2}{mc^2} = v. \tag{5.090}$$

Substantial insight into the nature of de Broglie waves was not forthcoming until the work of Schrödinger in 1925 and 1926, the story of which occupies the next chapter and much of the remainder of this book. However, a striking experimental confirmation of de Broglie's concept was supplied in 1927 by Davisson and Germer.[23] Even though this effort followed that of Schrödinger chronologically, it will be treated at this point because it certainly qualifies as a part of the experimental foundation upon which quantum theory stands. The experiment in question was performed by allowing monoenergetic electrons to impinge upon a nickel crystal and observing the angular dependence of the scattered electrons which emerged. The results showed that the electrons did indeed possess the wave characteristics attributed to them by de Broglie.

The scattering of any type of wave by a three-dimensional repetitive structure such as a crystal has been understood since the theoretical explanation of X-ray diffraction by von Laue[24] in 1912. A crystal can be thought of as a *space lattice* of mathematical points with each of which is associated a definite physical structure called a *basis*. A basis, in this context, consists of an atom (or ion) or a set of atoms (or ions) arranged in a definite pattern; all bases in a perfect crystal are identical and identically oriented.[†] When an incident monochromatic wave passes through the crystal, minute secondary wavelets originate at each basis, i.e. at each lattice point. If the crystal as a whole is auspiciously positioned with respect to the incident wave normal, there will be a direction of exit from the crystal along which all the secondary wavelets interfere constructively. In this direction, a diffracted wave will form. The intensity of the diffracted wave is influenced by such considerations as the polarization of the incident wave and the nature of the basis but this is not of great importance in the present discussion. Also in 1912, Bragg[25] pointed out that, when the phenomenon just described occurs, it is always possible to pass a set of imaginary "reflecting" planes through all the lattice points of the crystal so oriented that the incident wave normal e, the diffracted wave normal e', and the unit normal e'' of the imaginary planes are all coplanar and that the angles of incidence and of reflection are equal. If this statement is true, then the secondary wavelets from every lattice point in any one of the imaginary planes will interfere constructively. The planes are all parallel and are spaced a perpendicular distance d apart as shown in Fig. 5.14. It is also essential, of course, that the secondary wavelets originating in one plane interfere constructively with those originating in an adjacent plane and hence with those originating in any of the planes. Consider two lattice points A and B in adjacent planes and let the vector displacement of B with respect to A be r as shown in Fig. 5.14. Notice that r is not necessarily coplanar with e, e', and e''. The right triangle ACB is in the plane determined by e and r and:

$$\overline{CB} = e \cdot r. \tag{5.091}$$

Likewise, the right triangle ADB is in the plane determined by e' and r and:

$$\overline{AD} = e' \cdot r. \tag{5.092}$$

[†] If this statement appears to be contradicted in certain instances, it is only because too small a set is being visualized as the basis. For example, the basis of a sodium chloride crystal is a set of eight ions, four sodium and four chlorine, symmetrically disposed on the corners of a cube. The latter has an edge one-half as long as the edge of the unit cube of the space lattice.

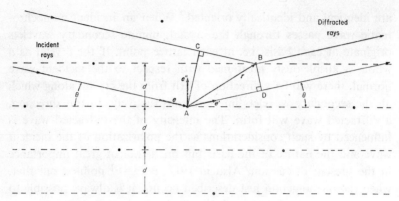

FIG. 5.14. Geometry for analysis of wave diffraction by a crystal. Dots are lattice points (bases not shown); dashed lines represent Bragg "reflecting" planes. Lattice point A is in the plane of the paper but B is not necessarily so.

The condition for constructive interference between wavelets originating at A and at B is simply that $\overline{AD} - \overline{CB}$ shall be an integral number of wavelengths. Thus:

$$e' \cdot r - e \cdot r = (e' - e) \cdot r = n\lambda, \qquad (5.093)$$

where n is an integer. The vector difference $e' - e$ lies in the direction e'' and is given by:

$$e' - e = 2e'' \sin \theta, \qquad (5.094)$$

where θ is the complement of the angle of incidence as the latter is usually understood. Combining (5.093) and (5.094), one obtains:

$$(2e'' \sin \theta) \cdot r = n\lambda. \qquad (5.095)$$

However, $e'' \cdot r = d$, the perpendicular distance between planes. Therefore:

$$2d \sin \theta = n\lambda. \qquad (5.096)$$

This relationship is known as *Bragg's law*.

Davisson and Germer employed an apparatus such as that illustrated schematically in Fig. 5.15 and invoked Bragg's law to identify the existence of waves associated with electrons emitted by the heated filament and projected toward the nickel crystal. In one of many trials, the electrons were accelerated through a potential difference of 54 volts and, consequently, arrived at the crystal with a momentum of 3.97×10^{-24} kg m/sec. According to equation (5.086), the wavelength

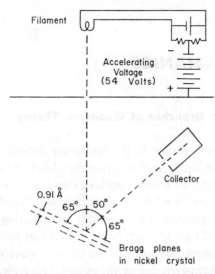

Fig. 5.15. Diagram of the Davisson–Germer apparatus.

h/p of the associated wave was 1.67 Å. The crystal was so oriented that a set of Bragg planes with a spacing (known from X-ray measurements) of 0.91 Å was utilized and, in the flux of scattered electrons, a pronounced maximum was detected at an angle of 50° from the direction of incidence. It follows that the angle θ was 65° and that Bragg's law was satisfied with the integer n equal to unity.

The data supplied by this experiment is now but a small part of the large body of evidence confirming the wave nature of material particles. This nature is possessed not only by electrons but also by other elementary particles and even by the centers of mass of aggregations of such particles.

CHAPTER 6

WAVE MECHANICS

6.01. The Two Branches of Quantum Theory

Modern quantum theory in its elementary formulation possesses two ostensibly different but nevertheless closely related branches known, respectively, as *matrix mechanics* and *wave mechanics*. The latter of these is considered, by a substantial majority of writers, to be less abstract and therefore more suitable as a starting point for beginners. In view of this, it is surprising to find that matrix mechanics was actually discovered a short time earlier than wave mechanics and was researched extensively before the essential equivalence of the two was appreciated.[†] The name principally associated with matrix mechanics is that of Heisenberg.[27] Born and Jordan, however, quickly became contributors[28] and the three together wrote a comprehensive paper[29] in which the foundations of modern quantum mechanics were first definitively set forth. The concepts of matrix mechanics were also developed in a somewhat different direction by Dirac.[30]

Since most of this book has been written to help the reader arrive at an understanding of quantum theory through the medium of waves, it would impair the coherence of the treatment to digress into an explanation of the matrix formulation merely because of the position of the latter in the history of physics. It will be sufficient, at this junc-

† For a useful guide to this era with a collection of original papers in English translation, see ref. 26.

ture, to mention that in matrix mechanics the dynamical conjugates q_j and p_j are represented by square matrices that are non-commutative under multiplication and that this feature is the avenue whereby specifically quantum manifestations make their entrance. A detailed discussion of matrix mechanics is reserved for Chapter 13; at that point the background built up in the intervening chapters can be used to advantage.

Attention is now directed to wave mechanics and the immediate objective is to derive the fundamentals of this branch of quantum theory in a way that takes inspiration from one of Schrödinger's lines of thought.[31] As a specific example, from which broader conclusions may be readily deduced, consider an electron moving in a prescribed field characterized by a scalar potential $\varphi(r, t)$ and at most a negligible vector potential $A(r, t)$.[†] The wave which, according to experimental evidence, is in some way associated with this electron is called the *wave function* and is denoted $\Psi(r, t)$. The program of derivation begins by assuming properties for the Ψ-wave such that, in a classical situation, a packet of these waves moves according to the laws of Newtonian mechanics and thereby "explains" the motion of the electron. This is in the spirit of the correspondence principle since it expects as a first requirement that the new mechanics should predict, in a classical context, behavior appropriate to that context. The hypotheses involved in this program are by no means gratuitous but are suggested by Hamilton–Jacobi theory and by de Broglie's results. Once the fundamental properties of the Ψ-wave have been determined in this way, it is an easy matter to derive the linear wave equation which Ψ must obey. This equation stands at the apex of wave mechanics; from it an enormous number of deductions, some within the domain of classical mechanics but most going far beyond

---

[†] In the present context: (i) it is not necessary to invoke the intrinsic angular momentum or "spin" of the electron; (ii) the assumption of negligible vector potential implies that the electric field is at most quasi-static and that the magnetic field is negligible. This assumption is realistic in a very large number of cases including most examples of electron motion in atoms and molecules.

that domain, can be made. It is, of course, in the agreement between such deductions and the results of experiment that the ultimate justification of the theory lies.

6.02. Waves and Wave Packets

An electron moves classically in the inter-electrode spaces of a cathode-ray tube or similar piece of vacuum apparatus where the electric field is typically of the order of 10^5 volts per meter or less. In such situations, the electron radiates only a negligible amount of energy as it pursues one of a family of trajectories determinable by methods studied earlier in this book. Letting $V(r, t)$ be the potential energy (charge times scalar potential) of the electron, one finds that the Hamiltonian is as given by (3.063) and that the whole discussion of Section 3.04, which may be re-read with profit at this point, becomes applicable. The family of trajectories is related to Hamilton's principal function $S(r, t)$ and the relations,

$$\left.\begin{array}{c} p = \nabla S, \\[2mm] \mathcal{H} = -\dfrac{\partial S}{\partial t}, \end{array}\right\} \tag{6.001}$$

should be remembered. Substitution of these relations into the expression for the Hamiltonian produces the Hamilton–Jacobi equation which, for the present case, is:

$$\frac{1}{2m}\,|\nabla S|^2 + V(r, t) + \frac{\partial S}{\partial t} = 0. \tag{6.002}$$

As mentioned in Section 4.05, a function such as the wave function $\Psi(r, t)$ can always be written in the following form:

$$\Psi(r, t) = A_0(r, t)\exp i\sigma(r, t). \tag{6.003}$$

If a number of these waves are to be superposed to form a packet, it is essential that they be (i) quasi-plane and (ii) very similar to one another in terms of wave vector and frequency. Thus any one of these wave components, especially the one having the average wave vector and the average frequency, has a pattern of wave fronts and rays that is reasonably typical of all. The following relations apply to any of the individual wave components but here they are intended to refer especially to the typical or average wave:

$$\left.\begin{array}{r} k = \nabla\sigma; \\[2mm] \omega = -\dfrac{\partial\sigma}{\partial t} \,. \end{array}\right\} \tag{6.004}$$

The similarity between these equations and those of (6.001) is striking; both sets play an important part in the subsequent steps.

It is natural to make an identification between the surfaces of constant S and their orthogonal trajectories on the one hand and the wave front and ray pattern of the typical wave component in the packet on the other. Such an identification, which constitutes the basic hypothesis of this derivation, implies that S is proportional to the eikonal σ of the typical wave since the surfaces of constant S then become surfaces of constant σ, i.e. wave fronts. The constant of proportionality between S and σ can now be selected so as to satisfy the experimentally verified de Broglie relation which states that the momentum of the particle must be equal to $\hbar$ times the wave vector of the wave. If one writes:

$$\boxed{S = \hbar\sigma} \tag{6.005}$$

it follows immediately that:

$$\boxed{\begin{array}{r} p = \hbar k \\[1mm] \mathscr{H} = T + V = \hbar\omega \end{array}} \left.\begin{array}{l} , \\[2mm] , \end{array}\right\} \tag{6.006}$$

and it is seen that the former of these fulfills the de Broglie wave vector relation. Unfortunately, as the alert reader may have noticed, the latter of (6.006) does not fulfill the de Broglie frequency relation

and there is, in fact, an enormous but constant difference between the
frequency ω of the Ψ-wave as it appears above and the frequency[†]
ω_B of the de Broglie phase wave which, with the scalar potential
present, must be written as:[‡]

$$\omega_B = \frac{mc^2 + q\varphi}{\hbar} = \frac{m_0 c^2 + T + V}{\hbar}. \qquad (6.007)$$

The difference between the two frequencies is related to the rest mass
energy of the electron:

$$\omega_B - \omega = m_0 c^2 / \hbar, \qquad (6.008)$$

and produces a comparably enormous difference between the phase
velocities of the two waves. In spite of this, both the de Broglie phase
wave and the Ψ-wave lead to useful formulations, the former being
relativistic and the latter non-relativistic. Thus Schrödinger's $\Psi(r, t)$
is not literally the de Broglie phase wave but is a wave of the same wavelength which serves equally well in the non-relativistic domain.

Since the surfaces of constant S are also the wave fronts of the typical
Ψ-wave component in the packet, it follows that the speed of advance
of these surfaces, which was commented upon in Chapter 3, is the
phase velocity of the Ψ-wave. Using e as the unit vector for both p
and k, one has for this phase velocity:

$$v_p = e\,\frac{\omega}{k} = e\,\frac{\mathscr{H}}{p} = e\,\frac{\mathscr{H}}{mv_{\text{particle}}}. \qquad (6.009)$$

[†] The symbol for the de Broglie frequency has been changed to ω_B here to avoid
confusion.

[‡] Equation (6.007) is one member of the Lorentz-covariant pair:

$$k_B = \frac{mv + q\mathbf{A}}{\hbar}\,; \qquad \frac{\omega_B}{c} = \frac{mc + q\varphi/c}{\hbar}.$$

The former of these contains the canonical momentum of a charged particle,
$mv + q\mathbf{A}$, which may be inferred from (1.096) after deleting the empty term. A
more complete derivation of wave mechanics would invoke both of the above
equations, i.e. it would include the vector potential; this, however, would lead to
an advanced formulation not suitable for present purposes.

This is clearly in agreement with (3.070). Using $p = [2m(\mathcal{H}-V)]^{\frac{1}{2}}$, the magnitude of this same quantity may be expressed in terms of $\mathcal{H}$, V, and m only:

$$v_p = \frac{\mathcal{H}}{[2m(\mathcal{H}-V)]^{\frac{1}{2}}} . \qquad (6.010)$$

It may be remarked that $\mathcal{H}-V$ is never negative in places where undulatory solutions and wave packets exist. One may therefore say that where V is large, i.e. where it has almost the same value as $\mathcal{H}$, the denominator is small and the phase velocity large; similarly, where V is small, the phase velocity is small. Thus the Ψ-wave is analogous to a macroscopic wave, e.g. an acoustic wave, propagating through an inhomogeneous medium in which the local phase velocity varies from place to place and perhaps from time to time.

One now turns to a consideration of the group velocity of the packet which, if the wave mechanics is to explain the classical motion, must agree with the velocity of the electron. To demonstrate this agreement, let the first and seldom-used set of Hamilton equations (2.007) be invoked:

$$\dot{q}_j = \frac{\partial \mathcal{H}}{\partial p_j} . \qquad (6.011)$$

In Cartesian coordinates, these become:

$$\left. \begin{aligned} \dot{x} &= \frac{\partial \mathcal{H}}{\partial p_x} = \frac{\partial \omega}{\partial k_x} ; \\[4pt] \dot{y} &= \frac{\partial \mathcal{H}}{\partial p_y} = \frac{\partial \omega}{\partial k_y} ; \\[4pt] \dot{z} &= \frac{\partial \mathcal{H}}{\partial p_z} = \frac{\partial \omega}{\partial k_z} . \end{aligned} \right\} \qquad (6.012)$$

These are clearly the Cartesian components of the velocity of the electron. When gathered together to form a vector, the latter is seen to be:

$$v_{\text{particle}} = \nabla_k \omega = v_g . \qquad (6.013)$$

This equation shows that the Ψ-wave which, so far, has been defined only through the relation $S = \hbar\sigma$, does indeed have the properties expected of it in that a packet composed of these waves has the velocity required by the laws of classical mechanics. Also contained in this treatment is the resolution of the speed paradox mentioned in Chapter 3 for it is now seen that the difference between the velocity of the surfaces of constant S and the velocity of the particle to which they are related is simply the difference between the phase velocity and the group velocity of a packet of waves.

Schrödinger, whose work was pivotal in the development of wave mechanics, found it expedient to think of the charge of the electron as distributed in space according to the magnitude squared of its wave function and, occasionally, the concept has been advanced that the wave packet actually constitutes the physical substance of the electron (or of whatever particle happens to be under consideration). On this attractively simple view, there is no wave-particle duality; all matter is composed of waves! Such an interpretation raises difficulties, however. These arise from such facts as:

(i) A wave mechanical packet can represent the center of mass of a system of particles rather than a single particle, in which case there may be no particle present at the site of the packet.

(ii) There are non-classical situations involving strong forces in which a packet is disrupted, one part going in one direction and one in another. The particle which the packet represents is not disrupted, however, but appears whole and entire in one place or the other with calculable probability.

For these and perhaps other reasons, any point of view which totally identifies particle and packet has been largely superseded by the statistical interpretation suggested by Born.[32] The latter is officially adopted in this book and is explained in Section 6.04 of this chapter. It is adopted, however, with reservations; the last word on the interpretation of quantum theory has not yet been spoken and between now and the time that it is spoken, many changes may be expected.

6.03. The Schrödinger Equation

In the previous section, a classical situation is identified as one in which a number (usually an infinite number) of quasi-plane Ψ-waves are superposed to form a packet which moves continuously through space according to the laws of classical mechanics. Such a situation is readily conceivable if the de Broglie wavelength is small compared with the relevant dimensions of the apparatus with which the particle represented by the packet interacts. Pursuant to this line of thought, one may review the calculations of Section 5.07 in which it is shown that the de Broglie wavelength for an electron with a very modest amount of energy is of the order of 1 Å; generally speaking, this wavelength is much smaller for heavier particles and is extremely small for the centers of mass of macroscopic objects. Thus it is easy to see why the macroscopic domain is classical; the de Broglie wavelengths of continuously moving objects are fantastically small compared with the distances within which significant changes in their potential energies occur.

The hydrogen atom itself provides a whole gamut of examples ranging from the classical to the obviously non-classical. To show this, one has only to calculate the de Broglie wavelength for hypothetical electron motion in the circular Bohr orbits. For the orbit of quantum number n:

$$\lambda_n = \frac{h}{p_n} = \frac{hr_n}{L_n} = \frac{2\pi hr_n}{nh} = \frac{2\pi r_n}{n}. \tag{6.014}$$

This interesting and often quoted result shows that there are n de Broglie wavelengths around the circumference of the nth Bohr orbit. In the 1000th such orbit, for example, one wavelength subtends only 0.36 degree of the circumference and a hypothetical displacement of one wavelength toward or away from the nucleus would occasion only a slight change in the potential energy. At this location, then, quasi-plane waves forming a continuously moving packet can be readily imagined. The situation is much different at the first Bohr orbit where the wavelength is equal to the circumference and is therefore significantly larger than the distance to the point where the potential energy drops (theoretically) to minus infinity. Although superposition is always possible, waves so close to the singularity can hardly be quasi-plane and it is difficult to picture the kind of motion an interference maximum would execute. This latter situation is, of course, the non-classical one. The difficulty inherent in the Bohr–Sommerfeld concept can now be appreciated; orbits of classical form are assumed to exist for all positive integer values of the quantum number n but this is not a realistic hypothesis at the lower values and, even at the higher, the cooperation of more than one wave (quantum state) is required to create the effect of orbital motion. In spite of this, the relation reproduced in (6.014) is a memorable one because it gave Schrödinger the inspiration to visualize a stationary quantum state as a sort of wave resonance phenomenon for which some parameter (in this case, the wavelength) can take only one of a set of particular values.

One may summarize this discussion by saying that the classical domain is, with insignificant exceptions, the domain of short wavelength, i.e. short with respect to the scale of spatial variations in the potential energy. This criterion is so overwhelmingly fulfilled in the macroscopic world that the short wavelength limit discussed in Section 4.05 is, for all practical purposes, attained with respect to particles in continuous motion. This means that, under classical conditions, it is possible to characterize a single Ψ-wave entirely by the eikonal function $\sigma(r, t)$. The eikonal equation for this function is not difficult to find; since σ is proportional to S and S is governed by the Hamilton–Jacobi equation, it follows that σ is governed by this same equation. Formally, then, the eikonal equation is simply (6.002) with $S = \hbar\sigma$ substituted:

$$\frac{\hbar^2}{2m} |\nabla\sigma|^2 + V(r, t) + \hbar \frac{\partial\sigma}{\partial t} = 0. \qquad (6.015)$$

This equation, which is the starting point in the derivation of the wave equation for Ψ, takes its form ultimately from the classical Hamiltonian; the great importance of the latter in wave mechanics is therefore understandable.

In Section 4.05, the acoustic wave equation was analyzed in the context of quasi-plane waves and the corresponding eikonal equation was derived. This process is now to be reversed. A wave equation which is linear (in order to accommodate superpositions of waves) is to be synthesized in such a way that (6.015) will be its eikonal equation. For this purpose, the following partial differentiations[†] of (6.003) will suffice:

$$\nabla^2 \Psi = \left[\frac{\nabla^2 A_0}{A_0} - |\nabla \sigma|^2 + \frac{2i}{A_0} \nabla A_0 \cdot \nabla \sigma + i \nabla^2 \sigma \right] \Psi. \quad (6.016)$$

$$\frac{\partial \Psi}{\partial t} = \left[\frac{1}{A_0} \frac{\partial A_0}{\partial t} + i \frac{\partial \sigma}{\partial t} \right] \Psi. \quad (6.017)$$

As in Section 4.05, the term $-|\nabla \sigma|^2$ dominates in (6.016) and the term $i \, \partial \sigma / \partial t$, in (6.017). In the approach to the short wavelength limit, the following become rigorous:

$$|\nabla \sigma|^2 = -\frac{\nabla^2 \Psi}{\Psi} ; \quad (6.018)$$

$$\frac{\partial \sigma}{\partial t} = \frac{-i}{\Psi} \frac{\partial \Psi}{\partial t}. \quad (6.019)$$

Substitution into (6.015) produces:

$$-\frac{\hbar^2}{2m} \frac{\nabla^2 \Psi}{\Psi} + V(r, t) - \frac{i\hbar}{\Psi} \frac{\partial \Psi}{\partial t} = 0. \quad (6.020)$$

It is usual to multiply this by Ψ and to rearrange. The result is the partial differential equation known as the SCHRÖDINGER EQUATION:

$$\boxed{-\frac{\hbar^2}{2m} \nabla^2 \Psi + V\Psi = i\hbar \frac{\partial \Psi}{\partial t}} \cdot \quad (6.021)$$

[†] See equations (4.062) through (4.066).

It follows that, in the short wavelength limit, a packet composed of quasi-plane wave solutions of this equation will move with the group velocity predicted by classical mechanics. Thus Schrödinger's wave mechanics, which is much deeper than Newton's laws, explains the macroscopic motion of particles as indeed it must do. More interestingly, it also explains a vast array of other phenomena, largely unimaginable in macroscopic terms, in which the wavelength is not short in comparison with the scale of spatial variations of the potential energy. These are the quantum phenomena *par excellence* and are generally exemplified by the behavior of electrons in crystals, in molecules, and in atoms. In this quantum microworld, many long-cherished concepts lose their utility and, to deal effectively with this world, new habits of thought must be acquired.

6.04. Interpretation of $\Psi^*\Psi$; Normalization and Probability Current

Although the case in which Ψ is a compact packet of quasi-plane waves near the short wavelength limit is vital in the inductive approach to the Schrödinger equation, it is but one specialized application in the vast field of wave mechanics, a field which from now on will be treated as a deductive discipline based upon the Schrödinger equation as a starting point. In the general case, when conditions associated with the short wavelength limit do not necessarily apply, all the terms in equations (6.016) and (6.017) must be retained. The Schrödinger equation, then, asserts the following:

$$-\frac{\hbar^2}{2m}\left[\frac{\nabla^2 A_0}{A_0}-|\nabla\sigma|^2+\frac{2i}{A_0}\nabla A_0\cdot\nabla\sigma+i\,\nabla^2\sigma\right]\Psi+V\Psi$$

$$= i\hbar\left[\frac{1}{A_0}\frac{\partial A_0}{\partial t}+i\frac{\partial\sigma}{\partial t}\right]\Psi. \tag{6.022}$$

After dividing out Ψ, one may equate reals to reals and imaginaries to imaginaries with the following results:

$$-\frac{\hbar^2}{2m}\left[\frac{\nabla^2 A_0}{A_0} - |\nabla\sigma|^2\right] + V = -\hbar\frac{\partial\sigma}{\partial t}. \qquad (6.023)$$

$$-\frac{\hbar^2}{2m}\left[\frac{2}{A_0}\nabla A_0\cdot\nabla\sigma + \nabla^2\sigma\right] = \frac{\hbar}{A_0}\frac{\partial A_0}{\partial t}. \qquad (6.024)$$

The second of these equations is important in the present context. Let it be multiplied by $2A_0^2/\hbar$ and rearranged:

$$\frac{\hbar}{m}\left[2A_0\nabla A_0\cdot\nabla\sigma + A_0^2\nabla^2\sigma\right] + 2A_0\frac{\partial A_0}{\partial t} = 0. \qquad (6.025)$$

This is equivalent to:

$$\frac{\hbar}{m}[\nabla A_0^2\cdot\nabla\sigma + A_0^2\nabla^2\sigma] + \frac{\partial A_0^2}{\partial t} = 0, \qquad (6.026)$$

which can be recognized as:

$$\nabla\cdot\left(A_0^2\frac{\hbar\nabla\sigma}{m}\right) + \frac{\partial A_0^2}{\partial t} = 0. \qquad (6.027)$$

This equation has the precise form of the equation of continuity,

$$\nabla\cdot(\varrho v) + \frac{\partial\varrho}{\partial t} = 0, \qquad (6.028)$$

which involves a "substance" of density ϱ conveyed from place to place by a current density $J = \varrho v$. In equation (6.027) it is easy to recognize the following:

$$\varrho = A_0^2 = \Psi^*\Psi; \qquad (6.029)$$

$$v = \frac{\hbar\nabla\sigma}{m}; \qquad (6.030)$$

$$J = \varrho v = \Psi^*\Psi\frac{\hbar\nabla\sigma}{m}. \qquad (6.031)$$

An important theorem relevant to the interpretation of the wave function will now be given:

THEOREM 6.01. *If $v = \hbar |\nabla\sigma|/m$ is bounded at infinity and if $\int \varrho \, d\tau$ over all of space converges, then it converges to a constant and never to a function of time.*

To begin the proof, let the divergence theorem be applied to (6.028). The result is:

$$\oint_{\substack{\text{spere} \\ \text{at } \infty}} \varrho v \cdot e_n \, da = -\frac{d}{dt} \int_{\substack{\text{all} \\ \text{space}}} \varrho \, d\tau. \qquad (6.032)$$

The proof depends upon the fact that if, as $r \to \infty$, ϱ decays rapidly enough for the integral on the right-hand side to converge, then the integral on the left-hand side must vanish. The convergence in question is secured if, as $r \to \infty$,

$$\varrho \sim f(\theta, \varphi, t) r^{-(3+\varepsilon)}, \qquad (6.033)$$

where $\varepsilon > 0$. Since da contains r^2 as a factor, the integral on the left-hand side must vanish as $r^{-(1+\varepsilon)}$ and the theorem is proved. Since an unbounded v (i.e. and unbounded transport velocity for the "substance" and, concomitantly, an unbounded wave packet momentum in case Ψ is in the form of a wave packet) can be imagined only in very unphysical situations, it becomes a working principle that if $\int \varrho \, d\tau$ over all of space converges, it converges to a constant.

The Born interpretation states that the mysterious "substance", whose density ϱ has been identified as $\Psi^*\Psi$, is the probability of finding the particle which has Ψ as its wave function. Thus $\Psi^*\Psi$ is a *probability density* and the probability that, at time t, the particle is in a given volume element $d\tau$ at position r is:

$$dP(r, t) = \Psi^*\Psi(r, t) \, d\tau. \qquad (6.034)$$

In order to fit this interpretation, Ψ, in addition to satisfying the Schrödinger equation and such boundary conditions as may be applicable, must therefore have the following property:

$$\int_{\substack{\text{all} \\ \text{space}}} \Psi^* \Psi d\tau = 1 \ .$$

(6.035)

This is called the *normalization condition* and a wave function which satisfies it is said to be *normalized*. A solution Ψ' of the Schrödinger equation is normalizable if the integral of $\Psi'^* \Psi'$ over all of space converges for then it converges to a constant. Since $\Psi'^* \Psi' > 0$ everywhere, this constant is a positive real, say a. The trivially modified wave function $\Psi = a^{-\frac{1}{2}} \Psi'$ then satisfies (6.035). It follows from (6.034) that the probability of finding the particle in a region of space that is neither infinite nor infinitesimal is simply the integral of $\Psi^* \Psi$ over the region in question. This and other ideas developed here may, in a straightforward way, be extended to include wave functions representing systems of more than one particle; an example of such an extension will be given in a later chapter.

If a wave function is such that the integral of $\Psi^* \Psi$ over all of space does not converge, then it is not normalizable and does not correspond to a physically realizable situation. It is noteworthy that, by this criterion, plane waves with non-vanishing amplitudes cannot literally be wave functions. Because of their simplicity, however, such waves are extensively used in quantum mechanical analyses. When so used, they simulate the behavior of physically large but normalizable wave packets whose sizes are very great compared with the relevant dimensions of the objects with which they interact.

The vector quantity J is called the *probability current density*; it exists whenever there is both a probability density and an eikonal gradient.[†] This current density is often written in an alternative form

---

[†] For economy of terminology, σ will continue to be called the eikonal even though there is some question about the propriety of doing this when the situation is not restricted to the quasi-plane wave case.

which does not mention the eikonal gradient explicitly:

$$J = \frac{\hbar}{2im} [\Psi^* \triangledown \Psi - \Psi \triangledown \Psi^*]. \tag{6.036}$$

Reconciliation of this form with (6.031) is left as an exercise.

6.05. Expectation Values

If a given quantity is measured experimentally N times, its average value is simply the sum of all the measured values divided by N but its expectation value (as that term is used in quantum theory) is the value which this average would have if the measurements were distributed exactly according to *a priori* probability. Thus one might assert, largely on experiential grounds, that the expectation value is the limit of the average value as N tends to infinity. For example, the expectation value of the score obtained by throwing two dice is seven; the average value of N such throws will usually not be precisely seven but is generally found to approach seven as N is made larger. It may be noticed that this definition of expectation value presupposes the existence of an *a priori* theory of probability applicable to the measurements in question. Quantum mechanics is supposed to be just such a theory.

In order to apply these remarks in the context of the present chapter, consider a one-particle system with wave function Ψ and an experiment X which is designed to measure some property of that system. Suppose that X can be duplicated N times under precisely the same conditions. This means that either:

(i) there are N similarly constructed systems, identical in the sense that for each the wave function is initially Ψ and that all N performances of the experiment occur simultaneously, or:

(ii) the experiment X is repeated N times on the same system, the latter being restored to its initial condition (given its initial wave function Ψ) before each repetition of the experiment.

As an illustration, suppose that the experiment is designed to measure the position vector r of the particle. Let space be divided into an infinite number of small volume elements $\Delta\tau_1$, $\Delta\tau_2$, $\Delta\tau_3$, ... located respectively at the vector displacements r_1, r_2, r_3, ... from the origin. Suppose that in n_1 trials the particle is found to be in $\Delta\tau_1$, in n_2 trials it is found to be in $\Delta\tau_2$, etc., where $n_1 + n_2 + n_3 + ... = N$. The sum of all the N measured values of r is therefore:

$$\sum_{i=1}^{N} r_i = \sum_{j=1}^{\infty} r_j n_j, \tag{6.037}$$

where the index j refers to the volume elements. The average value is:

$$\bar{r} = \frac{1}{N} \sum_{i=1}^{N} r_i = \sum_{j=1}^{\infty} r_j(n_j/N). \tag{6.038}$$

If the determinations are distributed in accordance with probability theory, then each of the ratios n_j/N must be equal to the corresponding *a priori* probability that the particle will be found in volume element $\Delta\tau_j$, i.e. to $\Psi^*\Psi(r_j, t)\,\Delta\tau_j$; in this particular example, it is not likely that this could be accomplished with finite N. In any event, the expectation value,[†] denoted $\langle r \rangle$, is given by:

$$\langle r \rangle = \sum_{j=1}^{\infty} r_j \Psi^*\Psi(r_j, t)\,\Delta\tau_j. \tag{6.039}$$

Notice that this value may be a function of time because of the possible dependence of $\Psi^*\Psi$ upon the time. If all the $\Delta\tau_j$ are allowed to go to zero, the sum on the right becomes an integral and one has:

$$\langle r \rangle = \int_{\substack{\text{all} \\ \text{space}}} r\Psi^*\Psi \, d\tau. \tag{6.040}$$

This entire development could apply equally well to a quantity such

---

[†] The difference between "average value" and "expectation value" is often slighted in spoken and even written discourse, the former term being used when, strictly speaking, the latter is meant.

as r^2. In lieu of equation (6.037), one would then write:

$$\sum_{i=1}^{N} r_i^2 = \sum_{j=1}^{\infty} r_j^2 n_j, \tag{6.041}$$

and proceed exactly as before. The result would be:

$$\langle r^2 \rangle = \int_{\substack{all \\ space}} r^2 \Psi^* \Psi \, d\tau. \tag{6.042}$$

By the same argument, it is seen that the development also applies to an arbitrary function of r and t. Thus:

$$\boxed{\langle f(r, t) \rangle = \int_{\substack{all \\ space}} \Psi^* f(r,t) \Psi \, d\tau} \,. \tag{6.043}$$

Notice that $f(r, t)$ has been written *between* Ψ^* and Ψ in the right-hand side of this equation. This peculiar style, which is conventional in wave mechanics, has a reason for existence; this reason will come to light in Chapter 8.

CHAPTER 7

THE TIME-INDEPENDENT SCHRÖDINGER EQUATION AND SOME OF ITS APPLICATIONS

7.01. Time-independent Potential Energy Functions and Stationary Quantum States

The Schrödinger equation for a system consisting of one particle of mass m moving in three dimensions and having a potential energy $V(r, t)$ was derived in Chapter 6. It is:

$$-\frac{\hbar^2}{2m}\nabla^2\Psi + V\Psi = i\hbar\frac{\partial\Psi}{\partial t}. \qquad (7.001)$$

The present chapter introduces the case in which the potential energy does not depend explicitly upon the time and can be written $V(r)$. This case has wide applicability; it permits solutions of (7.001) in which Ψ can be written as a function of r only times a complex exponential factor involving a single frequency ω. Thus:

$$\Psi(r, t) = \psi(r)\exp{-i\omega t}. \qquad (7.002)$$

The classical mechanical counterpart of this situation is the one in which $\mathcal{H}$ does not contain the time. For such a case, $\mathcal{H}$ is a constant of the motion (equal to $\mathcal{H}_0$) and Hamilton's principal function may be written in the separated form: $S(q_j, \zeta_j, t) = W(q_j, \zeta_j) - \mathcal{H}_0 t$.

Substitution of (7.002) into (7.001), with subsequent cancellation of the complex exponential factor, yields:

$$-\frac{\hbar^2}{2m}\nabla^2\psi + V\psi = \hbar\omega\psi. \tag{7.003}$$

According to equations (6.006), $\hbar\omega$ is to be identified as the total energy E. The wave function (7.002), characterized by a single definite frequency ω, is therefore also characterized by a single definite value of the total energy. Substitution of E for $\hbar\omega$ produces the TIME-INDE-PENDENT SCHRÖDINGER EQUATION:

$$\boxed{-\frac{\hbar^2}{2m}\nabla^2\psi + V\psi = E\psi}. \tag{7.004}$$

In the context of this equation, the function $\psi(r)$ is often loosely called the "wave function". Although this usage is widespread, it is well to keep in mind that the true wave function is the complete expression $\Psi(r, t)$ as given by (7.002).

When a system has a given wave function Ψ, it is often said to be "in the quantum state Ψ". A quantum state with a wave function like (7.002) has the interesting property that the probability density is not a function of time. Thus:

$$\Psi^*\Psi = (\psi^* \exp i\omega t)\,(\psi \exp -i\omega t) = \psi^*\psi. \tag{7.005}$$

A state of this type, which has a static probability density, is a *stationary state* in the wave mechanical sense. Although more fundamental than the stationary state as conceived by Bohr and Sommerfeld, it nevertheless has some features in common with the latter as will be seen in the context of the hydrogenic atom. It is noteworthy that a probability current density may exist even in a stationary state; if it does exist, it will have zero divergence as required by the equation of continuity.

Given a potential energy $V(r)$, there is in general more than one solution of (7.004) and each such solution ψ_n has an associated value E_n of E. For every ψ_n, there is a Ψ_n with frequency $\omega_n = E_n/\hbar$ and the totality of these solutions constitutes a "spectrum" of stationary states containing many, perhaps infinitely many, members. This matter will become clearer when, later in this chapter, the Schrödinger equation will be solved for a specific potential energy so as to exhibit the entire spectrum of stationary states. For the present, it is sufficient to observe that the solutions may be written:

$$\left.\begin{aligned}
\Psi_1(r, t) &= \psi_1(r) \exp{-i\omega_1 t}; \\
\Psi_2(r, t) &= \psi_2(r) \exp{-i\omega_2 t}; \\
&\ .\ .\ .\ .\ .\ .\ .\ .\ ; \\
\Psi_n(r, t) &= \psi_n(r) \exp{-i\omega_n t}; \\
&\ .\ .\ .\ .\ .\ .\ .\ .\ ..
\end{aligned}\right\} \qquad (7.006)$$

If the system has any one of these as its wave function, it is obviously in the particular stationary state in question and has the definite energy associated with that state. As intimated earlier, however, it is quite possible for the system to have a wave function Ψ which is a superposition of some or all of the Ψ_n listed above. Such a function, which is a solution of (7.001) but not of (7.004), would be expressed with the aid of coefficients α_n as follows:

$$\Psi = \sum_n \alpha_n \psi_n \exp{-i\omega_n t}. \qquad (7.007)$$

A state of this type has many different frequencies and, on the quantum level, many different energies. Its classical energy is the expectation value $\langle E \rangle$; a method for finding this value will be given in Chapter 8. A superposition of stationary states in which different frequencies (energies) are involved is, by its very nature, a non-stationary state, i.e. it has a probability density that changes with time. Suppose, for example, that the superposition consists of only two stationary states and that these have different energies:

$$\Psi = \alpha_1 \psi_1 \exp{-i\omega_1 t} + \alpha_2 \psi_2 \exp{-i\omega_2 t}. \qquad (7.008)$$

One then finds for the probability density:

$$\Psi^*\Psi = \alpha_1^*\alpha_1\psi_1^*\psi_1 + \alpha_2^*\alpha_2\psi_2^*\psi_2$$
$$+ \alpha_1^*\alpha_2\psi_1^*\psi_2 \exp -i(\omega_2-\omega_1)t$$
$$+ \alpha_2^*\alpha_1\psi_2^*\psi_1 \exp i(\omega_2-\omega_1)t. \tag{7.009}$$

It is seen that this density inevitably includes a function which is not only dependent upon coordinates but is also oscillatory in time at the difference frequency. Thus is emphasized the fact that the absolute frequency (energy) of a given quantum state is not important; only the differences between the frequency of this state and those of other states are important.

The ordinary plane wave is a simple example of a stationary state; it is a solution of the Schrödinger equation whenever the potential energy is constant in space as well as in time. One may write for such a wave:

$$\Psi = A \exp i(k \cdot r - \omega t) = A \exp ik \cdot r \exp -i\omega t. \tag{7.010}$$

Evidently the non-time-dependent part of the wave function is:

$$\psi = A \exp ik \cdot r \tag{7.011}$$

and

$$\left. \begin{array}{l} \varrho = A^*A; \\ J = A^*A\hbar k/m. \end{array} \right\} \tag{7.012}$$

As indicated in Chapter 6, plane waves with $|A| > 0$ are not normalizable and therefore not physically realizable; nevertheless, they are very useful for simulating the behavior of large packets. This use is clearly exemplified in the subsequent section.

7.02. The Rectangular Step; Transmission and Reflection

In this section, the quantum mechanics of a particle which passes (or attempts to pass) from one region of space in which $V = 0$ to another in which $V = V_0$, as illustrated in Fig. 7.01, will be studied.

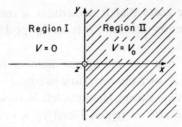

FIG. 7.01. Two regions of space, separated by a plane interface, in which the potential energy assumes different values.

If the microscopics of the transition region between the two media are examined, one might find a gradual increase in potential energy as in Fig. 7.02a or an abrupt (rectangular) step as in Fig. 7.02b. The former case is meaningful in both quantum and classical mechanics. For it, a wave packet incident from the left would surmount the acclivity and proceed into region II provided that it had sufficient energy; lacking sufficient energy, it would stop at some intermediate point and return into region I. A quantum packet and a classical particle will behave alike if the average wavelength in the former is very short,

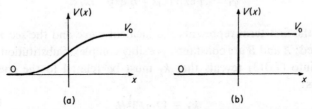

(a)   (b)

FIG. 7.02. A gradual transition and an abrupt (rectangular) step.

i.e. very small compared with the width of the transition between the two regions, a requirement which can be fulfilled if the transition is gradual. If, on the other hand, the step is rectangular as in Fig. 7.02b, it is impossible for the wavelength to be short compared with the width of the transition and the situation has no meaning in a classical context. In a purely quantum context, however, there is no necessary requirement about shortness of wavelength and a rectangular step has meaning in the sense that it constitutes a readily solvable and pedagogically interesting problem. It is on this problem that attention will now be focused.

The rectangular step has an infinite number of stationary-state solutions each with its own characteristic energy, E. A typical solution involves a plane wave incident from the left in region I accompanied by a reflected wave also in region I and by a transmitted wave in region II; for such a solution to be possible, it is necessary for E to be greater than zero.[†] The present discussion will treat only waves which travel in a direction normal to the interface (in a direction parallel to the x-axis) although extension to cases involving other angles of incidence is not difficult. The time-independent Schrödinger equation for region I is readily put into the following form:

$$\nabla^2 \psi_I + \frac{2mE}{\hbar^2}\, \psi_I = 0 \qquad (7.013)$$

and the solution of this equation, which takes the form of an incident and a reflected wave, may be written:

$$\psi_I = A \exp ik_I x + B \exp -ik_I x. \qquad (7.014)$$

Here, the first term represents the incident wave and the second, the reflected; A and B are constants, possibly complex. Substitution of this back into (7.013) reveals that k_I must be related to the energy as follows:

$$k_I = [2mE]^{\frac{1}{2}}/\hbar. \qquad (7.015)$$

The complete wave function is:

$$\Psi_I = A \exp i(k_I x - \omega t) + B \exp i(-k_I x - \omega t), \qquad (7.016)$$

[†] If $E > V_0$, it is possible to have a second independent solution with the same energy as the original solution. This second solution is characterized by a wave incident from the right in region II with a reflected wave also in region II and a transmitted wave in region I. Since no new principles are illustrated by this case, it will not be dealt with here.

where $\omega = E/\hbar$. In region II, the time-independent Schrödinger equation becomes:

$$\nabla^2\psi_{\text{II}} + \frac{2m(E-V_0)}{\hbar^2}\,\psi_{\text{II}} = 0 \qquad (7.017)$$

and the solution of interest is the transmitted wave:

$$\psi_{\text{II}} = C \exp ik_{\text{II}}x, \qquad (7.018)$$

where

$$k_{\text{II}} = [2m(E-V_0)]^{\frac{1}{2}}/\hbar \qquad (7.019)$$

and

$$\Psi_{\text{II}} = C \exp i(k_{\text{II}}x - \omega t) \qquad (7.020)$$

with $\omega = E/\hbar$ as before. One now faces the task of evaluating the coefficients B and C (which contain the amplitudes and phases of the reflected and transmitted waves) in terms of A, the amplitude and phase of the incident wave. Before this step can be taken, however, a digression must be made to investigate the conditions applicable in general at boundaries between regions in which V has different values and in which separate solutions of the Schrödinger equation have been obtained.

The boundary conditions sought are those applicable to the ordinary situation in which Ψ, $\nabla\Psi$, and $\partial\Psi/\partial t$ (and therefore ϱ and J) are finite at all points of interest. The finiteness of the normal component of $\nabla\Psi$ at the boundary implies the continuity of Ψ at the boundary; the continuity of Ψ, in turn, implies the continuity of the tangential components of $\nabla\Psi$. The continuity of the normal component of $\nabla\Psi$ at the boundary is, however, contingent upon the finiteness of the two values of V. To demonstrate this, let G be a closed surface in the form of a pill box so situated that the boundary passes midway between its two circular faces as in Fig. 7.03. The Schrödinger equation may be written as follows:

$$\nabla^2\Psi = \frac{2m}{\hbar^2}\left(V\Psi - i\hbar\frac{\partial\Psi}{\partial t}\right). \qquad (7.021)$$

If both sides of this equation are integrated over the volume of the

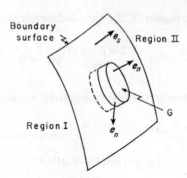

FIG. 7.03. A portion of a boundary surface upon which pill box G
has been constructed for derivational purposes.

pill box and if the divergence theorem is invoked, the result is:

$$\int_{\text{volume}} \nabla^2 \Psi \, d\tau = \oint_{\text{surface}} \nabla \Psi \cdot e_n \, da = \frac{2m}{\hbar^2} \int_{\text{volume}} \left(V \Psi - i\hbar \frac{\partial \Psi}{\partial t} \right) d\tau, \quad (7.022)$$

where e_n is the outward normal to the surface of the pill box. Allowing the pill box to collapse to zero thickness, the right-hand side of this equation vanishes if V is finite on both sides of the boundary. This implies the following result where e_s is a unit normal to the boundary pointing from region I into region II:

$$e_s \cdot \nabla \Psi_{\text{II}} - e_s \cdot \nabla \Psi_{\text{I}} = 0. \quad (7.023)$$

The result is obvious, namely that the normal component of $\nabla \Psi$ is also continuous. To summarize, one may say that at boundaries where V takes different values:

(i) Ψ and the tangential components of $\nabla \Psi$ are continuous regardless of the values of V.

(ii) If both of the two values of V are finite, the normal component of $\nabla \Psi$ also is continuous.

To return to the case at hand, namely the plane boundary with a plane wave normally incident, it is seen that the two ψ-functions of

(7.014) and (7.018) must obey the following conditions:

$$(\psi_I)_0 = (\psi_{II})_0;$$
$$\left(\frac{d\psi_I}{dx}\right)_0 = \left(\frac{d\psi_{II}}{dx}\right)_0 . \tag{7.024}$$

The zero subscripts indicate evaluation at $x = 0$. Equations (7.024) yield:

$$A + B = C;$$
$$ik_I A - ik_I B = ik_{II} C. \tag{7.025}$$

Since A is a known and B and C are unknowns, these equations may be written in more standard form as follows:

$$-B + C = A;$$
$$B + \frac{k_{II}}{k_I} C = A. \tag{7.026}$$

For convenience, let:

$$\frac{k_{II}}{k_I} = \beta = \left[\frac{E - V_0}{E}\right]^{\frac{1}{2}}. \tag{7.027}$$

Equations (7.026) may now be solved in terms of A with β appearing as a parameter:

$$B = \frac{1-\beta}{1+\beta} A;$$
$$C = \frac{2}{1+\beta} A. \tag{7.028}$$

Thus:

$$\psi_I = A\left[\exp ik_I x + \frac{1-\beta}{1+\beta} \exp - ik_I x\right];$$
$$\psi_{II} = \frac{2A}{1+\beta} \exp ik_{II} x. \tag{7.029}$$

Two cases may now be distinguished according to whether the energy E of the incident wave is greater than or less than the height V_0 of the step. For the case in which $E > V_0$, β is real and it is easily seen that the probability density in region I is given by:

$$\varrho = \Psi_I^* \Psi_I = A^* A \left[1 + \left(\frac{1-\beta}{1+\beta} \right)^2 + 2 \frac{1-\beta}{1+\beta} \cos 2k_I x \right]. \quad (7.030)$$

This function, which displays the interference between the incident and the reflected waves, is typical of standing wave phenomena and has minima separated by half a de Broglie wavelength. There is only a single (transmitted) wave in region II and therefore no interference; the probability density is simply a constant:

$$\varrho_{II} = \Psi_{II}^* \Psi_{II} = A^* A \left(\frac{2}{1+\beta} \right)^2. \quad (7.031)$$

These probability densities are plotted in Fig. 7.04 for two typical values of β. Notice that, paradoxically, there is greater average probability density (per unit length along the x-axis) on the side of the boundary where the potential energy is greater; this is in harmony with the classical observation that a particle spends more time in a region of higher potential energy where its motion is slower.

An interesting facet of this problem emerges when the probability current densities J_I and J_{II} are derived. If the wave function for region I is simply substituted into (6.036), one finds that:

$$J_I = e_x \frac{\hbar k_I}{m} [A^* A - B^* B] = e_x \frac{\hbar k_I}{m} A^* A \left[1 - \left(\frac{1-\beta}{1+\beta} \right)^2 \right]. \quad (7.032)$$

This same result could have been obtained, however, by treating the incident and reflected waves separately:

$$J_i = e_x \frac{\hbar k_I}{m} A^* A; \quad J_r = -e_x \frac{\hbar k_I}{m} B^* B; \quad J_I = J_i + J_r. \quad (7.033)$$

Thus, even though the incident and reflected waves interpenetrate one another and thereby create a standing wave pattern, they nevertheless retain their separate identities in the sense that each carries its own probability current independently of the other. This behavior is characteristic of any two waves which have *different* wave vectors. (Here the two wave vectors are $e_x k_I$ and $-e_x k_I$, respectively.)

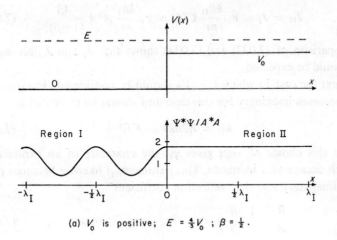

(a) V_0 is positive; $E = \frac{4}{3}V_0$; $\beta = \frac{1}{2}$.

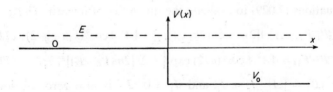

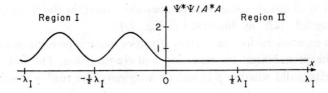

(b) V_0 is negative; $E = -\frac{1}{3}V_0$; $\beta = 2$.

FIG. 7.04. Plots of potential energy and of $\Psi^*\Psi/A^*A$ as functions of x for positive-going and negative-going rectangular steps. In both cases, $E > V_0$. The quantity λ_1, the de Broglie wavelength in region I, is used as a convenient length parameter in both regions.

The probability current density for region II is readily found to be:

$$J_{\mathrm{II}} = J_t = e_x \frac{\hbar k_{\mathrm{II}}}{m} C^*C = e_x \frac{\hbar k_{\mathrm{I}}}{m} A^*A \frac{4\beta}{(1+\beta)^2} . \qquad (7.034)$$

Comparison of (7.032) and (7.034) shows that J_{I} and J_{II} are equal, as would be expected.

Next the case in which $E < V_0$ should be considered. The quantity k_{II} becomes imaginary for this case and should be evaluated as

$$k_{\mathrm{II}} = i[2m(V_0-E)]^{\frac{1}{2}}/\hbar \qquad (7.035)$$

since this choice of sign gives ψ_{II} the character of an exponential which decays as x increases. The parameter β likewise becomes positive imaginary and the "reflection coefficient"

$$\frac{B}{A} = \frac{1-\beta}{1+\beta} = \exp-i\varphi, \quad \varphi = 2 \arctan |\beta|, \qquad (7.036)$$

is complex with magnitude unity as indicated. One must now go back to equations (7.029) to evaluate the quantities of interest. Thus:

$$\varrho_{\mathrm{I}} = \Psi_{\mathrm{I}}^*\Psi_{\mathrm{I}} = A^*A[2+2\cos(2k_{\mathrm{I}}x+\varphi)] = 4A^*A \cos^2(k_{\mathrm{I}}x+\varphi/2); \quad (7.037)$$

$$\varrho_{\mathrm{II}} = \Psi_{\mathrm{II}}^*\Psi_{\mathrm{II}} = 4A^*A \cos^2(\varphi/2) \exp\left\{-2x[2m(V_0-E)]^{\frac{1}{2}}/\hbar\right\}. \qquad (7.038)$$

Since $|B| = |A|$, $J_r = J_i$, and $J_{\mathrm{I}} = 0$. J_{II} is also zero, obviously, and there is no continuing transmission of probability density into region II although some density does exist there in the form of an exponential "tail" as illustrated in Fig. 7.05.

The expressions for the current densities make it possible to calculate the probabilities for reflection and transmission. The following formulas, valid whether β is real or imaginary, are readily obtained:

$$\text{Probability of reflection} = \frac{J_r}{J_i} = \left|\frac{1-\beta}{1+\beta}\right|^2 . \qquad (7.039)$$

$$\text{Probability of transmission} = \frac{J_t}{J_i} = \frac{4\mathrm{Re}\,\beta}{|1+\beta|^2} . \qquad (7.040)$$

A plane wave solution of non-zero amplitude, although not normalizable, is valuable because it simulates, in a very simple way, the

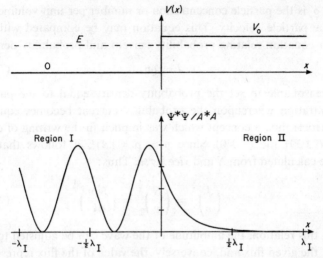

Fig. 7.05. Plots of potential energy and of $\Psi^*\Psi/A^*A$ as functions of x for a positive-going rectangular step with $E = \frac{3}{4}V_0$; $\beta = i3^{-\frac{1}{2}}$. The quantity λ_1, the de Broglie wavelength in region I, is used as a convenient length parameter in both regions.

behavior of a physically large wave packet. Thus the examples discussed here can be regarded as instances involving the incidence of a large packet on a rectangular step when the center of the packet is relatively close to the step. In this context there should also be mentioned another application for a non-normalizable plane wave solution of the Schrödinger equation with constant potential energy; such a wave can be used to study the quantum behavior of a *flux* of widely separated (non-interacting) identical particles having, to a very good approximation, a common momentum $p = \hbar k$. Such particle fluxes are widely used in atomic and nuclear experiments, i.e. in the very kind of situation in which quantum mechanics is eminently useful.

The flux N is defined as the number of particles crossing unit normal area per unit time and is given by:

$$N = \varrho' v, \tag{7.041}$$

where ϱ' is the particle concentration or number per unit volume and v is the particle velocity. This equation may be compared with the similar equation relating probability current and probability density:

$$J = \varrho v. \tag{7.042}$$

It is reasonable to set the probability density equal to the particle concentration whereupon the probability current becomes equal to the particle flux, a concept which was implicit in the writing of equations (7.039) and (7.040). Since $\varrho' = \varrho = |A|^2$, it follows that $|A|$ can be calculated from N and *vice versa*. Thus:

$$|A| = \left(\frac{J}{v}\right)^{\frac{1}{2}} = \left(\frac{N}{v}\right)^{\frac{1}{2}} = \left(\frac{Nm}{\hbar k}\right)^{\frac{1}{2}}. \tag{7.043}$$

Using this relation, the amplitude of the wave can be adjusted to represent the given flux and, conversely, the value of the flux represented by a wave of given amplitude can be obtained. It must be emphasized that the role of the plane wave in this case is only simulative of the true wave function. The latter is a composite of many broadly extended normalized packets, one for each particle.

7.03. The Rectangular Barrier and Tunneling

As a sequel to the rectangular step problem, it is interesting to consider a situation in which there are two semi-infinite regions in which the potential energy is zero separated by a third region, of finite width, in which the potential energy is V_0, as in Fig. 7.06. This central

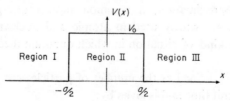

FIG. 7.06. Potential energy as a function of x for the rectangular barrier.

region, of total width a, is considered to be the "barrier". This appellation is especially appropriate when $V_0 > 0$ as will be assumed henceforth. Although stationary-state solutions exist for all $E > 0$, those for which $V_0 > E > 0$ are more informative for present purposes and are the only ones considered here.

As before, the Schrödinger equation takes the form of (7.013) in region I and a similar form (with ψ_{III} substituted) in region III. A wave of given amplitude A is assumed incident from the left in region I and, in order to permit the satisfaction of the boundary conditions, a transmitted wave must be present in region III. In region II, the Schrödinger equation is of the form of (7.017) and, since $V_0 > E$, the solutions are real exponentials rather than traveling waves. The respective solutions in the three regions, in terms of the unknown constants B, C, D, and F, are as follows:

$$\psi_{\text{I}} = A \exp ik_{\text{I}}x + B \exp -ik_{\text{I}}x; \qquad k_{\text{I}} = [2mE]^{\frac{1}{2}}/\hbar; \qquad (7.044)$$

$$\psi_{\text{II}} = C \exp \varkappa x + D \exp -\varkappa x; \qquad \varkappa = [2m(V_0-E)]^{\frac{1}{2}}/\hbar; \quad (7.045)$$

$$\psi_{\text{III}} = F \exp ik_{\text{III}}x; \qquad k_{\text{III}} = k_{\text{I}} = [2mE]^{\frac{1}{2}}/\hbar. \qquad (7.046)$$

Once more, the continuity requirements on ψ and on $d\psi/dx$ constitute the boundary conditions. Thus at $x = -a/2$:

$$\left. \begin{aligned} A \exp\left(-ik_{\text{I}}a/2\right) + B \exp\left(ik_{\text{I}}a/2\right) \\ = C \exp\left(-\varkappa a/2\right) + D \exp\left(\varkappa a/2\right); \\ ik_{\text{I}}A \exp\left(-ik_{\text{I}}a/2\right) - ik_{\text{I}}B \exp\left(ik_{\text{I}}a/2\right) \\ = \varkappa C \exp\left(-\varkappa a/2\right) - \varkappa D \exp\left(\varkappa a/2\right). \end{aligned} \right\} \quad (7.047)$$

At $x = a/2$:

$$\left. \begin{aligned} C \exp\left(\varkappa a/2\right) + D \exp\left(-\varkappa a/2\right) = F \exp\left(ik_{\text{I}}a/2\right); \\ \varkappa C \exp\left(\varkappa a/2\right) - \varkappa D \exp\left(-\varkappa a/2\right) = ik_{\text{I}}F \exp(ik_{\text{I}}a/2). \end{aligned} \right\} \quad (7.048)$$

These four equations make it possible to obtain B, C, D, and F in terms of A. The following temporary substitutions make this task easier:

$$\exp\left(ik_{\text{I}}a/2\right) = \mu; \qquad \exp\left(\varkappa a/2\right) = \eta. \qquad (7.049)$$

The equations, written in standard form with knowns on the right, become:

$$\left.\begin{array}{r} -\mu B+\eta^{-1}C+\eta D = \mu^{-1}A; \\ ik_1\mu B+\varkappa\eta^{-1}C-\varkappa\eta D = ik_1\mu^{-1}A; \\ \eta C+\eta^{-1}D-\mu F = 0; \\ \varkappa\eta C-\varkappa\eta^{-1}D-ik_1\mu F = 0. \end{array}\right\} \quad (7.050)$$

The determinant of the coefficients is:

$$\begin{aligned} \varDelta &= \mu^2[(k_1^2-\varkappa^2)(\eta^2-\eta^{-2})+2ik_1\varkappa(\eta^2+\eta^{-2})] \\ &= 2[(k_1^2-\varkappa^2)\sinh\varkappa a+2ik_1\varkappa\cosh\varkappa a]\exp ik_1 a. \end{aligned} \quad (7.051)$$

One readily finds that:

$$B = (k_1^2+\varkappa^2)(\eta^2-\eta^{-2})A/\varDelta, \quad (7.052)$$

which becomes:

$$\frac{B}{A} = \frac{(k_1^2+\varkappa^2)(\sinh\varkappa a)\exp-ik_1 a}{(k_1^2-\varkappa^2)\sinh\varkappa a+2ik_1\varkappa\cosh\varkappa a}. \quad (7.053)$$

Similarly:

$$F = 4ik_1\varkappa A/\varDelta, \quad (7.054)$$

which is equivalent to:

$$\frac{F}{A} = \frac{2ik_1\varkappa\exp-ik_1 a}{(k_1^2-\varkappa^2)\sinh\varkappa a+2ik_1\varkappa\cosh\varkappa a}. \quad (7.055)$$

As in equation (7.012):

$$\left.\begin{array}{l} J_i = A^*A\hbar k_1/m; \\ J_r = B^*B\hbar k_1/m; \\ J_t = F^*F\hbar k_{\mathrm{III}}/m = F^*F\hbar k_1/m. \end{array}\right\} \quad (7.056)$$

The probabilities of reflection and transmission therefore become:

$$P_r = \frac{J_r}{J_i} = \left|\frac{B}{A}\right|^2; \qquad P_t = \frac{J_t}{J_i} = \left|\frac{F}{A}\right|^2. \quad (7.057)$$

The denominator of $|B/A|^2$ and of $|F/A|^2$ can be transformed as follows:

$$(k_I^2 - \varkappa^2)^2 \sinh^2 \varkappa a + 4k_I^2 \varkappa^2 \cosh^2 \varkappa a$$

$$= (k_I^4 - 2k_I^2 \varkappa^2 + \varkappa^4) \sinh^2 \varkappa a + 4k_I^2 \varkappa^2 (1 + \sinh^2 \varkappa a)$$

$$= (k_I^2 + \varkappa^2) \sinh^2 \varkappa a + 4k_I^2 \varkappa^2$$

$$= 4m^2 V_0^2 \hbar^{-4} \sinh^2 \varkappa a + 16m^2 E(V_0 - E)\hbar^{-4}. \tag{7.058}$$

The final results are:

$$P_r = \frac{\sinh^2 \varkappa a}{\sinh^2 \varkappa a + 4E(V_0 - E)V_0^{-2}} \; ; \tag{7.059}$$

$$P_t = \frac{4E(V_0 - E)V_0^{-2}}{\sinh^2 \varkappa a + 4E(V_0 - E)V_0^{-2}} \; . \tag{7.060}$$

In particular, if conditions are such that $\varkappa a >> 1$, P_t is approximately proportional to a decaying exponential of argument $2\varkappa a$:

$$P_t \approx 16E(V_0 - E)V_0^{-2} \exp{-2\varkappa a}; \qquad \varkappa a \gg 1. \tag{7.061}$$

The remarkable feature of this analysis is that a non-zero probability of transmission exists even though the energy of the incident wave is less than the potential energy in the barrier region. The transmission of particles in this context is called "tunneling" and actually occurs in physical situations of practical interest. Examples are found in the radioactive decay of certain nuclei by alpha particle emission and in the passage of electrons through energetically forbidden regions as in field emission and in the operation of the tunnel diode. It is usual to contrast quantum mechanical tunneling with the classical case of a particle meeting a potential energy barrier which it cannot surmount and from which it is observed to be reflected with unit probability. In making this comparison, however, it must be remembered that the classical case always involves potential energy inclines which are very gradual compared with the de Broglie wavelength of the incident particle and that, under these same circumstances, a detailed quantum mechanical analysis would also yield an utterly negligible probability of transmission. The contrast, in other words, is not between the results of classical mechanics and of quantum mechanics as if these were two antagonistic theories. Rather the contrast is between two significantly different physical situations, one involving a wave packet incident upon a barrier with relatively abrupt inclines and the other involving a similar packet incident upon a barrier with relatively gradual inclines. The former must be treated by quantum methods and, when so treated, the possibility of tunneling becomes apparent. The latter may be treated by either quantum or classical methods and the same gross behavior, with essentially no possibility of tunneling, is predicted regardless of which discipline is invoked.

7.04. Stationary States of the Infinite Rectangular Well

Another very simple yet instructive quantum problem is provided by considering the motion of a particle (or the center of mass of a gross object) which is constrained to move along the x-axis within the range $-(a/2) < x < (a/2)$ but is otherwise unforced. Infinitely hard walls or bumpers must be postulated to provide the constraints at the ends of the range; this means that the potential energy is assumed to rise to infinity at $x = \pm a/2$ and to be equal to a finite constant at points between these limits. This constant is conventionally taken to be zero and the potential energy as a function of x is as illustrated in Fig. 7.07. The author readily acknowledges that the mathe-

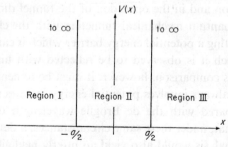

FIG. 7.07. Potential energy as a function of x for the infinite rectangular well.

matics of this problem is considerably easier if the origin of coordinates is taken at the left end of the range rather than at its center. With the origin at the center, however, the very important concept of *parity* can be introduced and a significant mathematical technique (also useful in later examples) can be illustrated. For these pedagogical reasons, then, the origin has been placed as in Fig. 7.07.

By the very nature of this problem, the particle cannot be found outside the range $-(a/2) < x < (a/2)$, hence one may say at the outset that $\psi = 0$ in $-\infty < x < -(a/2)$ and in $(a/2) < x < \infty$. In the central region where $V = 0$ and $\psi \neq 0$, the Schrödinger equation again reduces to (7.013) and the most general solution consists of two oppositely traveling waves:

$$\psi = B' \exp ikx + C' \exp -ikx; \qquad k = [2mE]^{\frac{1}{2}}/\hbar. \qquad (7.062)$$

An equivalent but more convenient form for this solution (and one that is readily obtained from the above) is given by:

$$\psi = B \cos kx + C \sin kx; \qquad k = [2mE]^{\frac{1}{2}}/\hbar. \qquad (7.063)$$

This solution is significantly different from the solutions of earlier examples in this chapter in that there is no "incident wave" of prescribed amplitude and wave number. The contrast is like that between a freely running physical system and a system that is driven by applied forces; the difference is manifested both in the process of analysis and in the final mathematical outcome.

The boundary conditions at $x = \pm a/2$ are also different from those found in earlier examples. At $x = -a/2$, for example, $V = \infty$ on the left side of the boundary and therefore $d\psi/dx$ need not be, and in fact is not, continuous. The function ψ itself must be continuous, however, and this leads to the requirement that $\psi = 0$ at $x = -a/2$. Similarly, $\psi = 0$ at $x = a/2$. These two conditions are both necessary and sufficient; from them, the following equations are obtained:

$$\left. \begin{array}{l} \cos(-ka/2)B + \sin(-ka/2)C = 0; \\ \cos(ka/2)B + \sin(ka/2)C = 0. \end{array} \right\} \qquad (7.064)$$

These have been deliberately written in the form of simultaneous linear equations with B and C as unknowns; the equations are unusual in that the two right-hand sides are equal to zero. If the determinant of the coefficients does not vanish, then the only solution possible is the trivial one, $B = C = 0$. If, on the other hand, the determinant of the coefficients does vanish, then the two equations are linearly dependent and each assigns the same value to the ratio B/C. Any B and C which have this ratio then constitute a non-trivial solution.

Only specific values of k permit the determinant to vanish. Thus one seeks k values for which:

$$\begin{vmatrix} \cos(-ka/2) & \sin(-ka/2) \\ \cos(ka/2) & \sin(ka/2) \end{vmatrix} = \sin ka = 0. \qquad (7.065)$$

Evidently it is necessary for ka to be an integral multiple of π. For an anticipated reason, this multiple is written $(n+1)\pi$, where n is an integer. Thus:

$$k = k_n = (n+1)\pi/a. \qquad (7.066)$$

Since k is dependent upon the value of n, both the stationary state wave function and the energy are related to n:

$$\psi_n = B_n \cos k_n x + C_n \sin k_n x; \qquad k_n = [2mE_n]^{\frac{1}{2}}/\hbar. \qquad (7.067)$$

Thus there is a specified wave function, a value of k, and a value of E for each value of the integral index n. The latter is clearly a quantum number and k and E are quantized.

It is now desirable to consider individually the cases $n = 0$ and $n = 1$. Suppose that $n = 0$; then $k_0 = \pi/a$ and the simultaneous equations (7.064) become:

$$\left. \begin{array}{l} 0B_0 - 1C_0 = 0; \\ 0B_0 + 1C_0 = 0. \end{array} \right\} \qquad (7.068)$$

These require that $(C_0/B_0) = 0$, i.e. that $C_0 = 0$, and that the ψ_0 function be of the form:

$$\psi_0 = B_0 \cos \pi x/a. \qquad (7.069)$$

The constant B_0 is determinable only by normalization. Thus:

$$\int\limits_{-\infty}^{\infty} \psi_0^* \psi_0 \, dx = |B_0|^2 \int\limits_{-a/2}^{a/2} \cos^2(\pi x/a) \, dx = |B_0|^2 \, a/2 = 1. \quad (7.070)$$

Evidently:

$$|B_0| = (2/a)^{\frac{1}{2}}. \quad (7.071)$$

The phase of B_0 is not determined at all and may be assigned arbitrarily; it is usual to assign the value zero, making B_0 positive real. The energy E_0 is $\hbar^2 k_0^2/2m$ or $\hbar^2\pi^2/2ma^2$ and the complete wave function for the stationary state corresponding to $n = 0$ is therefore given by:

$$\Psi_0 = (2/a)^{\frac{1}{2}} \cos(\pi x/a) \exp{-i\omega_0 t}; \qquad \omega_0 = \frac{E_0}{\hbar} = \frac{\hbar\pi^2}{2ma^2}. \quad (7.072)$$

One may next consider the case of $n = 1$. The quantity k_1 is $2\pi/a$ and the simultaneous equations (7.064) become:

$$\left.\begin{array}{l} 1B_1 + 0C_1 = 0; \\ 1B_1 + 0C_1 = 0. \end{array}\right\} \quad (7.073)$$

By reasoning similar to that used earlier, it is now found that $B_1 = 0$ and that the ψ_2 function is of the form:

$$\psi_1 = C_1 \sin(2\pi x/a). \quad (7.074)$$

Once more the process of normalization and the assignment of zero phase dictate that $C_1 = (2/a)^{\frac{1}{2}}$; the complete wave function for the stationary state corresponding to $n = 1$ is:

$$\Psi_1 = (2/a)^{\frac{1}{2}} \sin(2\pi x/a) \exp{-i\omega_1 t}; \qquad \omega_1 = \frac{E_1}{\hbar} = \frac{\hbar 4\pi^2}{2ma^2}. \quad (7.075)$$

It is not difficult to show that $n = -1$ creates a null function and that $n = -2, -3, -4$, etc., produce solutions which are not different from those associated with $n = 0, 1, 2$, etc., respectively. Thus nothing is lost by excluding negative n, and one may say generally:

$$\Psi_n = (2/a)^{\frac{1}{2}} \begin{Bmatrix} \cos{[(n+1)\pi x/a]} \\ \sin{[(n+1)\pi x/a]} \end{Bmatrix} \exp{-i\omega_n t}, \quad \begin{Bmatrix} n \text{ even} \\ n \text{ odd} \end{Bmatrix};$$

$$\omega_n = \frac{E_n}{\hbar} = \frac{\hbar(n+1)^2\pi^2}{2ma^2};$$

$$n = 0, 1, 2, 3, \ldots \infty.$$

(7.076)

The ψ_n functions for the states $n = 0, 1, 2,$ and 3 are illustrated in Fig. 7.08a and the corresponding probability densities are presented in Fig. 7.08b. Figure 7.08c is an energy-level diagram which shows the relative placement of the four respective values of E_n. The state associated with the lowest energy level, i.e. with E_0, is the ground state.

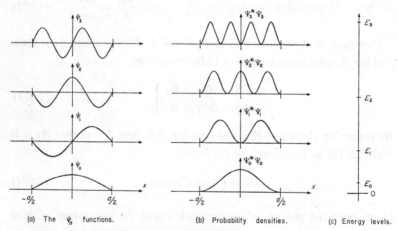

(a) The ψ_n functions.   (b) Probability densities.   (c) Energy levels.

FIG. 7.08. The four lowest-lying stationary states of the infinite rectangular well.

If the particle in the infinite rectangular well is to be in a stationary state, namely a state which involves only one frequency and only one particular value of energy, then it must have as its wave function one of the Ψ_n given in (7.076) and its energy must be equal to the corresponding quantized value E_n. Such a state would have a probability density for the location of the particle similar to one of those illustrated in Fig. 7.08b and would be rightly considered peculiar in the light of classical macroscopic experience. Actually there is no reason, in the material treated in this section or elsewhere, why the wave function of such a particle could not be a superposition of many, perhaps infinitely many, of the stationary states exhibited in (7.076). If it were such a superposition, the state would be non-stationary, the expectation value of its energy could be any value in the continuum $E_0 < \langle E \rangle < \infty$, and the probability density would vary with respect to time. A macroscopic body bouncing back and forth between two bumpers has such a superposition for its wave function and the corresponding probability density is a very narrow, pulse-like, structure which moves back and forth at uniform speed just as the particle itself would be expected to do under classical mechanics.[†]

As a final subject for discussion in this section, the concept of parity introduces itself quite naturally. Parity is an attribute of a function, e.g. a wave function $\Psi(r, t)$, involving a certain type of symmetry with respect to the origin of coordinates. In particular, if the symmetry is such that:

$$\Psi(-r, t) = \Psi(r, t), \qquad (7.077)$$

then Ψ is said to have *even parity*. On the other hand, if it is such that:

$$\Psi(-r, t) = -\Psi(r, t), \qquad (7.078)$$

then Ψ is said to have *odd parity*. In the one-dimensional problem of the infinite rectangular well, x may be substituted for r in these definitions. Thus one finds that Ψ_0, Ψ_2, Ψ_4, etc., all of which involve cosines of the argument $(n+1)\pi x/a$, have even parity and that Ψ_1, Ψ_3, Ψ_5, etc., all of which involve sines of the same argument, have

[†] In a macroscopic (and therefore necessarily approximate) realization of the system under discussion, a stationary state would be enormously improbable. If it were to exist for even an instant, it would be quickly transformed into a superposition of stationary states by the continual interaction between the system and the environment. Even the influence of light falling on the system would be amply sufficient to bring about this consequence.

odd parity. The reason for using $(n+1)$ rather than n is now apparent; with the use of $(n+1)$, an even quantum number is always associated with an even parity wave function and an odd with an odd.

It should not be supposed that (as in the examples just given) all wave functions have a definite parity. If a wave function is a superposition of, say, two functions, one of which is even and the other odd, then it will not have a definite parity; the definiteness of parity, like the definiteness of energy, exists only in special cases. The concept of parity, which acquires more significance in the context of operators (a subject to be discussed in Chapter 8), proves useful at all levels of quantum theory from the elementary to the very advanced.

7.05. Stationary States of the Finite Rectangular Well; Bound States and Continuum States

A natural sequel to the problem of the infinite rectangular well is that of a well in which the potential energy in the forbidden regions is finite rather than infinite. This example, like others in this chapter, is chosen because of its role in the illustration of new concepts. The potential energy as a function of x is shown in Fig. 7.09; as before, the origin of coordinates is placed at the center of the well for reasons related to the concept of parity.

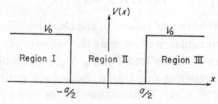

FIG. 7.09. Potential energy as a function of x for the finite rectangular well.

The stationary states in this problem may be divided into two broad classifications. In the first of these, $E < V_0$ and the solutions of the Schrödinger equation in the three regions have properties which are the exact reverse of those found in the rectangular barrier example. Thus ψ_I and ψ_{III} are real exponentials, chosen to decay to zero at $\pm \infty$, respectively, and ψ_{II} is equivalent to a solution composed of traveling waves. In terms of the unknown coefficients A, B, C, and A', the three functions are as follows:

$$\psi_I = A \exp \varkappa x; \qquad\qquad \varkappa = [2m(V_0-E)]^{\frac{1}{2}}/\hbar; \qquad (7.079)$$

$$\psi_{II} = B \cos kx + C \sin kx; \qquad k = [2mE]^{\frac{1}{2}}/\hbar; \qquad\qquad (7.080)$$

$$\psi_{III} = A' \exp -\varkappa x; \qquad\qquad \varkappa \text{ is as above.} \qquad\qquad (7.081)$$

It is important to realize that this solution has been constructed on the premise that both $\varkappa$ and k are positive real. (A negative value of k would yield nothing new and need not be considered; a negative value of $\varkappa$ would correspond to a non-normalizable solution and is not acceptable.)

By application of the usual boundary conditions (the continuity of ψ and of $d\psi/dx$) at $x = -a/2$ and at $x = a/2$, one obtains the following four equations:

$$A \exp(-\varkappa a/2) = B \cos(-ka/2) + C \sin(-ka/2); \qquad (7.082)$$

$$\varkappa A \exp(-\varkappa a/2) = -kB \sin(-ka/2) + kC \cos(-ka/2); \qquad (7.083)$$

$$A' \exp(-\varkappa a/2) = B \cos(ka/2) + C \sin(ka/2); \qquad (7.084)$$

$$-\varkappa A' \exp(-\varkappa a/2) = -kB \sin(ka/2) + kC \cos(ka/2). \qquad (7.085)$$

When written in standard form, these become four equations in the four unknowns A, B, C, and A' with the right-hand sides all equal to zero. In this respect, the problem is similar to that of the infinite rectangular well and quantization of some kind is to be expected. The labor of setting the determinant of the coefficients equal to zero can be lessened and new insights obtained by creating new equations through additions and subtractions of the old; it will be found that this is tantamount to an exploitation of parity. The four new equations and the methods of their generation are as follows:

Add   (7.082), (7.084): $(A+A') \exp(-\varkappa a/2) = 2B \cos(ka/2);$

$$(7.086)$$

Subtract (7.082), (7.084): $(-A+A') \exp(-\varkappa a/2) = 2C \sin(ka/2);$

$$(7.087)$$

Subtract (7.083), (7.085): $\varkappa(A+A') \exp(-\varkappa a/2) = 2kB \sin(ka/2);$

$$(7.088)$$

Add   (7.083), (7.085): $\varkappa(A-A') \exp(-\varkappa a/2) = 2kC \cos(ka/2).$

$$(7.089)$$

Let new variables be defined by:

$$F = A+A'; \qquad G = A-A'. \tag{7.090}$$

With the aid of these, the four equations can be written in the form of two completely disjoint systems:

$$\left. \begin{array}{l} \exp(-\varkappa a/2)F - 2\cos(ka/2)B = 0; \\ \varkappa \exp(-\varkappa a/2)F - 2k\sin(ka/2)B = 0; \end{array} \right\} \tag{7.091}$$

$$\left. \begin{array}{l} -\exp(-\varkappa a/2)G - 2\sin(ka/2)C = 0; \\ \varkappa \exp(-\varkappa a/2)G - 2k\cos(ka/2)C = 0. \end{array} \right\} \tag{7.092}$$

The condition for a non-trivial solution of (7.091) is that the following determinant, called $\varDelta^e$, shall be zero:

$$\varDelta^e = \begin{vmatrix} \exp(-\varkappa a/2) & -2\cos(ka/2) \\ \varkappa \exp(-\varkappa a/2) & -2k\sin(ka/2) \end{vmatrix}$$

$$= -2\exp(-\varkappa a/2)[k\sin(ka/2) - \varkappa\cos(ka/2)] = 0. \tag{7.093}$$

The following is readily seen to be true:

$$\varDelta^e = 0 \Leftrightarrow \varkappa = k \tan(ka/2). \tag{7.094}$$

Similarly, the condition for a non-trivial solution of (7.092) is that the following determinant, called Δ^o, be zero:

$$\Delta^o = \begin{vmatrix} -\exp(-\varkappa a/2) & -2\sin(ka/2) \\ \varkappa\exp(-\varkappa a/2) & -2k\cos(ka/2) \end{vmatrix}$$

$$= 2\exp(-\varkappa a/2)\,[k\cos(ka/2)+\varkappa\sin(ka/2)] = 0. \quad (7.095)$$

The corresponding conclusion is:

$$\Delta^o = 0 \Leftrightarrow \varkappa = -k\cot(ka/2). \quad (7.096)$$

It is impossible to satisfy both (7.094) and (7.096) with real values of k and $\varkappa$ since a multiplication of one of these equations by the other yields $\varkappa^2 = -k^2$. Assume then at first that $\Delta^e = 0$ and that $\Delta^o \neq 0$; it follows that (7.092) has the trivial solution $C = G = 0$, that $A = A'$, and that $F = 2A$. The solutions of the Schrödinger equation in the three regions become:

$$\psi_{\mathrm{I}} = A\exp\varkappa x; \quad \psi_{\mathrm{II}} = B\cos kx; \quad \psi_{\mathrm{III}} = A\exp-\varkappa x. \quad (7.097)$$

The parity of the complete function, considering all parts, is readily seen to be even. The next immediate task is to find k and $\varkappa$. From (7.094) one may write:

$$(\varkappa a/2) = (ka/2)\tan(ka/2). \quad (7.098)$$

Using the definitions of k and $\varkappa$ in (7.080) and (7.079), it is seen that:

$$(ka/2)^2+(\varkappa a/2)^2 = mV_0a^2/2\hbar^2. \quad (7.099)$$

Equations (7.098) and (7.099) must be solved simultaneously; this can be done graphically as in Fig. 7.10. Notice that (7.099) is the arc of a circle in the variables $X = ka/2$ and $Y = \varkappa a/2$ and that only first-quadrant intersections are acceptable since both k and $\varkappa$ must be positive real. The quantity k, the wave number in region II, is immediately relatable to the energy and larger values of k are associated with larger energies. It follows that the intersection labeled Ψ_0 in Fig. 7.10 denotes the even-parity state with the lowest k and the lowest energy, hence the designation. The even-parity state with the next higher k and the next higher energy is Ψ_2 and is so labeled in

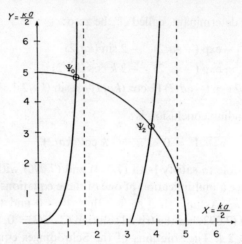

FIG. 7.10. Simultaneous solution of $Y = X \tan X$ and $X^2 + Y^2 = R^2$ to yield even parity bound stationary states of the finite rectangular well; $R = [mV_0 a^2/2\hbar^2]^{\frac{1}{2}}$. Situation illustrated is for $R = 5$.

Fig. 7.10. The number of states with $E < V_0$ is finite and is related to the radius of the circular arc, i.e. to the dimensionless constant $[mV_0 a^2/2\hbar^2]^{\frac{1}{2}}$.

Once k and $\varkappa$ have been found for a particular state, it becomes possible to relate A to B. Recalling that $F = 2A$ and using either of equations (7.091), it is found that:

$$A = B \exp (\varkappa a/2) \cos (ka/2). \qquad (7.100)$$

This can be used to eliminate A in favor of B in (7.097). The absolute determination of $|B|$ can then be accomplished by normalization.

Next, let it be assumed that $\varDelta^o = 0$ and that $\varDelta^e \neq 0$. Then (7.091) has the trivial solution $B = F = 0$; it follows that $A = -A'$ and that $G = -2A'$. The solutions of the Schrödinger equation in the three regions become:

$$\psi_{\text{I}} = -A' \exp \varkappa x; \qquad \psi_{\text{II}} = C \sin kx; \qquad \psi_{\text{III}} = A' \exp -\varkappa x. \qquad (7.101)$$

The parity of this complete function, considering all three parts, is seen to be odd. To find k and $\varkappa$ now, equation (7.096) must be invoked. From it:

$$(\varkappa a/2) = -(ka/2)\cot(ka/2). \qquad (7.102)$$

It is now a question of solving (7.102) and (7.099) simultaneously. This may be done graphically as in Fig. 7.11 and the intersections corresponding to the two odd-parity states Ψ_1 and Ψ_3 are indicated.

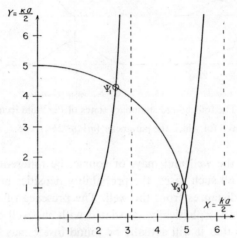

FIG. 7.11. Simultaneous solution of $Y = -X\cot X$ and $X^2 + Y^2 = R^2$ to yield odd parity bound stationary states of the finite rectangular well; $R = [mV_0 a^2/2\hbar^2]^{\frac{1}{2}}$. Situation illustrated is for $R = 5$.

It is now possible to relate A' to C; recalling that $G = -2A'$, one obtains from either of equations (7.092):

$$A' = C\exp(\varkappa a/2)\sin(ka/2). \qquad (7.103)$$

This enables the analyst to eliminate A' in favor of C in (7.101), whereupon $|C|$ may be found by normalization.

The ψ_n functions, the probability densities, and the energy levels for the four typical states Ψ_0, Ψ_1, Ψ_2, and Ψ_3 which have been discussed here are illustrated in Fig. 7.12a, b, and c, respectively. The number of such states, as has been pointed out, depends upon the physical

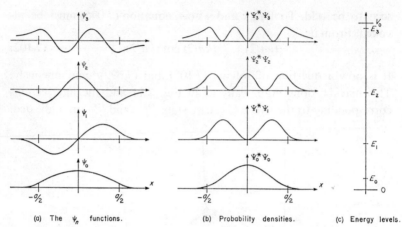

(a) The ψ_n functions.   (b) Probability densities.   (c) Energy levels.

FIG. 7.12. The four bound stationary states of the finite rectangular

well for which the parameter $[mV_0a^2/2\hbar^2]^{\frac{1}{2}} = 5$.

parameters of the well and may, of course, be different from four. For these and all such cases, the probability densities become negligible at great distances from the well. The presence of the particle, therefore, is always physically associated with the well even though the probability that it will actually be found inside may be less than one-half in some situations. Because of this association, the states in question are called *bound states*. Such states also have the common characteristic that $E < V_0$; like planets in elliptic orbits around the Sun, they relate to particles which have total energies less than the value of the potential energy at infinity. It is this circumstance that creates the exponential behavior of the wave function outside the well and is actually responsible for the "binding" of the particle. In retrospect, it can now be seen that all of the states of the infinite rectangular well are bound states; this feature is characteristic of systems in which the potential energy rises to infinity on both sides of some central "enclosed" region.

The second broad classification of the stationary states of the finite rectangular well is that for which $E > V_0$. In the context of this chapter, a suitable way of handling these states is to regard each one as composed of a wave incident from, say, the left in Fig. 7.09 accompanied by a reflected wave also in region I and by a transmitted wave in region III. (For each such state, there exists another independent state of the same energy involving a wave incident from the right with a reflected wave in region III and a transmitted wave in region I.) The solutions of the Schrödinger equation are of the traveling wave variety in all three regions and may be written:

$$\psi_{\mathrm{I}} = A \exp ik_{\mathrm{I}}x + B \exp -ik_{\mathrm{I}}x; \qquad k_{\mathrm{I}} = [2m(E-V_0)]^{\frac{1}{2}}/\hbar; \quad (7.104)$$

$$\psi_{\mathrm{II}} = C \exp ik_{\mathrm{II}}x + D \exp -ik_{\mathrm{II}}x; \quad k_{\mathrm{II}} = [2mE]^{\frac{1}{2}}/\hbar; \qquad (7.105)$$

$$\psi_{\mathrm{III}} = F \exp ik_{\mathrm{III}}x; \qquad k_{\mathrm{III}} = k_{\mathrm{I}} \qquad = [2m(E-V_0)]^{\frac{1}{2}}/\hbar. \quad (7.106)$$

Once more the enforcement of the boundary conditions at $x = \pm a/2$ creates four equations in four unknowns; these are in fact identical with the rectangular barrier equations (7.050) if $\varkappa$ is set equal to ik_{II} and if η is set equal to $\exp (ik_{\mathrm{II}}/2)$. These equations, unlike those related to the bound states, have right-hand sides which are not all zero but rather contain the amplitude A of the incident wave. Solutions are obtainable for B, C, D, and F in terms of A. Those for B and F may be adapted from (7.053) and (7.055) via the substitutions suggested above and the results are:

$$\frac{B}{A} = \frac{(k_{\mathrm{I}}^2 - k_{\mathrm{II}}^2)(\sin k_{\mathrm{II}}a) \exp -ik_{\mathrm{I}}a}{(k_{\mathrm{I}}^2 + k_{\mathrm{II}}^2) \sin k_{\mathrm{II}}a + 2ik_{\mathrm{I}}k_{\mathrm{II}} \cos k_{\mathrm{II}}a}; \qquad (7.107)$$

$$\frac{F}{A} = \frac{2ik_{\mathrm{I}}k_{\mathrm{II}} \exp -ik_{\mathrm{I}}a}{(k_{\mathrm{I}}^2 + k_{\mathrm{II}}^2) \sin k_{\mathrm{II}}a + 2ik_{\mathrm{I}}k_{\mathrm{II}} \cos k_{\mathrm{II}}a} \qquad (7.108)$$

Again the denominator of $|B/A|^2$ and of $|F/A|^2$ can be transformed as follows:

$$(k_{\mathrm{I}}^2 + k_{\mathrm{II}}^2) \sin^2 k_{\mathrm{II}}a + 4k_{\mathrm{I}}^2 k_{\mathrm{II}}^2 \cos^2 k_{\mathrm{II}}a$$
$$= (k_{\mathrm{I}}^2 - k_{\mathrm{II}}^2)^2 \sin^2 k_{\mathrm{II}}a + 4k_{\mathrm{I}}^2 k_{\mathrm{II}}^2$$
$$= 4m^2 V_0^2 \hbar^{-4} \sin^2 k_{\mathrm{II}}a + 16m^2 E(E-V_0)\hbar^{-4}. \qquad (7.109)$$

The final results for the probabilities of reflection and transmission
are:

$$P_r = \frac{\sin^2 k_{II}a}{\sin^2 k_{II}a + 4E(E-V_0)V_0^{-2}};$$ (7.110)

$$P_t = \frac{4E(E-V_0)V_0^{-2}}{\sin^2 k_{II}a + 4E(E-V_0)V_0^{-2}}.$$ (7.111)

The outcome of this analysis is interesting for at least two reasons.
First, there is no quantization of E or of the closely related quantities
k_I and k_{II}. In other words, the Schrödinger equation and the boundary
conditions permit the incident wave to have any wave number and
any energy greater than V_0. Since E may have any value in a semi-
infinite continuum, these states are called *continuum states*. It is
noteworthy that there is now no integral index which takes on discrete values and thereby counts these states. However, the energy
itself (or, alternatively, k_I or k_{II}) can be thought of as a "continuous
quantum number" which identifies them just as an index would do.
In fact, the energy can be regarded as an identifying number for all
the states of the finite rectangular well, taking on discrete values for
the bound states and continuous values for the continuum states.
The reader should keep this example in mind and be prepared for the
fact that the term "quantum number" is used rather loosely in theoretical discussions.

Just as in earlier examples, a non-normalizable wave is here used
as a convenient simulator for a normalizable wave packet. Such a
packet would spend only a finite time in the neighborhood of the well
but an infinite time in approaching and (perhaps after splitting in
two) in receding from, the well. Thus continuum states are non-bound
states; they represent the situation in which a particle interacts with
a given object only briefly and then leaves the scene never to return.

The second interesting feature of the results is that $P_r = 0$ and $P_t = 1$
for certain values of E, namely those for which $[2mE]^{\frac{1}{2}}a/\hbar$ or $k_{II}a$ is
an integer times π. This means that at certain energies the incident
wave "resonates" with the well so as to render the latter effectively

invisible. A similar effect occurs in optics when a plane wave is incident normally upon a thin lamina; if the number of half-wavelengths within the lamina is an integer, no reflection takes place. Again, a similar situation can exist in three dimensions with respect to electrons incident upon a rare gas atom. For a certain electron energy, the atom is invisible, i.e. the electron is not scattered by it. This phenomenon is called the *Ramsauer–Townsend effect*.

7.06. The Particle in a Box

There is beneficial experience to be gained by extending the example of the infinite rectangular well to three dimensions; the corresponding physical situation involves a central region in the form of a rectangular parallelepiped of dimensions a_1, a_2, a_3, in which the potential energy is zero, completely surrounded by a forbidden region in which the potential energy is infinite. An illustration is provided in Fig. 7.13.

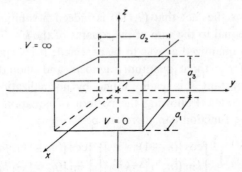

FIG. 7.13. Rectangular parallelepiped or "box" (with center at the origin) within which particle is confined.

Thus the particle in question is confined to the parallelepiped region, i.e. it is "in a box". Within this region, the particle has three degrees of freedom all of which come into play; the Schrödinger equation may be written as follows:

$$\nabla^2\psi + \frac{2mE}{\hbar^2}\,\psi = 0. \tag{7.112}$$

14*

The boundary conditions, like those of the infinite rectangular well in one dimension, are simply that $\psi = 0$ at the boundaries of the box. Under these conditions, the solution of (7.112) takes the form:

$$\psi = A \begin{Bmatrix} \cos k_x x \\ \sin k_x x \end{Bmatrix} \begin{Bmatrix} \cos k_y y \\ \sin k_y y \end{Bmatrix} \begin{Bmatrix} \cos k_z z \\ \sin k_z z \end{Bmatrix}, \qquad (7.113)$$

where the quantities k_x, k_y, and k_z are quantized as follows:

$$\left. \begin{array}{l} k_x = k_{x n_1} = (n_1 + 1)\pi/a_1; \\ k_y = k_{y n_2} = (n_2 + 1)\pi/a_2; \\ k_z = k_{z n_3} = (n_3 + 1)\pi/a_3. \end{array} \right\} \qquad (7.114)$$

It may be noticed that there are now three quantum numbers n_1, n_2, and n_3, and that the boundary conditions require that the upper function in (7.113) be chosen if the corresponding quantum number is even; the lower, if it is odd. Regardless of these choices, it is always true that:

$$\nabla^2 \psi = -(k_x^2 + k_y^2 + k_z^2)\psi, \qquad (7.115)$$

which confirms the fact that (7.113) is indeed a solution of (7.112) with $2mE/\hbar^2$ equal to the sum of the squares of the k's. The energy is therefore also quantized as was to be expected; it is dependent upon the values of the three quantum numbers and should be written $E_{n_1 n_2 n_3}$. The constant A is readily found by normalization and is the same for all states. Following the pattern of equation (7.076), the complete wave function can therefore be written:

$$\left. \begin{array}{l} \psi_{n_1 n_2 n_3} = \left(\dfrac{8}{a_1 a_2 a_3} \right)^{\frac{1}{2}} \begin{Bmatrix} \cos\left[(n_1+1)\pi x/a_1\right] \\ \sin\left[(n_1+1)\pi x/a_1\right] \end{Bmatrix} \begin{Bmatrix} \cos\left[(n_2+1)\pi y/a_2\right] \\ \sin\left[(n_2+1)\pi y/a_2\right] \end{Bmatrix} \\[2em] \begin{Bmatrix} \cos\left[(n_3+1)\pi z/a_3\right] \\ \sin\left[(n_3+1)\pi z/a_3\right] \end{Bmatrix} \exp - i\omega_{n_1 n_2 n_3} t, \quad \begin{Bmatrix} \text{quantum no. even} \\ \text{quantum no. odd} \end{Bmatrix}; \\[2em] \omega_{n_1 n_2 n_3} = \dfrac{E_{n_1 n_2 n_3}}{\hbar} = \dfrac{\hbar \pi^2}{2m} \left[\dfrac{(n_1+1)^2}{a_1^2} + \dfrac{(n_2+1)^2}{a_2^2} + \dfrac{(n_3+1)^2}{a_3^2} \right]; \\[2em] \left. \begin{array}{l} n_1 \\ n_2 \\ n_3 \end{array} \right\} = 0, 1, 2, \ldots \infty. \end{array} \right\} \qquad (7.116)$$

Each of the states displayed above has a definite parity, the latter being determined by the definition of parity as given in equations (7.077) and (7.078). If $+1$ is taken as the "value" of the parity of an even function and -1 as that of an odd function, then it is readily seen that the parity of the whole three-dimensional function is the product of the parities of its individual factors:

$$P = P_1 P_2 P_3. \tag{7.117}$$

The conclusion, expressed in words, is that if none or two of the individual one-dimensional factors is odd, then Ψ is even; if one or three of the individual one-dimensional factors is odd, then Ψ is odd.

Another feature of some quantum-wave functions which is readily illustrated by the particle in a box is *degeneracy*. Two or more independent wave functions are degenerate if they have exactly the same value of energy. For the general box problem, it is not at all impossible for two or more independent wave functions to have the same energy by sheer coincidence. If the three dimensions of the box are related in some especially simple way, however, the number of examples of degeneracy increases. The ultimate case is that of the cube in which $a_1 = a_2 = a_3$ and most of the stationary states are degenerate. (The states Ψ_{100}, Ψ_{010}, and Ψ_{001} of the cube, for example, all have the same energy and are said to be three-fold degenerate.) It should be noticed that the members of a degenerate set of stationary states can be superposed in infinitely many ways to form another state which is also stationary. Where degeneracy exists, then, the function $\Psi^*\Psi$ associated with a given energy is non-unique. In retrospect, it can now be appreciated that some of the continuum states of the rectangular step and all the continuum states of the finite rectangular well are two-fold degenerate. The concept of degeneracy will be encountered again in Chapter 8 where its significance will become more apparent.

7.07. The One-dimensional Harmonic Oscillator

The one-dimensional harmonic oscillator, when treated from the quantum point of view, is probably the best known and most basic example of the application of Schrödinger's wave mechanics. Although the techniques required to obtain the stationary-state solutions are interesting in themselves, the problem as a whole has a transcendent significance which is discernible in some of the very advanced phases of quantum theory. The analysis of this simple device is therefore eminently worthy of the reader's attention.

The system under consideration, illustrated in Fig. 1.01, is known to have the potential energy $Kx^2/2$ where K is the elastic constant of the spring. The ψ-function depends upon x alone and the Schrödinger equation takes the form:

$$-\frac{\hbar^2}{2M}\frac{d^2\psi}{dx^2} + \tfrac{1}{2}Kx^2\psi = E\psi. \tag{7.118}$$

Here M is the mass of the moving object; the classical radian frequency, $\Omega = (K/M)^{\frac{1}{2}}$, appears repeatedly in the derivation. As a first step, let the following dimensionless independent variable be defined:

$$\xi = (M\Omega/\hbar)^{\frac{1}{2}}x; \qquad x = (\hbar/M\Omega)^{\frac{1}{2}}\xi. \tag{7.119}$$

Evidently ψ can be regarded as $\psi(\xi)$ and, on substitution from (7.119), the Schrödinger equation becomes:

$$-\frac{\hbar^2}{2M}\frac{M\Omega}{\hbar}\frac{d^2\psi}{d\xi^2} + \frac{K\hbar}{2M\Omega}\xi^2\psi = E\psi. \tag{7.120}$$

Division by $\hbar\Omega$, which has the dimensions of energy, yields:

$$-\tfrac{1}{2}\frac{d^2\psi}{d\xi^2} + \tfrac{1}{2}\xi^2\psi = \frac{E}{\hbar\Omega}\,\psi. \tag{7.121}$$

It is found that the following substitution facilitates the solution. Let:

$$\psi(\xi) = u(\xi)\exp\left(-\xi^2/2\right). \tag{7.122}$$

It is readily seen that:

$$\frac{d^2\psi}{d\xi^2} = \left[\frac{d^2u}{d\xi^2} - 2\xi\frac{du}{d\xi} + (\xi^2 - 1)u\right]\exp\left(-\xi^2/2\right), \qquad (7.123)$$

whereupon the Schrödinger equation reduces to:

$$-\frac{1}{2}\frac{d^2u}{d\xi^2} + \xi\frac{du}{d\xi} - \varepsilon u = 0, \qquad (7.124)$$

with:

$$\varepsilon = \frac{E}{\hbar\Omega} - \frac{1}{2}. \qquad (7.125)$$

The power series method will now be used; it is sufficient to take the following series in non-negative integral powers of ξ:

$$u = \sum_{j=0}^{\infty} a_j\xi^j; \qquad (7.126)$$

$$\frac{du}{d\xi} = \sum_{j=0}^{\infty} ja_j\xi^{j-1}; \qquad (7.127)$$

$$\frac{d^2u}{d\xi^2} = \sum_{j=0}^{\infty} j(j-1)a_j\xi^{j-2}. \qquad (7.128)$$

When substituting into the second-derivative term of (7.124), one may let $j = k+2$ and begin the series with the $k = 0$ term since the first two terms of (7.128) vanish. In the other two terms of (7.124), let $j = k$. The result is:

$$-\frac{1}{2}\sum_{k=0}^{\infty}(k+2)(k+1)a_{k+2}\xi^k + \sum_{k=0}^{\infty}(k-\varepsilon)a_k\xi^k = 0. \qquad (7.129)$$

This is of the form:

$$\sum_{k=1}^{\infty}[-\tfrac{1}{2}(k+2)(k+1)a_{k+2} + (k-\varepsilon)a_k]\xi^k = 0, \qquad (7.130)$$

and can be satisfied for arbitrary ξ only if the coefficient in the square brackets vanishes for all k. The result is a recurrence relation for the coefficient a_k:

$$a_{k+2} = \frac{2(k-\varepsilon)}{(k+2)(k+1)} a_k. \tag{7.131}$$

It may be noticed that this relation generates two completely distinct series of coefficients and, therefore, two distinct types of functions. These types are distinguished by their parities. Thus, if one begins with the coefficient a_0 and applies (7.131), the result is the even function:

$$u^e(\xi) = a_0 + a_2\xi^2 + a_4\xi^4 + \dots. \tag{7.132}$$

On the other hand, if one begins with the coefficient a_1 and applies (7.131), and odd function is generated:

$$u^n(\xi) = a_1\xi + a_3\xi^3 + a_5\xi^5 + \dots. \tag{7.133}$$

In both of these series, any one of the coefficients may be chosen arbitrarily whereupon the others are determined. The resulting function satisfies (7.124) but the acceptability of the corresponding ψ as the ψ function of a quantum state must be scrutinized more closely.

The question of whether series (7.132) and (7.133) are finite or infinite is a crucial one. In general, both series will be infinite unless ε happens to be a non-negative integer of the appropriate parity for only then will k eventually become equal to ε with the result that subsequent coefficients vanish. The properties at large $|\xi|$ of an infinite series generated by (7.131) are relatively easy to deduce. Consider the ratio of adjacent coefficients at large k:

$$\frac{a_{k+2}}{a_k} \sim \frac{2}{k} \quad \text{as} \quad k \to \infty. \tag{7.134}$$

This is the same as the corresponding ratio for the function $\xi^N \exp \xi^2$ where N is an undetermined integer. The relevant series is:

$$\xi^N \exp \xi^2 = \xi^N + \xi^{N+2} + \frac{\xi^{N+4}}{2!} + \frac{\xi^{N+6}}{3!} + \dots.$$

$$= \sum_{k=N,\, N+2,\, \dots}^{\infty} b_k \xi^k. \tag{7.135}$$

Notice that k advances by two from one term to the next and that it always has the same parity as N. Evidently:

$$b_k = \frac{1}{[(k-N)/2]!} \tag{7.136}$$

and the ratio of adjacent coefficients at large k is:

$$\frac{b_{k+2}}{b_k} = \frac{[(k-N)/2]!}{[(k+2-N)/2]!} = \frac{1}{(k+2-N)/2} \sim \frac{2}{k} \quad \text{as} \quad k \to \infty. \tag{7.137}$$

The conclusion is that in the limit of large $|\xi|$, where the terms of very high degree become the most important, the functions $u(\xi)$ and $\xi^N \exp \xi^2$ become proportional to one another. This statement can be put into precise mathematical language which encompasses both parities; as $|\xi| \to \infty$,

$$u(\xi) \sim C\xi^N \exp \xi^2, \tag{7.138}$$

where C is a constant and the integer N is of the same parity[†] as $u(\xi)$. From (7.122), the following can be said of the corresponding ψ function in the same limit:

$$\psi(\xi) = u(\xi) \exp(-\xi^2/2) \sim C\xi^N \exp(\xi^2/2). \tag{7.139}$$

Obviously this asymptotic behavior at large $|\xi|$ is unacceptable in a quantum wave function. Such a function cannot be normalized nor can it simulate a normalizable packet as does a plane wave. The conclusion is that the series simply cannot be infinite and that the quantity ε must therefore equal a non-negative integer which will henceforth be called n. If n is even, the series $u^e(\xi)$ becomes an even polynomial of degree n; if odd, the series $u^o(\xi)$ becomes an odd polynomial, also of degree n. In either event, one obtains a normalizable wave function identified by the number n and the latter therefore becomes the quantum number for the system. Since the energy is related to ε through

[†] It would be usual to think of N as equal to zero or to unity. The argument under development would have the same conclusion, however, if N were equal to any finite integer of the proper parity.

equation (7.125) and since $\varepsilon = n$, it follows that E is quantized and that its stationary-state values are given by:

$$E_n = (n+\tfrac{1}{2})\hbar\Omega. \tag{7.140}$$

It is noteworthy that $E_{n+1} - E_n = \hbar\Omega$ which means that all the energy levels are equally spaced. This is a singular feature of the one-dimensional harmonic oscillator.

The polynomials alluded to above obviously have coefficients which obey recurrence relation (7.131) with ε replaced by n. Mathematical constructs with this property are called *Hermite polynomials* and are designated $H_n(\xi)$. A few of these (with the coefficient of the highest power of ξ set equal to 2^n according to convention) are displayed

TABLE 7.01. *Hermite Polynomials*

Even parity polynomials	Odd parity polynomials
$H_0 = 1$	$H_1 = 2\xi$
$H_2 = -2 + 4\xi^2$	$H_3 = -12\xi + 8\xi^3$
$H_4 = 12 - 48\xi^2 + 16\xi^4$	$H_5 = 120\xi - 160\xi^3 + 32\xi^5$

in Table 7.01. These polynomials have the following interesting properties which are stated here without proof:[†]

$$H_{n+1} = 2\xi H_n - 2nH_{n-1}; \tag{7.141}$$

$$\frac{\mathrm{d}H_n}{\mathrm{d}\xi} = 2nH_{n-1}; \tag{7.142}$$

$$\int_{-\infty}^{\infty} H_m(\xi)\,H_n(\xi)\,\exp{-\xi^2}\,\mathrm{d}\xi = 2^n n!\,\pi^{\frac{1}{2}}\delta_{mn}. \tag{7.143}$$

The quantity δ_{mn} is known as the "delta symbol"; it is equal to unity if $m = n$, to zero if $m \neq n$. Equation (7.143) is called an *orthogonality*

[†] For a detailed treatment of Hermite polynomials, see ref. 33.

relation. A set of functions, like a set of vectors, may be mutually orthogonal with respect to some prescribed rule of combination, i.e. it may be that when two different members of the set are combined according to the rule, the result is zero but when a member of the set is combined with itself, the result is positive definite. For ordinary vectors, where orthogonality connotes perpendicularity, the rule of combination is the dot product. For Hermite polynomials, the rule of combination is integration with the factor $\exp -\xi^2$ over the range $-\infty < \xi < \infty$. (The factor $\exp -\xi^2$ is called the *weight function*.) The concept of orthogonality is very important in quantum mechanics and numerous examples will be encountered.

In order to give the normalization integral the required value of unity, the Hermite polynomial $H_n(\xi)$ must be multiplied by a constant A_n before being incorporated into the wave function for the nth stationary state. The latter function, complete with time dependence, becomes:

$$\left.\begin{array}{l} \Psi_n(x, t) = A_n H_n(\xi) \exp(-\xi^2/2) \exp -i\omega_n t; \\[2mm] \xi = (M\Omega/\hbar)^{\frac{1}{2}} x; \\[2mm] \omega_n = \dfrac{E_n}{\hbar} = (n+\tfrac{1}{2})\Omega. \end{array}\right\} \qquad (7.144)$$

The normalization integral may be written as follows:

$$\int_{-\infty}^{\infty} \Psi_n^* \Psi_n \, dx = (\hbar/M\Omega)^{\frac{1}{2}} \int_{-\infty}^{\infty} \Psi_n^* \Psi_n \, d\xi = 1. \qquad (7.145)$$

Thus:

$$(\hbar/M\Omega)^{\frac{1}{2}} |A_n|^2 \, 2^n n! \pi^{\frac{1}{2}} = 1 \qquad (7.146)$$

and if the phase of A_n is arbitrarily set equal to zero, one obtains:

$$A_n = \left(\frac{M\Omega}{\pi\hbar}\right)^{\frac{1}{4}} (2^n n!)^{-\frac{1}{2}}. \qquad (7.147)$$

The orthogonality of the Hermite polynomials implies the orthogonality of the complete wave functions $\Psi_n(x, t)$. The latter is by no means an accidental property; it follows, rather, from a very general principle to be discussed in Chapter 8. The rule of combination which applies to the complete wave functions is simply the integration of the Hermitian product $\Psi_m^* \Psi_n$ over the whole of configuration space which, in this case, is over $-\infty < x < \infty$. Thus:

$$\int_{-\infty}^{\infty} \Psi_m^*(x, t)\, \Psi_n(x, t)\, \mathrm{d}x = \delta_{mn}. \tag{7.148}$$

The same orthogonality relation also applies to the ψ_n functions and one may say:

$$\int_{-\infty}^{\infty} \psi_m^*(x)\, \psi_n(x)\, \mathrm{d}x = \delta_{mn}. \tag{7.149}$$

Since the phases of the constants A_n were chosen to be zero, the ψ_n functions are all real in this case and the conjugation in (7.149) is without effect. It is important, however, that an expression like (7.149) be written with the first function conjugated since there are situations in which the reality of the ψ_n cannot be assumed. Functions like

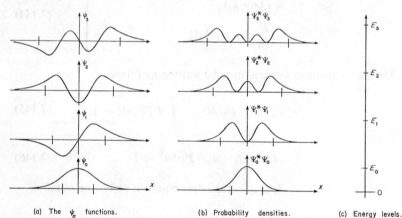

(a) The ψ_n functions.   (b) Probability densities.   (c) Energy levels.

FIG. 7.14. The four lowest-lying stationary states of the one-dimensional harmonic oscillator. The short vertical marks are points on the x-axis at which the potential energy becomes equal to the total energy E_n; these points would be the turning points of a hypothetical classical motion of total energy E_n.

$\Psi_n(x, t)$ and $\psi_n(x)$, which are both *orthogonal* and *normalized* are said to be *orthonormalized*, a term whose etymology is obvious.

The ψ_n functions for the four lowest-lying stationary states of the one-dimensional harmonic oscillator together with the corresponding probability densities and energy levels are illustrated in Fig. 7.14a, b, and c, respectively. All the states are bound states; there are no continuum states. The discussion concerning the possibility, in fact the overwhelming probability, that a macroscopic realization of the infinite rectangular well will have a superposition of stationary states as its wave function also applies to the harmonic oscillator. Under such conditions, the probability density may be put into the form of a spatially concentrated structure which executes the classical motion. This phenomenon will be discussed quantitatively in Chapter 9 after additional background material has been treated.

CHAPTER 8

OPERATORS, OBSERVABLES, AND THE QUANTIZATION OF A PHYSICAL SYSTEM

8.01. General Definition of Operators; Linear Operators

An *operator* is a mathematical entity whose presence in conjunction with a given function indicates that a particular process is to be performed on that function. The function in question is called the *operand*; it is in general changed by the action of the operator into another function called the *result*. In mathematical expressions, it is customary to place the symbol for the operator immediately before (to the left of) the symbol for the operand. In this book, all non-obvious symbols for operators will be identified by a circumflex (⌃); the following equation is illustrative:

$$\underset{\text{OPERATOR}}{\hat{A}} \quad \underset{\text{OPERAND}}{f(x)} \quad = \quad \underset{\text{RESULT}}{g(x).} \tag{8.001}$$

By their very nature, some operators can operate only on functions belonging to a certain restricted class. This matter will usually be clear from the context and should not cause undue difficulty.

An operator may be very simple as in $\hat{A} = x$. Multiplication is understood here and one has, for example:

$$\hat{A}(3x^2 + \sin x) = 3x^3 + x \sin x. \tag{8.002}$$

The operator $\hat{B} = x^2 + \mathrm{d}/\mathrm{d}x$ is more complicated; its action on the

210

function of the previous example is as follows:

$$\hat{B}(3x^2 + \sin x) = x^2(3x^2 + \sin x) + \frac{d}{dx}(3x^2 + \sin x)$$

$$= 3x^4 + x^2 \sin x + 6x + \cos x. \qquad (8.003)$$

Operators may be much more abstract. Examples are the parity operator $\hat{\Pi}$ and the exchange operator $\hat{X}_{mn}$. The action of the parity operator on a function of r and t is as follows:

$$\hat{\Pi}f(r, t) = f(-r, t). \qquad (8.004)$$

The exchange operator acts only on functions of the positions of two or more particles and has as its effect the *exchanging* of the coordinates of one particle with those of another within the function. Thus:

$$\hat{X}_{23}f(r_1, r_2, r_3, r_4, t) = f(r_1, r_3, r_2, r_4, t). \qquad (8.005)$$

The unit operator $\hat{I}$ and the null operator $\hat{0}$ can operate on any function and have the following obvious effects:

$$\left.\begin{aligned} \hat{I}f &= f. \\ \hat{0}f &= 0. \end{aligned}\right\} \qquad (8.006)$$

Some operators have inverse operators associated with them. Thus if $\hat{A}$ has an inverse, the latter is designated by the symbol $\hat{A}^{-1}$; $\hat{A}^{-1}$ undoes the operation performed by $\hat{A}$. Specifically, if:

$$\hat{A}f(x) = g(x), \qquad (8.007)$$

then:

$$\hat{A}^{-1}g(x) = f(x). \qquad (8.008)$$

An operator $\hat{A}$ is *linear* if and only if:

$$\hat{A}(\alpha_1 f_1 + \alpha_2 f_2 + \alpha_3 f_3 + \ldots) = \alpha_1 \hat{A} f_1 + \alpha_2 \hat{A} f_2 + \alpha_3 \hat{A} f_3 + \ldots, \qquad (8.009)$$

where the α_j are arbitrary constants and the f_j are arbitrary members of the class of functions upon which $\hat{A}$ can operate. In other words,

(i) it is immaterial whether a given constant immediately precedes or immediately follows a linear operator and (ii) a linear operator acting on an arbitrary superposition has the same effect as the superposition of the results of the same operator acting on the individual terms. All the specific examples of operators given so far have been linear operators. Two examples of non-linear operators are:

$$\hat{C}f(x) = f^2(x); \qquad (8.010)$$
$$\hat{D}f(x) = x + f(x). \qquad (8.011)$$

Notice that $\hat{D}$ in the above is not equal to $(x + \hat{1})$; $(x + \hat{1})$ is a linear operator and yields $xf(x) + f(x)$ rather than $x + f(x)$.

Subsequent discussion in this chapter will be concerned with only linear operators; such operators, in fact, will be the only types used in the remainder of this book.

8.02. The Non-commutative Algebra of Operators

The presence of an operator indicates the presence of a mathematical process and conversely; for every allowable process, a corresponding operator implicitly exists. Thus if operators $\hat{A}$ and $\hat{B}$ are permitted to act separately on the function $f(x)$ and the results added, yielding $g(x)$, the whole set of processes may be regarded as a single process which converts $f(x)$ into $g(x)$ and which can therefore be represented by a single operator $\hat{C}$:

$$\hat{A}f(x) + \hat{B}f(x) = g(x) = \hat{C}f(x). \qquad (8.012)$$

One may now consider the operators $\hat{A}$, $\hat{B}$, and $\hat{C}$ apart from the operand $f(x)$ and write:

$$\hat{A} + \hat{B} = \hat{C}, \qquad (8.013)$$

which is an equation in operators alone and expresses the fact that $\hat{A}$ and $\hat{B}$ have been added to obtain $\hat{C}$.

Another way to combine two operators is to allow them to act in sequence on a given function, say $f(x)$; once more the set of processes can be thought of as a single process with which is associated a single operator $\hat{D}$. Thus:

$$\hat{A}\hat{B}f(x) = \hat{A}[\hat{B}f(x)] = \hat{A}k(x) = l(x) = \hat{D}f(x). \qquad (8.014)$$

Again, one may consider the operators apart from the operand and write:

$$\hat{A}\hat{B} = \hat{D}. \qquad (8.015)$$

The sequential action of two operators, which is equivalent to the action of a third operator, is defined as *operator multiplication* and is succinctly expressed by an equation like (8.015). It is noteworthy that, whereas the addition of operators is a very natural extension of the addition of numbers, the multiplication of operators is derived from the concept of sequential operation and is less closely related to the multiplication of numbers.

With respect to operators, it is found that (i) addition is associative and commutative, (ii) multiplication is associative, and (iii) the two operations of addition and multiplication are, provided that the operators are linear, distributive. Thus a whole algebra of operators exists and this algebra has all the properties of the ordinary algebra of numbers except one. The one exception is very important, namely, the *non-commutativity of operator multiplication*. This exception occurs because, in general, the order in which two operators act in sequence has an influence upon the result, i.e. in general $\hat{A}\hat{B} \neq \hat{B}\hat{A}$. Perhaps the simplest example is that in which $\hat{A} = x$ and $\hat{B} = d/dx$. Then:

$$\left. \begin{array}{l} \hat{A}\hat{B}f(x) = x\,\dfrac{df}{dx}\,; \\[2mm] \hat{B}\hat{A}f(x) = \dfrac{d}{dx}[xf] = x\,\dfrac{df}{dx}+f(x). \end{array} \right\} \qquad (8.016)$$

Thus:

$$(\hat{A}\hat{B}-\hat{B}\hat{A})f(x) = x\,\frac{df}{dx} - \left[x\,\frac{df}{dx}+f(x) \right] = -f(x). \qquad (8.017)$$

It is seen that, in this case, $\hat{A}\hat{B} \neq \hat{B}\hat{A}$ and that $\hat{A}\hat{B}-\hat{A}\hat{B} = -\hat{1}$, not

$\hat{0}$; it is therefore said that $\hat{A}$ and $\hat{B}$ do not commute. The operator $(\hat{A}\hat{B} - \hat{B}\hat{A})$ is often designated by the special symbol $[\hat{A}, \hat{B}]$ and is called the *commutator* of $\hat{A}$ and $\hat{B}$. (Notice carefully that the first operator in the bracketed symbol is first in the first term of the commutator.) Evidently, one way of saying that two operators do not commute is to say that their commutator is not zero. Many pairs of operators employed in quantum mechanics commute but some do not; the latter have a special significance.

8.03. Eigenfunctions and Eigenvalues; the Operators for Momentum and Position

It has been mentioned that a given operator, say $\hat{A}$, may be able to accept as operands only functions belonging to a certain restricted class. Of this class of possible operands, however, there exists a generally much narrower subclass composed of functions which bear a very special relationship to $\hat{A}$. These special operands are called *eigenfunctions*;[†] they have the remarkable property that, when operated upon by $\hat{A}$, the result is the operand itself multiplied by a constant. Thus the functions $f_1, f_2, \ldots, f_n, \ldots$, which satisfy the following relationships:

$$\left.\begin{aligned} \hat{A}f_1 &= a_1 f_1, \\ \hat{A}f_2 &= a_2 f_2, \\ & \cdot \quad \cdot \quad \cdot \quad \cdot \quad \cdot, \\ \hat{A}f_n &= a_n f_n, \\ & \cdot \quad \cdot \quad \cdot \quad \cdot \quad \cdot, \end{aligned}\right\} \tag{8.018}$$

are eigenfunctions of $\hat{A}$ and the multiplying constants $a_1, a_2, \ldots, a_n, \ldots$ associated with them are called *eigenvalues*. Notice that these relationships are always of the following form:

$$(\text{OPERATOR})(\text{EIGENFUNCTION}) = (\text{EIGENVALUE}) \times (\text{EIGENFUNCTION}).$$
$$\tag{8.019}$$

---

† From the German word *eigen*, meaning "own" or "proper".

In (8.018) and in what follows, it is implicitly assumed that none of the eigenfunctions is trivial, i.e. none is equal to zero everywhere.

It is interesting to attempt to list all of the distinct eigenfunctions of a given operator $\hat{A}$. In so doing, an eigenfunction which has an eigenvalue different from all previously listed eigenvalues is obviously distinct and must be included. A set of eigenfunctions, each member of which has a unique eigenvalue, is said to be *non-degenerate*; by their very nature, the members of such a set are linearly independent which means that no one of them can be expressed as a linear super-position of others. (If an eigenfunction could be so expressed, then both it and the functions in terms of which it was expressed would have to have the same eigenvalue and this contradicts the hypothesis.) The matter can be stated in another way by saying that a non-trivial linear superposition of non-degenerate eigenfunctions of $\hat{A}$ cannot be an eigenfunction of $\hat{A}$. This principle is important in quantum me-chanics, as will be seen.

In continuing to examine the eigenfunctions of $\hat{A}$, one may come upon a set of functions which do have a common eigenvalue. This set may include some linearly dependent members but, after rejecting all of these as non-distinct and therefore non-interesting, there may still remain a subset of, say, N linearly independent eigenfunctions with the same eigenvalue. Such a subset is characterized as *N-fold degenerate*, a designation already applied to wave functions with the same quantized energy level, and special considerations apply to it.

As a concrete example, consider the operator d/dx. Any exponential function $\exp ax$ is an eigenfunction with eigenvalue a because:

$$\frac{d}{dx} \exp ax = a \exp ax. \tag{8.020}$$

There is no degeneracy because there are no subsets of linearly inde-pendent exponentials with the same value of a. As another example, consider the operator d^2/dx^2. Again the eigenfunctions are the expo-nentials but the eigenvalues are now a^2 rather than a, as is evident from the following:

$$\frac{d^2}{dx^2} \exp ax = a^2 \exp ax. \tag{8.021}$$

An important difference exists between this case and the previous one; the functions exp ax and exp $-ax$ are linearly independent eigenfunctions but they nevertheless have the same eigenvalue. In this second case, then, there are an infinite number of two-fold degenerate subsets of eigenfunctions.

It is interesting that, in an N-fold degenerate subset of eigenfunctions, the N original functions can be combined with one another in infinitely many ways to produce N new functions which are also linearly independent degenerate eigenfunctions having the original common eigenvalue. Thus the two degenerate functions exp ax and exp $-ax$ of the previous example can be combined with one another to yield cosh ax and sinh ax; these two functions are also linearly independent two-fold degenerate eigenfunctions of d^2/dx^2 with eigenvalue a^2.

An example taken directly from quantum mechanics will illustrate one concept already proposed and, in addition, suggest new ones. Consider the parity operator $\hat{\Pi}$ as applied to the stationary-state wave functions of the infinite rectangular well. Notice that the ground state Ψ_0, which has even parity, is an eigenfunction of $\hat{\Pi}$ with an eigenvalue $+1$:

$$\hat{\Pi}\left[(2/a)^{\frac{1}{2}}\cos(\pi x/a)\exp-i\omega_0 t\right] = (2/a)^{\frac{1}{2}}\cos(-\pi x/a)\exp-i\omega_0 t$$
$$= +1\left[(2/a)^{\frac{1}{2}}\cos(\pi x/a)\exp-i\omega_0 t\right].$$
$$(8.022)$$

Similarly, the first excited state Ψ_1, which has odd parity, is an eigenfunction of $\hat{\Pi}$ with eigenvalue -1:

$$\hat{\Pi}\left[(2/a)^{\frac{1}{2}}\sin(2\pi x/a)\exp-i\omega_1 t\right] = (2/a)^{\frac{1}{2}}\sin(-2\pi x/a)\exp-i\omega_1 t$$
$$= -1\left[(2/a)^{\frac{1}{2}}\sin(2\pi x/a)\exp-i\omega_1 t\right].$$
$$(8.023)$$

As a matter of fact, all the stationary states of the infinite rectangular well, as analyzed in Section 7.04, have definite parities and are eigenfunctions of $\hat{\Pi}$ with eigenvalues which are alternately $+1$ and -1. Any two of these which have different parities, e.g. Ψ_0 and Ψ_1, are non-degenerate under the operator $\hat{\Pi}$ since they have different eigenvalues. When two such functions are superposed with the aid of two non-vanishing coefficients α_0 and α_1, the function thus produced cannot be an eigenfunction of $\hat{\Pi}$ as is evident from the following equation:

$$\hat{\Pi}[\alpha_0\Psi_0+\alpha_1\Psi_1] = \alpha_0\Psi_0-\alpha_1\Psi_1$$
$$\neq (\text{constant})\,[\alpha_0\Psi_0+\alpha_1\Psi_1]. \qquad (8.024)$$

This superposition, it may be observed, does not have a definite parity.

It was seen in Section 7.06 that, with some justification, the "value" of the parity of an even function can be considered as $+1$ and the "value" of the parity of an odd function, as -1; no function having a definite parity can have a value of parity other than ± 1. Thus parity is a sort of variable which is connected with the operator $\hat{\Pi}$ in such a way that if a given function has associated with it a definite value of this variable, then that function will be an eigenfunction of $\hat{\Pi}$ with the definite value cast in the role of the eigenvalue. The pattern involved here suggests a technique to be used in the remainder of this section for obtaining the operators corresponding to the momentum p and the position r of a particle, quantities which are often called the momentum and position *observables*. This pattern is a characteristic feature of quantum mechanics; its significance goes far beyond any particular example.

To implement the suggestion just mentioned with regard to the operator for the vector momentum p, the following steps must be carried out:

   (i) Find the wave function Ψ_p for a particle which has a definite momentum p_0.
   (ii) Find the operator, to be designated $\hat{p}$, which has Ψ_p as an eigenfunction with the vector momentum p_0 as the corresponding eigenvalue.

It is not difficult to find a function which incorporates a single definite value k_0 of the wave vector and, by equations (6.006), a single definite value of momentum, namely $p_0 = \hbar k_0$. The prime example of such a function is the plane wave; the latter is a solution of the Schrödinger equation when the potential energy is constant, i.e. when the particle is unforced. Of course, when one attempts to identify a one-particle quantum state with a plane wave of non-vanishing amplitude, a problem arises because of the impossibility of normalization. What is needed is a plane wave of zero amplitude! Bizarre though such a function is, it is eminently useful for conceptual purposes. On the practical level, it is unattainable but it may be regarded as a limiting case toward which a sequence of wave packets tends as the members of the sequence become spatially broader and increasingly mono-directional and monoenergetic.

In order to express the function just described, the subscript (↓) will be annexed to the symbol for the wave amplitude. This is intended to convey the message that the amplitude approaches the limit zero as the function approaches plane wave form. A plane wave quantum state will therefore be represented by the following expression:

$$\Psi_p(r, t) = A_{\downarrow} \exp i(k_0 \cdot r - \omega_0 t). \tag{8.025}$$

Step (ii) may be carried out by inspection. It is seen at once that Ψ_p above is an eigenfunction of the operators $\partial^l / \partial x^l$, $\partial^m / \partial y^m$, $\partial^n / \partial z^n$, and $\partial^q / \partial t^q$ with respective eigenvalues $(ik_{0x})^l$, $(ik_{0y})^m$, $(ik_{0z})^n$, and $(-i\omega_0)^q$ where l, m, n, and q are non-negative integers. Since the required eigenvalue p_0 is equal to $\hbar(e_x k_{0x} + e_y k_{0y} + e_z k_{0z})$, it is clear that only the first powers of k_{0x}, k_{0y}, and k_{0z} and the zeroth power of ω_0 are acceptable. The operator sought must therefore include only the partial first derivatives $\partial / \partial x$, $\partial / \partial y$, and $\partial / \partial z$. The vector gradient immediately suggests itself and it is found that Ψ_p is indeed an eigenfunction of ∇ with eigenvalue ik_0. All that is necessary is to multiply ∇ by a factor which will convert ik_0 into $\hbar k_0$, namely by $-i\hbar$. The desired operator is:

$$\boxed{\hat{p} = -i\hbar \nabla} \; . \tag{8.026}$$

It is seen that:

$$\hat{p}A_{\downarrow} \exp i(k_0 \cdot r - \omega_0 t) = \hbar k_0 A_{\downarrow} \exp i(k_0 \cdot r - \omega_0 t)$$
$$= p_0 A_{\downarrow} \exp i(k_0 \cdot r - \omega_0 t), \qquad (8.027)$$

and the operator $\hat{p}$ fulfills the postulated requirements. The individual components of $\hat{p}$ are frequently used; for these, one writes:

$$\begin{cases} \hat{p}_x = -i\hbar\partial/\partial x \quad ; \\ \hat{p}_y = -i\hbar\partial/\partial y \quad ; \\ \hat{p}_z = -i\hbar\partial/\partial z \quad . \end{cases} \qquad (8.028)$$

In retrospect, it may be remarked that (8.025) is not the most general example of a state of definite momentum. Any Ψ function with a spatial dependence of $\exp ik_0 \cdot r$ is an eigenfunction of $\hat{p}$ even though, in an environment where forces act, it can exist as such for only an instant of time. If this instant is called t_0, the function in question can be written:

$$\Psi_p(r, t_0) = B_{\downarrow} \exp ik_0 \cdot r. \qquad (8.029)$$

It is now desirable to find the operator corresponding to the vector position r of a particle. Following the same program, one begins by constructing a wave function for a state of definite position. As in the previous example, this state is also impossible to achieve practically and is so for even more cogent reasons. Nevertheless, it is just as useful for conceptual purposes as is the state of definite momentum. Both states are freely discussed by physicists as if they were actually realizable.

If a situation involving infinite containing forces and a situation involving no forces are both excluded, a Ψ-function of definite position and a Ψ-function of definite momentum are equally ephemeral; like (8.029) above, the state of definite position can exist for only an instant of time. Calling the definite position r_0 and the instant of time t_0, the wave function sought must imply unit probability that the particle will be located at r_0 and zero probability that it will be located anywhere else. This is expressed by the following:

$$\Psi_r^* \Psi_r(r, t_0) = \begin{cases} \infty & \text{if } r = r_0 \\ 0 & \text{if } r \neq r_0 \end{cases}; \qquad \int\limits_{\substack{\text{all} \\ \text{space}}} \Psi_r^* \Psi_r \, d\tau = 1. \quad (8.030)$$

A function with these unusual features is called a *Dirac delta function* and is written $\delta(r-r_0)$. Even from a purely mathematical point of view, it exists only as the hypothetical limit of a sequence of functions designed to more nearly approximate its properties as some particular parameter is varied. Thus for (8.030), one writes:

$$\Psi_r^* \Psi_r(r, t_0) = \delta(r-r_0). \quad (8.031)$$

Evidently Ψ itself must be given by:

$$\Psi_r(r, t_0) = \delta^{\frac{1}{2}}(r-r_0) \exp i\varphi, \quad (8.032)$$

where φ is arbitrary. The function $\delta^{\frac{1}{2}}(r-r_0)$ is also infinite at $r = r_0$ and zero at all other r but it has the property that the integral of its *square* over all of space is unity. It exists under the same limitations as the delta function itself since it is also the hypothetical limit of a specially designed sequence of functions. Setting the arbitrary φ in (8.032) equal to zero, one may write for the wave function of a state of definite position:

$$\Psi_r(r, t_0) = \delta^{\frac{1}{2}}(r-r_0). \quad (8.033)$$

Although this is quite different from (8.029), the two functions are subtly related; they are members of a duality as can be appreciated if both are seen from the viewpoint of "momentum space" since, in that space, their forms are interchanged. Momentum space is treated in Chapter 10 and a suitable explanation of this important duality is provided therein.

It is now necessary to find an operator such that the Ψ_r of (8.033) is one of its eigenfunctions with the definite vector position r_0 as the corresponding eigenvalue, i.e. an operator for which the following is true:

$$(\text{OPERATOR}) \, \delta^{\frac{1}{2}}(r-r_0) = r_0 \delta^{\frac{1}{2}}(r-r_0). \quad (8.034)$$

The operator which fits this situation is none other than r itself as is shown by the fact that:

$$r\delta^{\frac{1}{2}}(r-r_0) = r_0\delta^{\frac{1}{2}}(r-r_0).$$ (8.035)

Thus the operator for the vector position and the operators for the individual components thereof (for the coordinates) are simply given by:

$$\left.\begin{array}{l} \hat{r} = r \quad ; \\ \hat{x} = x \quad ; \\ \hat{y} = y \quad ; \\ \hat{z} = z \quad . \end{array}\right\}$$ (8.036)

The commutators for the operators for coordinates and momenta are very interesting. Specifically, one finds that the commutator of a given coordinate and its conjugate momentum is equal to $i\hbar\hat{1}$ but the commutator of a given coordinate and a non-conjugate momentum is zero. Thus:

$$\left.\begin{array}{lll} [\hat{x}, \hat{p}_x] = i\hbar\hat{1}; & [\hat{x}, \hat{p}_y] = \hat{0}; & [\hat{x}, \hat{p}_z] = \hat{0}; \\ [\hat{y}, \hat{p}_x] = \hat{0}; & [\hat{y}, \hat{p}_y] = i\hbar\hat{1}; & [\hat{y}, \hat{p}_z] = \hat{0}; \\ [\hat{z}, \hat{p}_x] = \hat{0}; & [\hat{z}, \hat{p}_y] = \hat{0}; & [\hat{z}, \hat{p}_z] = i\hbar\hat{1}. \end{array}\right\}$$ (8.037)

The information in this table of commutators (which should be memorized) plays a very important role in quantum mechanics.

8.04. The Association of an Operator with an Observable and the Calculation of Expectation Values

Two observables, namely r and p, have been considered in some detail in the previous section. To these, there correspond the operators $\hat{r} = r$ and $\hat{p} = -i\hbar\nabla$. As was intimated earlier, this pattern is universal in quantum theory and, to a given observable A, there always corresponds an operator $\hat{A}$. Moreover, if the quantum state Ψ is such that A has a definite value, i.e. if A is not an average over a distribution of many values, then Ψ is an eigenfunction of $\hat{A}$ with the definite value appearing as the eigenvalue. It is only in certain cases, of course, that Ψ will be an eigenfunction of a particular given operator; in general it is not such an eigenfunction and the corresponding observable is distributed over a range of values. In that event, the quantity that lends itself to calculation is the average value of the observable, e.g. of A, on the assumption that all probabilities are fulfilled; this is none other than the expectation value, $\langle A \rangle$.

From the wave function Ψ, one can immediately derive the probability density ϱ and the probability current density J. In Chapter 6, the expectation value of r was seen to be calculable from ϱ:

$$\langle r \rangle = \int_{\substack{\text{all} \\ \text{space}}} r\varrho \, d\tau = \int_{\substack{\text{all} \\ \text{space}}} \Psi^* r \Psi \, d\tau = \int_{\substack{\text{all} \\ \text{space}}} \Psi^* \hat{r} \Psi \, d\tau. \quad (8.038)$$

The calculation of the expectation value of p is only a little less obvious. If m is the mass of the particle in question, $m\varrho$ can be regarded as the average local mass density and mJ as the average local momentum density if all probabilities are fulfilled. The expectation value of the momentum will then be simply the integral of this latter density:

$$\langle p \rangle = \int_{\substack{\text{all} \\ \text{space}}} mJ \, d\tau = \frac{\hbar}{2i} \int_{\substack{\text{all} \\ \text{space}}} [\Psi^* \nabla \Psi - \Psi \nabla \Psi^*] \, d\tau. \quad (8.039)$$

In order to simplify the right-hand side of (8.039), consider the gradient of the probability density:

$$\nabla \varrho = \nabla(\Psi^* \Psi) = \Psi^* \nabla \Psi + \Psi \nabla \Psi^*. \quad (8.040)$$

By one of the standard theorems of vector analysis, the volume integral of a gradient such as this within a sphere K of radius r is transformable into an integral on the surface of the same sphere:

$$\int \nabla \varrho \, d\tau = \oint_K \varrho e_n \, da, \quad (8.041)$$

where e_n is the outward normal unit vector on the spherical surface. In (8.041), one may consider any one of the rectangular components. Choosing the z component for convenience, one has:

$$\left| \int \frac{\partial \varrho}{\partial z} \, d\tau \right| = \left| \oint_K \varrho \cos \theta \, da \right| \leqslant \oint_K \varrho \, da. \qquad (8.042)$$

Here θ is the familiar spherical polar coordinate; the final inequality is valid because ϱ is positive definite. It can also be easily seen that the final integral is at most a function of r and t, where r is the radius of sphere K:

$$\oint_K \varrho \, da = r^2 \int_0^{2\pi} \int_0^\pi \varrho(r, \theta, \varphi, t) \sin \theta \, d\theta \, d\varphi = g(r, t). \qquad (8.043)$$

Since ψ is normalized:

$$\int_{\substack{\text{all} \\ \text{space}}} \varrho \, d\tau = \int_0^\infty \left[\oint_K \varrho \, da \right] dr = \int_0^\infty g(r, t) \, dr = 1. \qquad (8.044)$$

The time dependence of $g(r, t)$ must be of such a nature that the integral of this function from zero to infinity is equal to unity as stated. This implies that:

$$\oint_K \varrho \, da \sim G r^{-(1+\varepsilon)} \quad \text{as} \quad r \to \infty, \qquad (8.045)$$

where both G and ε are positive real and G may even be a function of time. It follows at once that the integral of $\partial \varrho / \partial z$ in (8.042) must vanish when r is infinite; this is also true of the other two rectangular components of $\nabla \varrho$ since, on reorientation of the coordinate system, any rectangular component could become the z component. Thus:

$$\int_{\substack{\text{all} \\ \text{space}}} \nabla \varrho \, d\tau = 0 \quad \text{and} \quad \int_{\substack{\text{all} \\ \text{space}}} \Psi \nabla \Psi^* \, d\tau = - \int_{\substack{\text{all} \\ \text{space}}} \Psi^* \nabla \Psi \, d\tau, \qquad (8.046)$$

according to (8.040). Substituting this into (8.039), one obtains for the expectation value of the momentum:

$$\langle p \rangle = \frac{\hbar}{2i} \int_{\substack{\text{all} \\ \text{space}}} 2\Psi^* \nabla \Psi \, d\tau = \int_{\substack{\text{all} \\ \text{space}}} \Psi^*(-i\hbar \nabla \Psi) \, d\tau. \quad (8.047)$$

The momentum operator is easily identified and the above may be written:

$$\langle p \rangle = \int_{\substack{\text{all} \\ \text{space}}} \Psi^* \hat{p} \Psi \, d\tau. \quad (8.048)$$

If the result just obtained is compared with (8.038), a remarkable similarity is noticed. In (8.038), the expectation value of r is found by integrating Ψ^* times the function $\hat{r}\Psi$ over all of space; in (8.048), the expectation value of p is found by integrating Ψ^* times the function $\hat{p}\Psi$ over all of space. The format in which the operator is placed between Ψ^* and Ψ is essential in the second case and obviously permissible in the first. One sees here the exemplification of a principle that is found to be universal in quantum mechanics; the expectation value of any observable A in the quantum state Ψ is given by:

$$\boxed{\langle A \rangle = \int_{\substack{\text{all} \\ \text{space}}} \Psi^* \hat{A} \Psi \, d\tau} \, , \quad (8.049)$$

where $\hat{A}$ is the operator corresponding to the observable A. In particular, if Ψ happens to be an eigenfunction of $\hat{A}$ with eigenvalue a, then $\hat{A}\Psi = a\Psi$ and:

$$\langle A \rangle = \int_{\substack{\text{all} \\ \text{space}}} \Psi^* a \Psi \, d\tau = a \int_{\substack{\text{all} \\ \text{space}}} \Psi^* \Psi \, d\tau = a. \quad (8.050)$$

Thus the expectation value of A reduces to the definite value a in this special case. Applications of these results are very important, as will be seen.

8.05. The Hamiltonian Operator and the Generalized Derivation of the Schrödinger Equation

The operator $\hat{p} = -i\hbar \nabla$ was discovered by first writing a Ψ function to represent a state of definite momentum. It was then relatively easy to identify $-i\hbar \nabla$ as the operator of which Ψ was an eigenfunction with $p = \hbar k$ as the corresponding eigenvalue. The same procedure can be used to find the operator for the Hamiltonian which, in a sense, is a canonical momentum conjugate (with a minus sign) to the time t. As suggested by equations (6.006), a quantum state with a definite value of the Hamiltonian is a quantum state with a definite value of ω where $\mathcal{H} = \hbar\omega$. Such a state is a stationary state and has the form:

$$\Psi(r, t) = \psi(r) \exp{-i\omega t}. \tag{8.051}$$

The Hamiltonian operator must have the property that a function of this type is one of its eigenfunctions with $\hbar\omega$ as the corresponding eigenvalue. In seeking this operator, one quickly arrives at the partial derivative $\partial/\partial t$, of which the above Ψ is an eigenfunction with eigenvalue $-i\omega$. Multiplication of $\partial/\partial t$ by $i\hbar$ changes the eigenvalue to $\hbar\omega$ and therefore produces the operator desired. One has:

$$\boxed{\hat{\mathcal{H}} = i\hbar \frac{\partial}{\partial t}} \ . \tag{8.052}$$

This result is eminently reasonable. The canonical momentum p_x, for example, is conjugate to the coordinate x and the operator $\hat{p}_x$ is $-i\hbar\partial/\partial x$. Similarly, the canonical momentum $-\mathcal{H}$ is conjugate to t and therefore the operator for $-\mathcal{H}$ should be $-i\hbar\partial/\partial t$. This is in accord with (8.052).

In classical mechanics, when one speaks of the Hamiltonian, one means the function $\mathcal{H}(q_j, p_j, t)$. A similar usage obtains in quantum mechanics and "the Hamiltonian" is almost invariably understood to mean the operator $\hat{\mathcal{H}}$ expressed not as $i\hbar \, \partial/\partial t$ but rather as a function of the operators $\hat{q}_j$, $\hat{p}_j$, and, possibly, of t. Taken in this sense, $\hat{\mathcal{H}}$ will be a different operator, i. e. a different function of $\hat{q}_j$, $\hat{p}_j$, and t, for every different system which comes under consideration. Thus a single particle moving in three dimensions with potential energy $V(\mathbf{r}, t)$ has as its classical Hamiltonian:

$$\mathcal{H} = \frac{1}{2m}(p_x^2 + p_y^2 + p_z^2) + V(\mathbf{r}, t) \tag{8.053}$$

and, if one converts every observable in this expression to its corresponding operator, the result is:[†]

$$\hat{\mathcal{H}} = \frac{-\hbar^2}{2m}\nabla^2 + V(\mathbf{r}, t). \tag{8.054}$$

This is regarded as "the Hamiltonian" for the one-particle system under consideration.

It is important to appreciate that $\hat{\mathcal{H}}$ as expressed in (8.054) is not identical with $\hat{\mathcal{H}}$ as expressed in (8.052); it merely has the same effect as (8.052) for the particular system at hand. Thus, when both forms are allowed to act upon the system wave function Ψ and the results equated, there is created a non-trivial condition which Ψ must satisfy. This condition is none other than the Schrödinger equation itself, as is evident on comparison with (6.021). Extending the concepts involved here, one may write for any system whose classical Hamiltonian is known:

$$\boxed{\hat{\mathcal{H}}(\hat{q}_j, \hat{p}_j, t)\Psi = i\hbar\frac{\partial \Psi}{\partial t}} \tag{8.055}$$

and this constitutes the Schrödinger equation applicable to the system in question. The use of the form $\hat{\mathcal{H}}(\hat{q}_j, \hat{p}_j, t)$ becomes so habitual that

[†] In quantum mechanics it is usual to consider time as a parameter rather than as an observable, much as is done in classical mechanics. There is never any need to speak of the operator for time although the latter would presumably be $\hat{t} = t$. Even if this point of view were adopted, it would have no effect in, for example, the passage from (8.053) to (8.054).

it scarcely needs to be specified explicitly. By the use of (8.055), then, a system whose classical features are given can be analyzed quantum mechanically; in other words, it can be "quantized".

If, as explained in Section 7.01, $\mathcal{H}$ does not contain the time, then Ψ can be a stationary state as displayed in (8.051). Such a Ψ, which is an eigenfunction of $i\hbar\,\partial/\partial t$ with eigenvalue $\mathcal{H} = E = \hbar\omega$, yields on substitution into (8.055):

$$\boxed{\mathcal{H}(\hat{q}_j,\,\hat{p}_j)\psi = E\psi} \ . \tag{8.056}$$

Evidently ψ (or Ψ) is also an eigenfunction of $\mathcal{H}(\hat{q}_j,\,\hat{p}_j)$ with the eigenvalue E. Equation (8.056) is a generalized version of the time-independent Schrödinger equation considered in Section 7.01.

As a non-trivial example of the method outlined above whereby the Schrödinger equation for a given system may be found, one may consider a helium or helium-like[†] atom with nuclear charge Ze and, for simplicity, infinite nuclear mass. There are six degrees of freedom as evidenced by the fact that the configuration, illustrated in Fig. 8.01, is described by the vector displacements r_a and r_b of the two

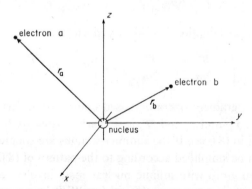

FIG. 8.01. A helium or helium-like atom (with infinite nuclear mass) and an inertial Cartesian reference frame.

† Singly ionized lithium and doubly ionized beryllium are examples of helium-like atoms.

electrons, i.e. by the six coordinates x_a, y_a, z_a, x_b, y_b, and z_b. The classical Hamiltonian is readily found to be:

$$\mathcal{H} = \frac{1}{2m}(p_{xa}^2 + p_{ya}^2 + p_{za}^2) + \frac{1}{2m}(p_{xb}^2 + p_{yb}^2 + p_{zb}^2)$$
$$+ V(x_a, y_a, z_a, x_b, y_b, z_b), \tag{8.057}$$

where m is the mass of an electron and the potential energy V is a rather complicated expression in the Cartesian coordinates† used here:

$$V = \frac{e^2}{4\pi\varepsilon_0}\left\{-Z(x_a^2 + y_a^2 + z_a^2)^{-\frac{1}{2}} - Z(x_b^2 + y_b^2 + z_b^2)^{-\frac{1}{2}}\right.$$
$$\left. + [(x_a - x_b)^2 + (y_a - y_b)^2 + (z_a - z_b)^2]^{-\frac{1}{2}}\right\}. \tag{8.058}$$

The Hamiltonian does not contain the time and stationary states are therefore possible. To construct the Hamiltonian operator, one replaces p_{xa} by $-i\hbar\partial/\partial x_a$, p_{xa}^2 by $-\hbar^2\partial^2/\partial x_x^2$, etc. Differentiation with respect to each of the six coordinates is required and one may infer correctly that the wave function Ψ is a function of all six of the coordinates and of time:

$$\Psi = \Psi(x_a, y_a, z_a, x_b, y_b, z_b, t). \tag{8.059}$$

The Schrödinger equation is evidently given by:

$$\left[-\frac{\hbar^2}{2m}\nabla_a^2 - \frac{\hbar^2}{2m}\nabla_b^2 + V\right]\Psi = i\hbar\frac{\partial\Psi}{\partial t}, \tag{8.060}$$

where ∇_a is a gradient operator based upon partial derivatives with respect to x_a, y_a, and z_a and ∇_b is similarly related to x_b, y_b, and z_b; V is as given in (8.058). If the stationary states are sought, the above equation can be simplified according to the pattern of (8.056).

The helium atom with infinite nuclear mass involves a six-dimensional configuration space and the product $\Psi^*\Psi$ is a probability den-

---

† Since this problem is being discussed only to illustrate the concept of the generalized derivation of the Schrödinger equation, no attempt has been made to select the most appropriate coordinate system.

sity in this space. Thus $\Psi^*\Psi(r_a, r_b, t)\,\Delta\tau_a\,\Delta\tau_b$ is the probability at time t that electron (a) will be in volume element $\Delta\tau_a$ located at r_a and electron (b) will be in volume element $\Delta\tau_b$ located at r_b as illustrated in Fig. 8.02. It may be remarked in passing that, since electrons are indistiguishable particles, a necessary condition for the admissibility

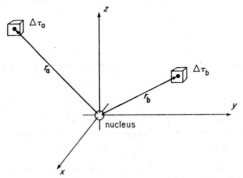

FIG. 8.02. A configuration for the helium atom showing volume elements $\Delta\tau_b$ and $\Delta\tau_a$ within which this and immediately neighboring configurations lie.

of a given wave function is that $\Psi^*\Psi(r_a, r_b, t) = \Psi^*\Psi(r_b, r_a, t)$. This, in turn, implies that Ψ must be an eigenfunction of $\hat{X}_{ab}$ with an eigenvalue of magnitude unity.[†] The normalization of Ψ requires that an integration of the probability density over all possible configurations shall yield unity. This is expressed mathematically as follows:

$$\int_{-\infty}^{\infty} \int_{-\infty}^{\infty} \int_{-\infty}^{\infty} \int_{-\infty}^{\infty} \int_{-\infty}^{\infty} \int_{-\infty}^{\infty} \Psi^*\Psi \, dx_a \, dy_a \, dz_a \, dx_b \, dy_b \, dz_b = 1. \quad (8.061)$$

Other integrals which might occur in conjunction with this example, such as those required in the evaluation of expectation values, must

[†] Since $\hat{X}_{ab}^2 = \hat{1}$, the only possible eigenvalues of $\hat{X}_{ab}$ are $+1$ and -1. The eigenfunctions corresponding to these respective values have substantially different properties and the one to be used in a given situation is determined by the *Pauli exclusion principle*. The latter has implications going considerably beyond the present brief treatment.

also be taken over the total ranges of all six of the coordinates, i.e. over all of configuration space. The many three-dimensional integrals in this book which are designated to be taken over "all space" should be construed in a typical sense. When configuration space has more than three dimensions, it is implicitly assumed that such integrals (and the associated theory) are extended to cover the additional dimensions.

8.06. Hermitian Operators and Expansion in Eigenfunctions

In this section, the mathematical properties of a particular type of operator known as an *Hermitian* operator will be surveyed in general terms. A certain economy of thought is achieved by first discussing these mathematical properties apart from their many applications to quantum mechanics; later, when the applications are considered, the theory will acquire additional vitality and meaning. Complete mathematical rigor would require an inordinately long treatment and will not be attempted. It is sufficient if the reader acquires a working knowledge of Hermitian operators and their eigenfunctions and is able to recognize the many situations in quantum mechanics where such knowledge is applicable.

With a given space of interest and a given operator, say $\hat{A}$, there is associated a class of functions, defined on that space, upon which $\hat{A}$ can operate and to which the results of its operation also belong. For the purposes of this section, this class is further restricted to include only functions which are normalizable (but not necessarily normalized in every instance) and, when so restricted, will be called the class C. In this context, the operator $\hat{A}$ is said to be Hermitian if and only if:

$$\int f^* \hat{A} f \, d\tau \quad \text{is real,} \quad (8.062)$$

where f is an *arbitrarily chosen* member of C. (This and all integrals in what follows are taken over the space of interest mentioned above.) Notice that the integrand in (8.062) is the product of two functions,

f^* and $\hat{A}f$. The conjugate of the integrand is therefore $f(\hat{A}f)^*$ and the stipulation of (8.062) is often expressed alternatively as:

$$\int f^*\hat{A}f\, d\tau = \int f(\hat{A}f)^*\, d\tau. \tag{8.063}$$

If this, then, is true for arbitrary f, it follows that $\hat{A}$ is Hermitian. A corollary to this definition is that the eigenvalues of an Hermitian operator are real. Suppose that f is an eigenfunction of $\hat{A}$ with eigenvalue a. The integral in (8.062) is then given by:

$$\int f^*\hat{A}f\, d\tau = \int f^*af\, d\tau = a \int f^*f\, d\tau \tag{8.064}$$

and if $\hat{A}$ is Hermitian, all three of the forms displayed above are real. Since the factor $\int f^*f\, d\tau$ must be real, it follows that a is also real.

It will now be desirable to prove some theorems about Hermitian operators.

THEOREM 8.01. *If f and g are two arbitrarily chosen members of C and if $\hat{A}$ is Hermitian, then:*

$$\int f^*\hat{A}g\, d\tau = \int g(\hat{A}f)^*\, d\tau. \tag{8.065}$$

Notice that if f and g are identical, this statement reduces to the definition of Hermiticity already given; if they are different, it embodienew content. As proof, consider the two superpositions:

$$R = f+g; \qquad S = f+ig. \tag{8.066}$$

By the Hermiticity of $\hat{A}$:

$$\int (f+g)^*\hat{A}(f+g)\, d\tau = \int (f+g)\,[\hat{A}(f+g)]^*\, d\tau \tag{8.067}$$

and therefore:

$$\int [f^*\hat{A}f+f^*\hat{A}g+g^*\hat{A}f+g^*\hat{A}g]\, d\tau$$
$$= \int [f(\hat{A}f)^*+f(\hat{A}g)^*+g(\hat{A}f)^*+g(\hat{A}g)^*]\, d\tau. \tag{8.068}$$

Again, because of the Hermiticity of $\hat{A}$, two terms on the left-hand side of this cancel two terms on the right and one has:

$$\int [f^*\hat{A}g+g^*\hat{A}f]\, d\tau = \int [f(\hat{A}g)^*+g(\hat{A}f)^*]\, d\tau. \tag{8.069}$$

16*

Using the function S in the same way, it follows that:

$$\int (f+ig)^* \, \hat{A}(f+ig) \, d\tau = \int (f+ig) \, [\hat{A}(f+ig)]^* \, d\tau. \quad (8.070)$$

With omission of the intermediate step, this becomes:

$$\int [if^* \hat{A}g - ig^* \hat{A}f] \, d\tau = \int [-if(\hat{A}g)^* + ig(\hat{A}f)^*] \, d\tau. \quad (8.071)$$

Finally, if (8.069) is multiplied by $\frac{1}{2}$ and (8.071) is multiplied by $-i/2$ and the results added, one obtains (8.065) which was to be proved.

It should be mentioned that equation (8.065) is sometimes used as the definition of Hermiticity rather than (8.062) and, when so used, (8.062) readily follows as a corollary. Thus either (8.062) or (8.065) can be deduced from the other; the two statements are logically equivalent and it is a matter of taste as to which should be employed as the primary definition. In this book, (8.062) is favored because it is a simple statement with a clear connotation. Moreover, it is interesting to find that the more elaborately structured (8.065) is a logical consequence of the simpler (8.062).

Two distinct members f_m and f_n of C are said to be orthogonal if:

$$\int f_m^* f_n \, d\tau = 0. \quad (8.072)$$

The discussion of orthogonality preceding equation (7.148) in the previous chapter may be recalled here. In conjunction with this concept, one is lead to the following theorem:

THEOREM 8.02. *Any two eigenfunctions of an Hermitian operator are orthogonal if they belong to different eigenvalues.*

Suppose that f_m and f_n are two eigenfunctions of Hermitian operator $\hat{A}$ and that they belong to different eigenvalues. One may then write:

$$\hat{A}f_m = a_m f_m; \quad (8.073)$$

$$\hat{A}f_n = a_n f_n. \quad (8.074)$$

Now multiply f_m^* onto both sides of (8.074) and integrate:

$$\int f_m^* \hat{A}f_n \, d\tau = a_n \int f_m^* f_n \, d\tau. \quad (8.075)$$

Likewise, multiply f_n^* onto both sides of (8.073) and integrate:

$$\int f_n^* \hat{A} f_m \, d\tau = a_m \int f_n^* f_m \, d\tau. \qquad (8.076)$$

Since $\hat{A}$ is Hermitian, its eigenvalues are real and the conjugate of the above equation can be written:

$$\int f_n (\hat{A} f_m)^* \, d\tau = a_m \int f_m^* f_n \, d\tau. \qquad (8.077)$$

By Theorem 8.01, the left-hand sides of (8.075) and (8.077) are equal. Subtraction of these equations therefore yields:

$$0 = (a_m - a_n) \int f_m^* f_n^* \, d\tau. \qquad (8.078)$$

The factor $(a_m - a_n)$ cannot be zero because, by hypothesis, the two eigenvalues are different; it follows that the integral must be zero and that the functions are orthogonal.

If all the eigenfunctions of a given Hermitian operator are non-degenerate, then by Theorem 8.02 they must all be orthogonal and the situation is extremely simple. On the other hand, there may exist one or more degenerate subsets of eigenfunctions and these, temporarily at least, introduce complications. The members of an N-fold degenerate subset of eigenfunctions are, by hypothesis, linearly independent and they must (because of differing eigenvalues) be orthogonal to all eigenfunctions which do not belong to their own subset; they may not, however, be orthogonal to one another. In an earlier discussion, it was pointed out that the members of such a subset can be combined with one another by linear superposition so as to form a new subset of N linearly independent eigenfunctions. It is now asserted that this task can be performed in such a way that the N new functions are mutually orthogonal. This process can be accomplished in infinitely many ways, however, and the resulting N-fold degenerate orthogonal subset of eigenfunctions does not have a unique form. This matter suggests a theorem which, because of the non-uniqueness, must be stated with unusual care:

THEOREM 8.03. *For a given Hermitian operator, it is possible to construct a set of eigenfunctions, from which no linearly independent eigenfunction is excluded, in such a way that all functions in the set are orthogonal.*

A set of eigenfunctions from which no linearly independent eigenfunction is excluded will, in this book, be called a *whole* set. In simple language, then, it is possible to construct a whole orthogonal set of eigenfunctions for any given Hermitian operator.

To prove Theorem 8.03, it is merely necessary to show that the members of an N-fold degenerate subset of eigenfunctions can be orthogonalized as asserted above. There is a standard method for doing this, called the *Schmidt orthogonalization procedure*, which will now be outlined. Let g_1, g_2, $\ldots$, g_N be the original members of the subset in question; these are normalized and linearly independent but not necessarily orthogonal. The new functions, $f_1, f_2, \ldots, f_N$, are to be linear combinations of the g_n and are to be orthogonalized; in the course of the Schmidt procedure, they also become normalized. One may begin by arbitrarily letting $f_1 = g_1$, whereupon f_2 should be written:

$$f_2 = \beta_2[g_2 - f_1 \int f_1^* g_2 \, d\tau]. \tag{8.079}$$

The constant β_2 may be adjusted so as to normalize f_2; it has no bearing upon the matter of orthogonality. Now multiply f_1^* onto both sides of (8.079) and integrate:

$$\int f_1^* f_2 \, d\tau = \beta_2[\int f_1^* g_2 \, d\tau - \int f_1^* g_2 \, d\tau] = 0. \tag{8.080}$$

Evidently f_2 is orthogonal to f_1. Continuing, one writes for f_3:

$$f_3 = \beta_3[g_3 - f_1 \int f_1^* g_3 \, d\tau - f_2 \int f_2^* g_3 \, d\tau]. \tag{8.081}$$

As before, β_3 is to be used for normalization. Now multiply f_1^* onto both sides of (8.081) and integrate; repeat with f_2^*. The results are:

$$\int f_1^* f_3 \, d\tau = \beta_3[\int f_1^* g_3 \, d\tau - \int f_1^* g_3 \, d\tau - 0] = 0; \tag{8.082}$$

$$\int f_2^* f_3 \, d\tau = \beta_3[\int f_2^* g_3 \, d\tau - 0 - \int f_2^* g_3 \, d\tau] = 0. \tag{8.083}$$

Evidently f_3 is orthogonal to both f_1 and f_2 This procedure can be continued as far as necessary and the theorem is therefore proved.

To summarize, one may observe that, out of the class C of functions related to Hermitian operator $\hat{A}$, there exists a whole set of functions with members f_n which are eigenfunctions of $\hat{A}$. Even though there may be instances of degeneracy, this set (which by hypothesis is normalizable) can be orthogonalized to form a *whole orthonormal* set of eigenfunctions, i.e. a set from which no linearly independent eigenfunction of $\hat{A}$ has been excluded and whose members have the property that:

$$\int f_m^* f_n \, d\tau = \delta_{mn} \, . \qquad (8.084)$$

The next matter for consideration in this section is the representation of an arbitrarily chosen member f of C as a linear superposition of the orthonormal eigenfunctions of $\hat{A}$ described above. When f is so represented, it is said to be *expanded* in the eigenfunctions in question; the expansion takes the following form where the α_n are coefficients:

$$f = \sum_n \alpha_n f_n \, . \qquad (8.085)$$

If the whole orthonormal set of eigenfunctions is capable of being employed in this fashion, it is called a *complete* set. For mathematical rigor, completeness should be proved in individual cases and proofs have been devised for all the standard examples of such sets. In textbooks such as this one, it is customary to omit proofs of this type; the utility of a given orthonormal set in physical applications is usually accepted as satisfactory evidence of completeness.

The method for calculating the coefficients α_n in (8.085) is extremely simple. One has only to multiply f_m^* onto both sides and integrate:

$$\int f_m^* f \, d\tau = \sum_n \alpha_n \int f_m^* f_n \, d\tau = \sum_n \alpha_n \delta_{mn} = \alpha_m . \qquad (8.086)$$

The coefficient α_m is clearly equal to the integral on the left. It is convenient to rewrite this with the index m changed to n so that the result takes the form:

$$\alpha_n = \int f_n^* f \, d\tau \quad . \tag{8.087}$$

Equations (8.085) and (8.087) find many applications in quantum mechanics.

If the function f is normalized, a condition is automatically imposed upon the coefficients α_n. The nature of this condition may be investigated as follows:

$$\begin{aligned}
\int f^* f \, d\tau &= \int \left(\sum_n \alpha_n^* f_n^* \right) \left(\sum_m \alpha_m f_m \right) d\tau \\
&= \sum_n \sum_m \alpha_n^* \alpha_m \int f_n^* f_m \, d\tau \\
&= \sum_n \sum_m \alpha_n^* \alpha_m \delta_{nm} = \sum_n \alpha_n^* \alpha_n .
\end{aligned} \tag{8.088}$$

Since the integral on the left is equal to unity, so is the final sum on the right; one therefore has:

$$\sum_n \alpha_n^* \alpha_n = 1 \quad . \tag{8.089}$$

This is typical of the many simple results which can be obtained with the aid of a complete set of orthonormal functions.

A final theorem concerning the status of the commutator of two Hermitian operators will now be given:

THEOREM 8.04. *If $\hat{A}$ and $\hat{B}$ are two Hermitian operators related to the same class C of functions, then $-i[\hat{A}, \hat{B}]$ is also Hermitian.*

As proof, let f be an arbitrarily chosen member of C. Then:

$$\int f^* \hat{A} \hat{B} f \, d\tau = \int (\hat{B} f)(\hat{A} f)^* \, d\tau = Y. \tag{8.090}$$

This is true because the function $\hat{B} f$ can be identified with the function g of (8.065). The symbol Y is introduced as a convenient shorthand to represent the integral. One may now invert the order of $\hat{A}$ and $\hat{B}$ and say:

$$\int f^* \hat{B} \hat{A} f \, d\tau = \int (\hat{A}f)(\hat{B}f)^* \, d\tau = Y^*. \qquad (8.091)$$

It follows that:

$$\int f^*(-i[\hat{A}, \hat{B}])f \, d\tau = -i(Y - Y^*). \qquad (8.092)$$

This quantity is real and therefore $-i[\hat{A}, \hat{B}]$ is Hermitian.

It is interesting to apply this theorem to the commutators listed in (8.037). One sees, for example, that $-i[\hat{x}, \hat{p}_x] = \hbar \hat{1}$; this is obviously Hermitian.

8.07. The Role of Hermitian Operators and their Eigenfunctions in Quantum Mechanics

It is easy to apply the theory of the preceding section to the wave functions of quantum states and to the operators which act upon them. In particular, if an arbitrarily selected wave function Ψ is substituted for f in (8.062), it is seen that operator $\hat{A}$ will be Hermitian if the expectation value $\langle A \rangle$ of the corresponding observable is real and vice versa. The expectation value of a physical observable must be real since it is the value one would expect for the average of a set of actual physical measurements if all probabilities were fulfilled. It follows that physical observables can be represented only by Hermitian operators.

Some simple operators representing physical observables (such as $\hat{r}$ and $\hat{p}$, for example) were derived in earlier pages of this chapter. Although there is no reason to question the validity of the forms given to these operators, it is nevertheless instructive to demonstrate their Hermiticity. For this purpose, it is merely necessary to show that they generate real expectation values for arbitrary Ψ. Since Ψ is a quantum wave function, it is, of course, normalized. As a first example, consider the observable $F(r, t)$ where F is a real function of r and t. The operator for this observable is simply the function itself and:

$$\langle F \rangle = \int_{\substack{\text{all} \\ \text{space}}} \Psi^* F(\mathbf{r}, t)\Psi \, d\tau = \int_{\substack{\text{all} \\ \text{space}}} F(\mathbf{r}, t)\Psi^*\Psi \, d\tau. \qquad (8.093)$$

The integrand is seen to be the product of two real functions, hence $\langle F \rangle$ is real for arbitrary Ψ and $\hat{F}$ is Hermitian. As a second example, consider the operator $\hat{p} = -i\hbar\nabla$ for the momentum of a particle. The algebraic steps beginning with equation (8.039) and ending with (8.048) show that, for normalized but otherwise arbitrary Ψ, the following two integrals are to be identified:

$$\int_{\substack{\text{all} \\ \text{space}}} \Psi^* \hat{p}\Psi \, d\tau = \int_{\substack{\text{all} \\ \text{space}}} m\mathbf{J} \, d\tau. \qquad (8.094)$$

Both of these are real since $\mathbf{J}$ is real and it follows that $\hat{p}$ is Hermitian. As a third example, consider the operator for the kinetic energy of a particle, namely $\hat{T} = (-\hbar^2/2m)\nabla^2$. The expectation value of T becomes:

$$\langle T \rangle = \frac{-\hbar^2}{2m} \int_{\substack{\text{all} \\ \text{space}}} \Psi^* \nabla^2\Psi \, d\tau. \qquad (8.095)$$

The proof that this is real begins by invoking Green's theorem with respect to all of space and the sphere at infinity:

$$\int_{\substack{\text{all} \\ \text{space}}} (\Psi^* \nabla^2\Psi - \Psi \nabla^2\Psi^*) \, d\tau = \oint_{\substack{\text{sphere} \\ \text{at } \infty}} (\Psi^* \nabla\Psi - \Psi \nabla\Psi^*) \cdot \mathbf{e}_n \, da. \qquad (8.096)$$

Multiplication by $\hbar/2im$ yields:

$$\frac{\hbar}{2im} \int_{\substack{\text{all} \\ \text{space}}} (\Psi^* \nabla^2\Psi - \Psi \nabla^2\Psi^*) \, d\tau = \oint_{\substack{\text{sphere} \\ \text{at } \infty}} \mathbf{J} \cdot \mathbf{e}_n \, da. \qquad (8.097)$$

As in the discussion associated with Theorem 6.01, the integral in the right-hand side above must vanish for normalized Ψ. Therefore:

$$\int_{\substack{\text{all}\\\text{space}}} \Psi^* \nabla^2 \Psi \, d\tau = \int_{\substack{\text{all}\\\text{space}}} \Psi \nabla^2 \Psi^* \, d\tau. \tag{8.098}$$

This equation shows that the integral in (8.095) is equal to its own conjugate and is therefore real. It follows that $\hat{T}$ is Hermitian. It is clear from the first example discussed above that the operator $\hat{V}$ for the potential energy is Hermitian and, since $\hat{T}$ has been shown to be Hermitian, the Hamiltonian $\hat{T} + \hat{V}$ for a one-particle system must be Hermitian. Similar results for more general systems are readily obtainable.

In attempting to construct the quantum mechanical counterpart of the classical Hamiltonian, a dilemma arises if a term such as xp_x occurs in the latter. Two problems immediately present themselves: (i) since the sequence of the classical variables is immaterial, the combination xp_x does not indicate which of the two different combinations $\hat{x}\hat{p}_x$ or $\hat{p}_x\hat{x}$ is supposedly to be chosen and (ii) neither $\hat{x}\hat{p}_x$ nor $\hat{p}_x\hat{x}$ is Hermitian. Fortunately the resolution of this dilemma is quite simple; one chooses as the operator for xp_x the following *symmetrized combination* of the two possible sequences:

$$\widehat{(xp_x)} = \tfrac{1}{2}[\hat{x}\hat{p}_x + \hat{p}_x\hat{x}]. \tag{8.099}$$

It is easily shown that this operator is Hermitian, hence the symmetrized combination is also a Hermitized combination and both difficulties are overcome.

In the light of Section 8.06, the reasons for some results obtained much earlier can now be appreciated. For example, it is easy to see why the stationary states of the harmonic oscillator are orthogonal. The Hamiltonian for the system

$$\hat{\mathscr{H}} = -\frac{\hbar^2}{2M} \frac{\partial^2}{\partial x^2} + \tfrac{1}{2}Kx^2 \tag{8.100}$$

is an Hermitian operator and the stationary states $\Psi_n(x, t)$ are its non-degenerate eigenfunctions since they obey the relationship:

$$\hat{\mathscr{H}}\Psi_n = E_n\Psi_n, \tag{8.101}$$

where the E_n, given by $(n+\frac{1}{2})\hbar\Omega$, are all different. Thus the states in question have to be orthogonal; the same conclusion is true for the stationary states of the infinite rectangular well and for the bound stationary states of the finite rectangular well. For the particle in a box, the members of the numerous degenerate subsets do not have to be mutually orthogonal although, as expressed in (7.116), they happen to be because of the systematic way in which the wave functions are constructed.

It should not be supposed that the theory of Hermitian operators and their eigenfunctions is necessarily applied to complete wave functions $\Psi_n(r, t)$ in all situations. As one of many possibilities, the theory can be applied to the partial functions $\psi_n(r)$. By definition, these functions are eigenfunctions of a time-independent Hamiltonian and, in most cases, can be readily formed into an orthonormal set and used as the basis for the expansion of an arbitrary function $f(r)$. Sometimes even less extensive portions (i.e. factors) of wave functions are eigenfunctions of Hermitian operators and constitute orthonormal systems of considerable interest and importance. The operators for angular momentum and their eigenfunctions are cases in point; their relevance will become apparent when they are considered in Chapter 12.

The typical task of the science of mechanics is to predict the future development of a system, given that the latter is in some prescribed initial state at, say, time t_0. In a quantum context, the initial state is specified by a statement such as:

$$\Psi(r, t_0) = f(r), \tag{8.102}$$

where $f(r)$ is a known function. This is a complete specification in the sense that, from it, one can in principle predict the form of the wave function at all future times provided that the system remains isolated. It is not a complete specification, however, from a point of view which contemplates the values of all the relevant physical observables since the fact that the wave function has a definite form does not imply the existence of precisely determined values for all of these observables. This is a matter to be taken up in Chapter 11; for the present it is interesting to see how the theory of orthonormal functions provides a means, in cases in which the Hamiltonian does not contain the time, for actually predicting the future form of the wave function given (8.102) as a starting point.

Let the functions $\Psi_n(r, t) = \psi_n(r)\exp-i\omega_n t$ be the stationary states of the system. As such, they are eigenfunctions of the Hamiltonian and can be formed into a set which is orthonormal at every instant of time; these functions and any linear superposition of them are also, obviously, solutions of the Schrödinger equation. Assuming that this set is complete, which it almost invariably is, one may proceed to represent the system wave function with the aid of the cofficients α_n as follows:

$$\Psi(r, t) = \sum_n \alpha_n \Psi_n(r, t). \tag{8.103}$$

Once these coefficients are determined, the future form of the wave function is predicted since the latter, obviously, can then be calculated for any r and t. At $t = t_0$, one has:

$$\Psi(r, t_0) = f(r) = \sum_n \alpha_n \Psi_n(r, t_0) \tag{8.104}$$

and, according to (8.087), the coefficients are given by:

$$\alpha_n = \int_{\substack{\text{all} \\ \text{space}}} \Psi_n^*(r, t_0)f(r)\,d\tau = \exp i\omega_n t_0 \int_{\substack{\text{all} \\ \text{space}}} \psi_n^*(r)f(r)\,d\tau. \tag{8.105}$$

It is seen that the coefficients can be calculated from integrals involving the known function $f(r)$; the objective in question is therefore accomplished.

When a wave function is represented in eigenfunctions of the Hamiltonian, as in (8.103), it is a very simple matter to calculate the expectation value of the Hamiltonian, i.e. of the total energy. Since $\hat{\mathcal{H}}\Psi_n = E_n\Psi_n$, operation of the Hamiltonian on Ψ has the following effect:

$$\hat{\mathcal{H}}\Psi = \sum_m \alpha_m E_m \Psi_m. \tag{8.106}$$

The expectation value is therefore:

$$\langle \mathcal{H} \rangle = \langle E \rangle = \int_{\substack{\text{all} \\ \text{space}}} \sum_n \alpha_n^* \Psi_n^* \sum_m \alpha_m E_m \Psi_m \, d\tau$$

$$= \sum_n \sum_m \alpha_n^* \alpha_m E_m \int_{\substack{\text{all} \\ \text{space}}} \Psi_n^* \Psi_m \, d\tau$$

$$= \sum_n \sum_m \alpha_n^* \alpha_m E_m \delta_{nm}. \qquad (8.107)$$

This becomes:

$$\langle E \rangle = \sum_n \alpha_n^* \alpha_n E_n \qquad (8.108)$$

and it is seen that the quantity sought can be obtained directly from the coefficients α_n.

A similar result, of course, would follow for the expectation value of any observable if the wave function were represented in eigenfunctions of the operator corresponding to that observable. Thus if $\hat{A}\Psi_n = a_n \Psi_n$, and if wave function Ψ is represented in terms of the Ψ_n with coefficients α_n, then:

$$\langle A \rangle = \sum_n \alpha_n^* \alpha_n a_n. \qquad (8.109)$$

CHAPTER 9

THE SIGNIFICANCE OF EXPECTATION VALUES

9.01. Time Derivatives of Expectation Values

The derivation of quantum mechanics employed in this book made use of a wave group (i.e. a packet) composed of quasi-plane waves and identified — in some sense — with the presence of a particle. The properties of the action function S in conjunction with the de Broglie relations suggested a proportionality between S and the eikonal of the typical wave of the packet. This proportionality yielded the correct group velocity and allowed the eikonal equation to be deduced from the Hamilton–Jacobi equation. A linear wave equation was then constructed to be consistent with this known eikonal equation in the short wavelength limit. The equation thus obtained was the Schrödinger equation of wave mechanics.

Since this derivation, the Schrödinger equation has been applied to a number of situations. In many of these, the criteria for quasi-plane waves were not fulfilled in all relevant regions of configuration space and the phenomena encountered were of a type never contemplated in classical mechanics. Subsequently, operators were introduced, bringing with them a whole new world of concepts and techniques. Against this background, it is now time to re-examine certain aspects of both classical and quantum mechanics in order to arrive at a better understanding of the relationship between the two. This will be done through a more detailed study of the expectation values of observables.

Most observables originate as functions of the classical dynamical variables, the q_j, the p_j, and possibly the time; a typical one would be written $A = A(q_j, p_j, t)$. For any specific motion of the system, the q_j and the p_j are specific functions of time and, in this situation, the observable A itself is thus at most a function of time only. If the nature of A is such that it remains constant as the motion proceeds, it is known as a constant of the motion and occupies a unique position on the theoretical scene. In quantum mechanics, an operator $\hat{A} = \hat{A}(\hat{q}_j, \hat{p}_j, t)$ is associated with A, time being regarded as a parameter. (It is assumed that this operator has been Hermitized, if necessary, according to the procedure proposed in Section 8.07.) Just as the observable A becomes important in the context of a specific dynamical state of the system, so the expectation value $\langle A \rangle$ is defined with reference to a specific quantum state Ψ. This is clearly seen in (8.049). Notice that the two factors of the integrand, Ψ^* and $\hat{A}\Psi$, are functions of coordinates and time. Integration is performed over the coordinates and therefore the quantity $\langle A \rangle$, like the observable A, is at most a function of time only. It is reasonable, then, to speak of the total derivative $d\langle A \rangle/dt$; an investigation of this derivative occupies the remainder of this section.

Realizing that the time derivatives of the entities under the integral sign must be partial derivatives, one may differentiate (8.049) as follows:

$$\frac{d\langle A \rangle}{d\tau} = \int_{\substack{\text{all} \\ \text{space}}} \left[\frac{\partial \Psi^*}{\partial t} \hat{A}\Psi + \Psi^* \frac{\partial \hat{A}}{\partial t}\Psi + \Psi^* \hat{A} \frac{\partial \Psi}{\partial t} \right] d\tau. \quad (9.001)$$

In this expression, $\partial \hat{A}/\partial t$ is a new operator obtained by taking the partial derivative of $\hat{A}$ with respect to time; since, by (2.003), $\partial q_j/\partial t$ and $\partial p_j/\partial t$ are identically zero, it is easily seen that this is the same entity which would be obtained by first forming the observable $\partial A/\partial t$ and then converting it into an operator. Thus the derivative $d\langle A \rangle/dt$

becomes:

$$\frac{\mathrm{d}\langle A \rangle}{\mathrm{d}t} = \langle \partial A/\partial t \rangle + \int\limits_{\substack{\text{all}\\\text{space}}} \left[\frac{\partial \Psi^*}{\partial t} \hat{A}\Psi + \Psi^* \hat{A} \frac{\partial \Psi}{\partial t} \right] \mathrm{d}\tau. \quad (9.002)$$

For a quantum state Ψ, one may say by (8.055) that:

$$\frac{\partial \Psi}{\partial t} = \frac{-i}{\hbar} [\hat{\mathcal{H}}(\hat{q}_j, \hat{p}_j, t)\Psi]. \quad (9.003)$$

The conjugate of this is:

$$\frac{\partial \Psi^*}{\partial t} = \frac{i}{\hbar} [\hat{\mathcal{H}}(\hat{q}_j, \hat{p}_j, t)\Psi]^*. \quad (9.004)$$

With the aid of these expressions, the bracketed quantity in (9.002) can be rewritten and:

$$\frac{\mathrm{d}\langle A \rangle}{\mathrm{d}t} = \langle \partial A/\partial t \rangle + \frac{i}{\hbar} \int\limits_{\substack{\text{all}\\\text{space}}} [(\hat{\mathcal{H}}\Psi)^* \hat{A}\Psi - \Psi^* \hat{A}\hat{\mathcal{H}}\Psi] \mathrm{d}\tau. \quad (9.005)$$

Now consider Theorem 8.01 and let f, $\hat{A}$, and g of that theorem represent Ψ, $\hat{\mathcal{H}}$, and $(\hat{A}\Psi)$, respectively. One obtains;

$$\int\limits_{\substack{\text{all}\\\text{space}}} \Psi^*\hat{\mathcal{H}}\hat{A}\Psi \, \mathrm{d}\tau = \int\limits_{\substack{\text{all}\\\text{space}}} \hat{A}\Psi(\hat{\mathcal{H}}\Psi)^* \, \mathrm{d}\tau. \quad (9.006)$$

This permits a substitution for the first term in the integral of (9.005). After making this substitution and inverting the order of the terms, one has:

$$\frac{\mathrm{d}\langle A \rangle}{\mathrm{d}t} = \langle \partial A/\partial t \rangle - \frac{i}{\hbar} \int\limits_{\substack{\text{all}\\\text{space}}} \Psi^*[\hat{A}\hat{\mathcal{H}} - \hat{\mathcal{H}}\hat{A}]\Psi \, \mathrm{d}\tau. \quad (9.007)$$

Notice that $\hat{A}\hat{\mathcal{H}} - \hat{\mathcal{H}}\hat{A}$ is the commutator $[\hat{A}, \hat{\mathcal{H}}]$. It is interesting that, by Theorem 8.04, the operator $-i[\hat{A}, \hat{\mathcal{H}}]$ is Hermitian and that the

T.M.C.Q. 17

final form of the time derivative may be written in terms of this operator as follows:

$$\frac{d\langle A \rangle}{dt} = \langle \partial A/\partial t \rangle + \frac{1}{\hbar} \int\limits_{\substack{\text{all}\\\text{space}}} \Psi^*(-i[\hat{A}, \hat{\mathcal{H}}])\Psi \, d\tau \quad . \qquad (9.008)$$

It is informative to examine the implications of expression (9.008) with respect to the constancy of some simple expectation values. As a first example, consider the observable $A = p_x$. Since $\partial p_x/\partial t$ is zero, it follows that $\langle p_x \rangle$ is constant if and only if $\hat{p}_x$ commutes with the Hamiltonian. Since $\hat{p}_x = -i\hbar\partial/\partial x$, it will commute with the Hamiltonian if and only if the latter does not contain x. If the Hamiltonian does not contain x, however, then p_x is a constant of the motion by the usual classical criterion. Thus $\langle p_x \rangle$ is constant if and only if p_x, according to the classical point of view, is a constant of the motion. As a second example, consider the Hamiltonian itself as the observable; let $A = \mathcal{H}$. Obviously $\mathcal{H}$ commutes with itself and the second term in (9.008) is identically zero. Thus $\mathcal{H}$ is constant if and only if $\partial\mathcal{H}/\partial t$ is zero, i.e. if the Hamiltonian does not contain the time. This, of course, is precisely the classical criterion, as expressed in (2.014), for the constancy of $\mathcal{H}$. Again, then, $\langle \mathcal{H} \rangle$ is constant if and only if $\mathcal{H}$, in the classical way of speaking, is a constant of the motion.

9.02. Ehrenfest's Theorem

The fact that the constancy of a quantum expectation value, e.g. of $\langle p_x \rangle$, is correlated with the constancy of the observable p_x in the corresponding classical context suggests a strong affinity between the two quantities. This suggestion is sharpened by Ehrenfest's theorem[34] which states that the quantum mechanical expectation values of observables which have classical counterparts do indeed obey the laws of classical mechanics. A completely general proof of this theorem will not be attempted; the proof which will be given, involving the position and momentum of a one-particle system, is convincing and usually sufficient.

The proof in question will be presented in two parts which complement one another. The first part begins with a consideration of the time derivative of $\langle r \rangle$:

$$\frac{d\langle r \rangle}{dt} = \langle \partial r/\partial t \rangle + \frac{1}{\hbar} \int_{\substack{\text{all} \\ \text{space}}} \Psi^*(-i[\hat{r}, \hat{\mathscr{H}}])\Psi \, d\tau. \qquad (9.009)$$

As emphasized in the previous section, the partial derivative $\partial r/\partial t$ is zero. Recalling that $\hat{\mathscr{H}} = \hat{T} + \hat{V}$ and that $\hat{r}$ commutes with $\hat{V}$, it is seen that:

$$\frac{d\langle r \rangle}{dt} = \frac{1}{\hbar} \int_{\substack{\text{all} \\ \text{space}}} \Psi^*(-i[\hat{r}, \hat{T}])\Psi \, d\tau$$

$$= \frac{i\hbar}{2m} \left[\int_{\substack{\text{all} \\ \text{space}}} \Psi^* r \, \nabla^2 \Psi \, d\tau - \int_{\substack{\text{all} \\ \text{space}}} \Psi^* \, \nabla^2(r\Psi) \, d\tau \right]. \qquad (9.010)$$

In preparing to evaluate $\nabla^2(r\psi)$, the simpler $\nabla^2(x \Psi)$ should first be considered:

$$\nabla^2(x\Psi) = x \, \nabla^2\Psi + 2 \, \nabla x \cdot \nabla \Psi + \Psi \, \nabla^2 x$$
$$= x \, \nabla^2\Psi + 2e_x \cdot \nabla \Psi + 0$$
$$= x \, \nabla^2\Psi + 2 \, \partial \Psi/\partial x. \qquad (9.011)$$

It follows that:

$$\nabla^2(r\Psi) = r \, \nabla^2\Psi + 2 \, \nabla \Psi. \qquad (9.012)$$

Substitution of this into (9.010) produces a cancellation; only the term involving $2 \nabla \Psi$ survives:

$$\frac{d\langle r \rangle}{dt} = \frac{-i\hbar}{2m} \int_{\substack{\text{all} \\ \text{space}}} \Psi^* 2 \, \nabla \Psi \, d\tau. \qquad (9.013)$$

It is easy to identify the momentum operator here. The final result may be written:

$$m \frac{\mathrm{d}\langle r \rangle}{\mathrm{d}t} = \langle p \rangle .$$

(9.014)

This completes the first part of the proof and confirms that the expectation values $\langle r \rangle$ and $\langle p \rangle$ are related according to the classical definition of momentum.

The second part of the proof investigates the time derivative of $\langle p \rangle$. It begins as follows:

$$\frac{\mathrm{d}\langle p \rangle}{\mathrm{d}t} = \langle \mathrm{d}p/\mathrm{d}t \rangle + \frac{1}{\hbar} \int_{\substack{\text{all} \\ \text{space}}} \psi^*(-i[\hat{p}, \hat{\mathcal{H}}])\Psi \, \mathrm{d}\tau .$$

(9.015)

Continuing as before, it may be recalled that $\partial p/\partial t$ is zero. Also, $\hat{p}$ commutes with $\hat{T}$ but, generally, not with $\hat{V}$. Therefore:

$$\frac{\mathrm{d}\langle p \rangle}{\mathrm{d}t} = \frac{1}{\hbar} \int_{\substack{\text{all} \\ \text{space}}} \Psi^*(-i[\hat{p}, \hat{V}])\Psi \, \mathrm{d}\tau$$

$$= -\left[\int_{\substack{\text{all} \\ \text{space}}} \Psi^* \, \nabla(V\Psi) \, \mathrm{d}\tau - \int_{\substack{\text{all} \\ \text{space}}} \Psi^*V \, \nabla\Psi \, \mathrm{d}\tau \right] .$$

(9.016)

The gradient of the product $V\Psi$ is given by:

$$\nabla(V\Psi) = V \, \nabla\Psi + \Psi \, \nabla V .$$

(9.017)

Therefore:

$$\frac{\mathrm{d}\langle p \rangle}{\mathrm{d}t} = - \int_{\substack{\text{all} \\ \text{space}}} \Psi^*\Psi \, \nabla V \, \mathrm{d}\tau .$$

(9.018)

Since $-\nabla V$ is equal to the force F, the above may be written as:

$$\frac{\mathrm{d}\langle p \rangle}{\mathrm{d}t} = \langle -\nabla V \rangle = \langle F \rangle .$$

(9.019)

This completes the second part of the demonstration and shows that the expectation values of momentum and of force are related according to Newton's second law.

Even though quantum expectation values formally exhibit classical behavior in all situations in which the observables have classical counterparts, the significance of this behavior can vary between wide limits. Thus in some cases this behavior is so trivial that it is devoid of physical interest whereas in others it is a salient feature of the system and constitutes such a satisfactory description of the motion that, for many purposes, the underlying wave mechanism can be ignored. The latter case, of course, is the quintessentially classical one. As an example of the former type of situation, consider a one-dimensional harmonic oscillator in a single stationary state $\Psi_n(x, t)$. Since Ψ_n is an eigenfunction of parity, $\Psi_n^* \Psi_n$ must have even parity and the expectation value of the position x is zero at all times:

$$\langle x \rangle = \int_{-\infty}^{\infty} \Psi_n^* x \Psi_n \, dx = 0. \tag{9.020}$$

Also, since the parities of Ψ_n^* and of $\partial \Psi_n / \partial x$ must be opposite, the expectation value of p_x must also be zero at all times:

$$\langle p_x \rangle = -i\hbar \int_{-\infty}^{\infty} \Psi_n^* (\partial \Psi_n / \partial x) \, dx = 0. \tag{9.021}$$

Finally, the force $F_x = -(\partial V / \partial x) = -Kx$ is simply the elastic constant times x and its expectation value must be zero for the same reason that $\langle x \rangle$ is zero:

$$\langle F_x \rangle = -K\langle x \rangle = 0. \tag{9.022}$$

It follows that Ehrenfest's theorem is trivially satisfied for this stationary state regardless of the value of the quantum number n. It should be added, of course, that a single stationary state of very high n is extremely improbable.

In general, an entirely different type of situation presents itself if the oscillator has a wave function which is a superposition of a number of stationary states. Suppose that Ψ is as follows:

$$\Psi(x, t) = \sum_{n=0}^{\infty} \alpha_n \Psi_n(x, t). \tag{9.023}$$

The expectation value of x is then given by:

$$\langle x \rangle = \int_{-\infty}^{\infty} \left[\sum_{m=0}^{\infty} \alpha_m^* \psi_m^* \exp i\omega_m t \right] x \left[\sum_{n=0}^{\infty} \alpha_n \psi_n \exp -i\omega_n t \right] dx. \tag{9.024}$$

Since $\omega_m - \omega_n = (m-n)\Omega$, where Ω is the classical frequency of the oscillator, one has:

$$\langle x \rangle = \sum_{m=0}^{\infty} \sum_{n=0}^{\infty} \alpha_m^* \alpha_n \int_{-\infty}^{\infty} \psi_m^* x \psi_n \, dx \exp i(m-n)\Omega t. \tag{9.025}$$

Here it is convenient to denote the integral by a special symbol:

$$\int_{-\infty}^{\infty} \psi_m^* x \psi_n \, dx = C_{mn}. \tag{9.026}$$

With the aid of this symbol, the quantity $\langle x \rangle$ can be displayed as follows:

$$\langle x \rangle =$$

$$\begin{aligned}
&\alpha_0^* \alpha_0 C_{00} &&+ \alpha_0^* \alpha_1 C_{01} \exp -i\Omega t + \alpha_0^* \alpha_2 C_{02} \exp -i2\Omega t + \ldots \\
&+ \alpha_1^* \alpha_0 C_{10} \exp i\Omega t &&+ \alpha_1^* \alpha_1 C_{11} &&+ \alpha_1^* \alpha_2 C_{12} \exp - i\Omega t + \ldots \\
&+ \alpha_2^* \alpha_0 C_{20} \exp i2\Omega t &&+ \alpha_2^* \alpha_1 C_{21} \exp i\Omega t &&+ \alpha_2^* \alpha_2 C_{22} &&+ \ldots \\
&+ \ldots
\end{aligned} \tag{9.027}$$

The quantities C_{mn} can be evaluated from information presented in Section 7.07. Details are left as an exercise; the result is given by:

$$C_{mn} = \begin{cases} \left[\dfrac{\hbar}{2M\Omega} \text{ (larger of } m \text{ and } n) \right]^{\frac{1}{2}} & \text{if } |m-n| = 1 \\[2mm] 0 & \text{if } |m-n| \neq 1 \end{cases}. \tag{9.028}$$

Thus, of the terms in (9.027), only two diagonal rows, one immediately above and the other immediately below the main diagonal, are non-zero. (The main diagonal extends from upper left to lower right and bisects the array of terms.) The expectation value $\langle x \rangle$ therefore reduces to:

$$\langle x \rangle = \left(\frac{\hbar}{2M\Omega}\right)^{\frac{1}{2}} \left(\left[\alpha_0^*\alpha_1 1^{\frac{1}{2}} + \alpha_1^*\alpha_2 2^{\frac{1}{2}} + \alpha_2^*\alpha_3 3^{\frac{1}{2}} + \ldots\right] \exp -i\Omega t\right.$$

$$\left. + \left[\alpha_1^*\alpha_0 1^{\frac{1}{2}} + \alpha_2^*\alpha_1 2^{\frac{1}{2}} + \alpha_3^*\alpha_2 3^{\frac{1}{2}} + \ldots\right] \exp i\Omega t\right). \qquad (9.029)$$

The quantities in square brackets above are, in general, complex numbers and are easily seen to be conjugates of one another. One may introduce the real quantities A_0 and θ by saying:

$$\left.\begin{array}{l} \left(\dfrac{\hbar}{2M\Omega}\right)^{\frac{1}{2}} \displaystyle\sum_{n=0}^{\infty} \alpha_n^*\alpha_{n+1}(n+1)^{\frac{1}{2}} = \tfrac{1}{2}A_0 \exp -i\theta; \\[3mm] \left(\dfrac{\hbar}{2M\Omega}\right)^{\frac{1}{2}} \displaystyle\sum_{n=0}^{\infty} \alpha_{n+1}^*\alpha_n(n+1)^{\frac{1}{2}} = \tfrac{1}{2}A_0 \exp i\theta. \end{array}\right\} \qquad (9.030)$$

On substitution of these into (9.029), the latter becomes:

$$\langle x \rangle = A_0 \cos(\Omega t + \theta). \qquad (9.031)$$

This shows that, provided A_0 is not zero, $\langle x \rangle$ does actually execute sinusoidal motion at the classical frequency Ω. (A_0 will be non-zero if there are at least two non-zero α_n coefficients with adjacent values of n.)

Although it is true that $\langle x \rangle$ oscillates at the classical frequency with an amplitude related to the distribution of the α_n coefficients, macroscopic classical behavior in which $\Psi^*\Psi$ is in the form of an extremely narrow packet which moves back and forth is not inevitably achieved. A typical situation in which it is achieved, however, may be described as follows:

(i) There is a very large number of states in the superposition. These have consecutive values of n and are clustered in a relatively compact set of width Δn about a central quantum number n_c as illustrated in Fig. 9.01. The relationships $n_c \gg \Delta n \gg 1$ obtain; reasonable values for n_c and Δn might be 10^{30} and 10^{26} respectively.

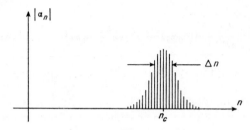

FIG. 9.01. A possible plot of $|\alpha_n|$ with respect to n for the harmonic oscillator. Width of the distribution is indicated qualitatively by Δn which, for present purposes, need not be defined precisely. If n_c and Δn are both large, classical behavior can be achieved.

(ii) The coefficients α_n, which can be written in the form $|\alpha_n| \exp i\varphi_n$, have the properties that $|\alpha_n|$ is a relatively smooth function of n and that φ_n is a linear function of n, i.e. that $\varphi_n = \beta - \gamma n$.

Under these conditions:

$$\alpha_n^* \alpha_{n+1} = |\alpha_n| \, |\alpha_{n+1}| \exp -i\gamma$$

$$\approx |\alpha_n|^2 \exp -i\gamma, \qquad (9.032)$$

and:

$$\alpha_{n+1}^* \alpha_n \approx |\alpha_n|^2 \exp i\gamma. \qquad (9.033)$$

Thus:

$$\tfrac{1}{2} A_0 \exp i\theta \approx \left(\frac{\hbar}{2M\Omega}\right)^{\frac{1}{2}} \left[\sum_{n=0}^{\infty} |\alpha_n|^2 (n+1)^{\frac{1}{2}} \right] \exp i\gamma. \qquad (9.034)$$

Because of the relative compactness of the cluster, $(n+1)^{\frac{1}{2}}$ is essentially constant over the summation and can be replaced by $n_c^{\frac{1}{2}}$. The sum of $|\alpha_n|^2$ is unity because of normalization and the results can be summarized by:

$$\left. \begin{array}{c} \theta = \gamma; \\[2mm] A_0 \approx \left(\dfrac{2\hbar n_c}{M\Omega} \right)^{\frac{1}{2}}. \end{array} \right\}$$ (9.035)

According to (8.108), the expectation value of the energy of the oscillator is:

$$\langle E \rangle = \sum_{n=0}^{\infty} \alpha_n^* \alpha_n (n + \tfrac{1}{2}) \hbar \Omega.$$ (9.036)

Again, because of the compactness of the cluster, this reduces to:

$$\langle E \rangle \approx n_c \hbar \Omega.$$ (9.037)

It is interesting to reconcile A_0 and $\langle E \rangle$. Solving (9.035) for $\hbar n_c$, one obtains:

$$\hbar n_c \approx \tfrac{1}{2} M \Omega A_0^2.$$ (9.038)

This may be multiplied on both sides by Ω and substituted into (9.037); $M\Omega^2 = K$, the elastic constant, and one has:

$$\langle E \rangle \approx \tfrac{1}{2} K A_0^2,$$ (9.039)

which is the correct relationship for a macroscopic oscillator.

9.03. A More Precise View of the Correspondence Principle and of the Nature of Classical Mechanics

Bohr's correspondence principle states that, in situations which can be considered classical, quantum mechanics must yield predictions in harmony with those of classical mechanics, i.e. the results of quantum mechanics must *correspond* with those of classical mechanics whenever the latter is relevant. In the light of Ehrenfest's theorem and the preceding discussion of the harmonic oscillator, this principle can now be appreciated with new insight. For classical mechanics to be meaningful, certain conditions must be fulfilled. Using a one-particle system as an example, these conditions may be stated as follows:

(i) It must be possible for quasi-plane waves to exist in all relevant regions of configuration space. This, in turn, requires that changes in the potential energy must be relatively gradual in terms of the longest de Broglie wavelength of interest, which is tantamount to saying that the force on the particle must be relatively weak.[†]

(ii) The quasi-plane waves which exist under the aegis of condition (i) above must be superposed to form a relatively compact packet.

A wave mechanical packet, formed under the conditions just stated, will move through three-dimensional configuration space in such a way that it can be said to trace out an orbit. The velocity of this packet, which is none other than the group velocity, agrees with the classically predicted velocity of the associated particle. The Schrödinger equation insures this; it was, in fact, derived with the understanding that such agreement should exist.

Since a wave packet incorporates many component waves each with a slightly different local wave vector, it has a collection of many slightly different momenta rather than a single definite momentum. Likewise, because of its physical extension, it embodies many closely juxtaposed positions rather than a single definite position. The packet does have a definite centroid, however, and the displacement of this centroid from the origin of coordinates is simply the expectation value $\langle r \rangle$. Also, there is associated with the packet a definite quantity called $\langle p \rangle$ which constitutes an average of all the individual momenta represented in its composition. As Ehrenfest's theorem states, this $\langle p \rangle$ is

---

[†] If an infinite potential energy step could actually exist, it would constitute an exception to this statement; for if a wave packet possessed classical characteristics prior to incidence upon such a step, it would continue to possess these characteristics after the perfect reflection that would occur. The infinite rectangular well illustrates this point since, for it, one can assemble a superposition of stationary states which duplicates the classical motion.

equal to the particle mass times the velocity of the centroid and both $\langle r \rangle$ and $\langle p \rangle$ obey the laws of classical mechanics. One is readily led to the conclusion, already stated in this book, that there is but one mechanics and that that is quantum mechanics. Classical mechanics is but a collection of relatively simple rules which codify the behavior of wave packet properties such as $\langle r \rangle$ and $\langle p \rangle$, behavior which is really determined by the wave mechanical infrastructure. Thus Bohr's correspondence principle, which was a heuristic principle from the beginning, is seen to be eminently well satisfied by Schrödinger's wave mechanics.

From what has been said, it follows that the motions described by classical mechanics are really not the motions of particles after all but only the motions of centroids of probability distributions. The matter may be summarized by saying that neither wave mechanics, which deals directly with the continuous wave function, nor classical mechanics, which provides a description (often eminently useful) of the gross motions of wave packets, gives specific information about the "true behavior" of the supposedly singular particles. Some physicists believe that the true motion of a particle is a meaningful concept and that a complete theory of such motions will some day explain wave mechanics in much the same way that statistical mechanics explains thermodynamics. Such is the format of so-called "hidden variable" theories which, in the present state of development of physics, are controversial. Other physicists maintain that the entities called particles are so abstract that any attempt to describe their "true motions" is inevitably naïve. Only the future holds the answer to this intriguing question.

CHAPTER 10

THE MOMENTUM REPRESENTATION

10.01. Fourier Series

An exponential function of $ik_0 \cdot r$ with vanishing amplitude, such as (8.029), is an attempt to express a normalized eigenfunction of the momentum operator even though such an entity is an unrealizable limiting case. An integral superposition of eigenfunctions of this type is realizable, however, and is useful for many purposes. When cast in the form of such a superposition, a quantum wave function is said to be in the *momentum representation*. This chapter begins with a review study of Fourier series, then passes on to the Fourier integral which is the basic tool needed to work with this representation.

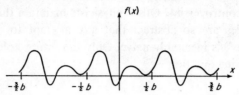

FIG. 10.01. A periodic function of x with period b.

Let $f(x)$ be a periodic function of x with period b as illustrated in Fig. 10.01. Whether real or complex, $f(x)$ can be represented by a complex exponential Fourier series as follows:

$$f(x) = \sum_{n=-\infty}^{\infty} \alpha_n \exp\left(in2\pi x/b\right). \qquad (10.001)$$

This series obviously has the necessary periodic property. To find the coefficients α_n, one may multiply by $\exp{(-im2\pi x/b)}$ on both sides and integrate over one period:

$$\int_{-b/2}^{b/2} f(x)\exp{(-im2\pi x/b)}\,\mathrm{d}x = \sum_{n=-\infty}^{\infty} \alpha_n \int_{-b/2}^{b/2} \exp{[i(n-m)2\pi x/b]}\,\mathrm{d}x.$$
(10.002)

Let $(2\pi x/b) = \theta$. Then $\mathrm{d}x = b\,\mathrm{d}\theta/2\pi$ and:

$$\int_{-b/2}^{b/2} \exp{[i(n-m)2\pi x/b]}\,\mathrm{d}x = \frac{b}{2\pi}\int_{-\pi}^{\pi} \exp{i(n-m)\theta}\,\mathrm{d}\theta. \quad (10.003)$$

It is easily seen that the integral on the right is equal to 2π if $n = m$ and to zero otherwise, i.e. it is given by $2\pi\delta_{nm}$. Therefore:

$$\int_{-b/2}^{b/2} f(x)\exp{(-im2\pi x/b)}\,\mathrm{d}x = \sum_{n=-\infty}^{\infty} \alpha_n b\delta_{nm} = b\alpha_m. \quad (10.004)$$

One may now replace m by n and write for the coefficient:

$$\alpha_n = \frac{1}{b}\int_{-b/2}^{b/2} f(x)\exp{(-in2\pi x/b)}\,\mathrm{d}x. \quad (10.005)$$

Equations (10.001) and (10.005) form the basis for the study of Fourier series.

Still another important result, having to do with the average value of f^*f over one period, will now be obtained. Notice that:

$$f^*f = \left[\sum_{n=-\infty}^{\infty} \alpha_n^* \exp{(-in2\pi x/b)}\right]\left[\sum_{m=-\infty}^{\infty} \alpha_m \exp{(im2\pi x/b)}\right]$$

$$= \sum_{n=-\infty}^{\infty}\sum_{m=-\infty}^{\infty} \alpha_n^*\alpha_m \exp{[i(m-n)2\pi x/b]}. \quad (10.006)$$

Integrating this over one period, one finds that:

$$\int_{-b/2}^{b/2} f^*f\,\mathrm{d}x = \sum_{n=-\infty}^{\infty}\sum_{m=-\infty}^{\infty} \alpha_n^*\alpha_m b\delta_{mn}. \quad (10.007)$$

After dividing by b and simplifying, this becomes:

$$\frac{1}{b} \int_{-b/2}^{b/2} f^*f \, dx = \sum_{n=-\infty}^{\infty} \alpha_n^* \alpha_n. \tag{10.008}$$

This equation is known as the *Parseval formula*. It relates the average value in question to the sum of the squared magnitudes of the coefficients.

10.02. Fourier Transforms and their Application to Quantum Mechanics

The subject of Fourier transforms is easily developed by substituting a single symbol for the combination $n2\pi/b$ which occurs in both (10.001) and (10.005). This symbol will be called k_n; it takes on discrete values as indicated:

$$k_n = n2\pi/b; \qquad \Delta k = k_{n+1} - k_n = 2\pi/b. \tag{10.009}$$

Since α_n can be regarded as a function of n, it can also be regarded as a function of k_n and one may rewrite (10.001) as follows:

$$f(x) = \sum_{n=-\infty}^{\infty} \alpha(k_n) \exp ik_n x. \tag{10.010}$$

Since $(b \, \Delta k/2\pi) = 1$ by (10.009), the above may be written:

$$f(x) = \sum_{n=-\infty}^{\infty} \frac{b}{2\pi} \alpha(k_n) \exp ik_n x \, \Delta k. \tag{10.011}$$

It is now convenient to define a new function of k_n related to $\alpha(k_n)$ as follows:

$$g(k_n) = (2\pi)^{-\frac{1}{2}} b\alpha(k_n), \tag{10.012}$$

whereupon (10.011) becomes:

$$f(x) = (2\pi)^{-\frac{1}{2}} \sum_{n=-\infty}^{\infty} g(k_n) \exp ik_n x \, \Delta k. \tag{10.013}$$

The same function, $g(k_n)$, can be introduced into (10.005) by multiplying both sides by $(2\pi)^{-\frac{1}{2}} b$. Thus:

$$g(k_n) = (2\pi)^{-\frac{1}{2}} \int_{-b/2}^{b/2} f(x) \exp{-ik_n x}\, dx. \qquad (10.014)$$

These two formulas, (10.013) and (10.014), are still only trivially different from (10.001) and (10.005), respectively, since they were obtained merely by defining new quantities and substituting. The crucial step, which will now be taken, is to allow the period b of the function $f(x)$ to approach infinity. As this happens, Δk approaches zero, k can be regarded as a continuous variable, and the designation k_n is no longer necessary. The sum in (10.013) becomes an integral and the two formulas exhibit a remarkable degree of symmetry:

$$f(x) = (2\pi)^{-\frac{1}{2}} \int_{-\infty}^{\infty} g(k) \exp{ikx}\, dk; \qquad (10.015)$$

$$g(k) = (2\pi)^{-\frac{1}{2}} \int_{-\infty}^{\infty} f(x) \exp{-ikx}\, dx. \qquad (10.016)$$

The function $g(k)$ is said to be the *Fourier transform* of $f(x)$; like the set of coefficients α_n it contains the same information contained in $f(x)$ since the latter can be obtained from it by (10.015). Conversely, $f(x)$ is called the inverse Fourier transform of $g(k)$, and $g(k)$ can be obtained from $f(x)$ by (10.016). The two formulas, (10.015) and (10.016) are extremely useful in mathematical physics and will soon be employed to develop the momentum representation.

The Parseval formula in the limit of infinite period is easily derived in terms similar to those just employed. Let the right-hand side of (10.008) be multiplied by $b\, \Delta k/2\pi$:

$$\frac{1}{b} \int_{-b/2}^{b/2} f^* f\, dx = \sum_{n=-\infty}^{\infty} \frac{b\alpha_n^* \alpha_n}{2\pi}\, \Delta k. \qquad (10.017)$$

The resulting equation may now be re-expressed as follows:

$$\int_{-b/2}^{b/2} f^*f \, dx = \sum_{n=-\infty}^{\infty} g^*(k_n) \, g(k_n) \, \Delta k. \qquad (10.018)$$

Once more the period is allowed to approach infinity, and Δk to approach zero. The result is the very symmetrical expression:

$$\int_{-\infty}^{\infty} f^*f \, dx = \int_{-\infty}^{\infty} g^*g \, dk. \qquad (10.019)$$

If $f(x)$ changes with time, i.e. if it is really $f(x, t)$, then $g(k)$ which, at every instant, is derivable from $f(x)$ by (10.016), becomes $g(k, t)$ and the Fourier transform pair may be written:

$$f(x, t) = (2\pi)^{-\frac{1}{2}} \int_{-\infty}^{\infty} g(k, t) \exp ikx \, dk; \qquad (10.020)$$

$$g(k, t) = (2\pi)^{-\frac{1}{2}} \int_{-\infty}^{\infty} f(x, t) \exp -ikx \, dx. \qquad (10.021)$$

In this form, the transforms are readily applicable to quantum mechanics. To make this application, let $f(x, t)$ be the wave function $\Psi(x, t)$ for a system with one degree of freedom described by the inertial Cartesian coordinate x such as, for example, the one-dimensional harmonic oscillator. To further emphasize that a one-dimensional system is being dealt with, the quantity k will be written as k_x, and, in order to employ a Greek capital notation in harmony with $\Psi(x, t)$, the function $g(k, t)$ will be written $\Phi(k_x, t)$. Thus one has:

$$\Psi(x, t) = (2\pi)^{-\frac{1}{2}} \int_{-\infty}^{\infty} \Phi(k_x, t) \exp ik_x x \, dk_x \quad ; \qquad (10.022)$$

$$\Phi(k_x, t) = (2\pi)^{-\frac{1}{2}} \int_{-\infty}^{\infty} \Psi(x, t) \exp -ik_x x \, dx \quad . \qquad (10.023)$$

Expressed as in (10.022), Ψ is an infinite superposition of the complex exponentials $\exp ik_x x$. Each of these is an eigenfunction of the operator $\hat{p}_x$ with eigenvalue $p_x = \hbar k_x$, hence the construction of the (one-dimensional) momentum representation has been accomplished. The quantity k_x here plays the part of a continuous quantum number, each value of which signifies the presence of the corresponding value of the actual momentum component p_x. The fact that p_x is proportional to k_x means that there is an unusually simple and direct relationship between eigenvalue and quantum number, so simple in fact that the two are often identified[†] and the continuum $-\infty < k_x < \infty$ is regarded as a one-dimensional momentum space. The function $\Phi(k_x, t)$ is a distribution function on this continuum and shows the degree, at time t, to which each of the various values of k_x, and therefore of p_x, participate in the composition of the wave function $\Psi(x, t)$. In this connection, it is interesting to observe that the normalization of $\Psi(x, t)$ on configuration space implies the normalization of $\Phi(k_x, t)$ on momentum space. Thus:

$$\int_{-\infty}^{\infty} \Phi^*\Phi(k_x, t)\,\mathrm{d}k_x = \int_{-\infty}^{\infty} \Psi^*\Psi(x, t)\,\mathrm{d}x = 1 \quad . \qquad (10.024)$$

On inspection of (10.024) above, the inference that $\Phi^*\Phi\,\mathrm{d}k_x$ is the probability that k_x will be found to have a value in the infinitesimal interval $\mathrm{d}k_x$ suggests itself. In the light of this inference, which is correct, it follows that the expectation value of k_x is given by the integral of $\Phi^*k_x\Phi\,\mathrm{d}k_x$ over the aforementioned continuum and, more importantly, that the expectation value of p_x is:

$$\langle p_x \rangle = \int_{-\infty}^{\infty} \Phi^*\hbar k_x\Phi\,\mathrm{d}k_x. \qquad (10.025)$$

Momentum space may be regarded as dual to configuration space and the function $\Phi(k_x, t)$ as a dual form of the wave function which, in conjunction with the appropriate operators, may be used in much the same way that $\Psi(x, t)$ is employed on configuration space. Thus,

[†] Theoreticians often employ a unit system in which $\hbar = 1$. This brings about a complete identification of p_x with k_x.

T.M.C.Q. 18

in the context of momentum space, it follows from (10.025) that:

$$\hat{p}_x = \hbar k_x \qquad (10.026)$$

and, in the same context, it can be shown[†] that:

$$\hat{x} = i\frac{\partial}{\partial k_x}. \qquad (10.027)$$

The momentum operator is Hermitian and one may rightly expect strong analogies between formulas developed here and some of the results obtained in Chapter 8. The fact that k_x acts like a continuous quantum number has been pointed out. In the momentum representation, therefore, it performs the same function that the discrete index n performs in the representation scheme discussed in Chapter 8. It follows that $\Phi(k_x, t)$ is analogous to the set of coefficients α_x (which, in the general case, may also be functions of time) affecting the individual eigenfunctions in which $\Psi(x, t)$ is represented. The normalization condition (10.024) is analogous to the result that the sum of $\alpha_n^* \alpha_n$ must be equal to unity and, finally, (10.025) is strikingly similar to (8.109).

10.03. Extension to Three Dimensions

A one-particle wave function $\Psi(r, t)$, defined on three-dimensional configuration space, possesses a distribution function $\Phi(k, t)$ defined upon the space of the wave vector k. The latter therefore becomes three-dimensional momentum space. The formulas connecting $\Psi(r, t)$ with $\Phi(k, t)$, which are relatively simple extensions of (10.022) and (10.023), will now be developed.

Equation (10.022) is taken as a model. In $\Phi(k_x, k_y, k_z, t)$, k_y, k_z, and t may be looked upon as parameters which must also appear in

---

[†] By equation (10.056) below, the wave function, which — at time t_0 — represents a quantum state of definite position, is expressed by $\Phi_x(k_x, t_0) = C_{\downarrow} \exp{-ik_x x_0}$. The operator which has this as an eigenfunction with x_0 as the corresponding eigenvalue is $i\partial/\partial k_x$.

Ψ. The latter, for an anticipated reason, will be called Ψ''. Thus:

$$\Psi''(x, k_y, k_z, t) = (2\pi)^{-\frac{1}{2}} \int_{-\infty}^{\infty} \Phi(k_x, k_y, k_z, t) \exp ik_x x \, dk_x. \quad (10.028)$$

The function $\Psi''(x, k_y, k_z, t)$ may now be regarded as a function of k_y with x, k_z, and t as parameters and, as such, subjected to a similar transformation involving the pair of variables y and k_y. The result is:

$$\Psi'(x, y, k_z, t) = (2\pi)^{-\frac{1}{2}} \int_{-\infty}^{\infty} \Psi''(x, k_y, k_z, t) \exp ik_y y \, dk_y. \quad (10.029)$$

The function $\Psi'(x, y, k_z, t)$ is now to be regarded as a function of k_z and, as such, subjected to a transformation involving the pair of variables z and k_z. This yields:

$$\Psi(x, y, z, t) = (2\pi)^{-\frac{1}{2}} \int_{-\infty}^{\infty} \Psi'(x, y, k_z, t) \exp ik_z z \, dk_z. \quad (10.030)$$

Chain substitution produces the result:

$$\Psi(\mathbf{r}, t) = (2\pi)^{-\frac{3}{2}} \int_{\substack{\text{all } \mathbf{k} \\ \text{space}}} \Phi(\mathbf{k}, t) \exp i\mathbf{k}\cdot\mathbf{r} \, d\tau_k \quad , \quad (10.031)$$

where $d\tau_k = dk_x \, dk_y \, dk_z$ is a volume element in $\mathbf{k}$ space. A similar sequence of steps with (10.023) as a model gives the opposite transformation:

$$\Phi(\mathbf{k}, t) = (2\pi)^{-\frac{3}{2}} \int_{\substack{\text{all} \\ \text{space}}} \Psi(\mathbf{r}, t) \exp -i\mathbf{k}\cdot\mathbf{r} \, d\tau \quad . \quad (10.032)$$

If the normalization condition (10.024) is applied in the context of (10.028), one has:

$$\int_{-\infty}^{\infty} \Phi^*\Phi(k_x, k_y, k_z, t) \, dk_x = \int_{-\infty}^{\infty} \Psi''^*\Psi''(x, k_y, k_z, t) \, dx. \quad (10.033)$$

Both sides may now be integrated with respect to k_y and k_z with the following result:

$$\int_{-\infty}^{\infty} \int_{-\infty}^{\infty} \int_{-\infty}^{\infty} \Phi^*\Phi(k_x, k_y, k_z, t)\, dk_x\, dk_y\, dk_z$$
$$= \int_{-\infty}^{\infty} \int_{-\infty}^{\infty} \int_{-\infty}^{\infty} \Psi'''^*\Psi'''(x, k_y, k_z, t)\, dx\, dk_y\, dk_z. \quad (10.034)$$

Similarly, the normalization formula can be applied to $\Psi'''(x, k_y, k_z, t)$ and $\Psi'(x, y, k_z, t)$ since these are Fourier transforms of one another through the variables y and k_y. The result may be subsequently integrated over x and k_z to yield:

$$\int_{-\infty}^{\infty} \int_{-\infty}^{\infty} \int_{-\infty}^{\infty} \Psi'''^*\Psi'''(x, k_y, k_z, t)\, dx\, dk_y\, dk_z$$
$$= \int_{-\infty}^{\infty} \int_{-\infty}^{\infty} \Psi'^*\Psi'(x, y, k_z, t)\, dx\, dy\, dk_z. \quad (10.035)$$

Finally, it can be applied to $\Psi'(x, y, k_z, t)$ and $\Psi(x, y, z, t)$ in the context of the variables z and k_z followed by integrations over x and y:

$$\int_{-\infty}^{\infty} \int_{-\infty}^{\infty} \int_{-\infty}^{\infty} \Psi'^*\Psi'(x, y, k_z, t)\, dx\, dy\, dk_z$$
$$= \int_{-\infty}^{\infty} \int_{-\infty}^{\infty} \int_{-\infty}^{\infty} \Psi^*\Psi(x, y, z, t)\, dx\, dy\, dz. \quad (10.036)$$

Once more, a chain substitution yields:

$$\boxed{\int_{\substack{\text{all } k \\ \text{space}}} \Phi^*\Phi(k, t)\, d\tau_k = \int_{\substack{\text{all} \\ \text{space}}} \Psi^*\Psi(r, t)\, d\tau = 1}. \quad (10.037)$$

Finally, the expressions for the two familiar operators, by extension of (10.026) and (10.027), take the following forms in the three-dimensional momentum representation:

$$\left. \boxed{\begin{aligned} \hat{p} &= \hbar k \quad ; \\ \hat{r} &= i\, \nabla_k \quad . \end{aligned}} \right\} \quad (10.038)$$

10.04. Eigenfunctions of Position and of Momentum

In Chapter 8, it was pointed out that, at least at a particular instant of time, it is possible to imagine a normalized one-particle quantum wave function which is an eigenfunction of $\hat{r}$ and of another such function which is an eigenfunction of $\hat{p}$. In each case, the forms envisioned can be approached arbitrarily closely but never actually attained. The object of this section is to show, in a quantitative way, how this approach can be implemented and how, concomitantly, the duality between the coordinate (conventional) representation and the momentum representation is emphasized.

As was done earlier in this chapter, an analysis in one dimension will be made first and, later, an extension to three dimensions will be indicated. Thus, analogously with the results of Section 8.03, an eigenfunction of $\hat{x}$ can be written as the square root of a Dirac delta function times an arbitrary phase factor:

$$\Psi_x(x, t_0) = \delta^{\frac{1}{2}}(x - x_0) \exp i\varphi. \qquad (10.039)$$

This indicates that, at t_0, the coordinate x has the definite value x_0. Similarly, an eigenfunction of $\hat{p}_x$ can be written:

$$\Psi_p(x, t_0) = B_1 \exp ik_{0x}x \qquad (10.040)$$

and this states that, at t_0, the momentum p_x has the definite value $\hbar k_{0x}$. The system wave function can, as of time t_0, approach either (10.039) or (10.040) but not both at once.

In carrying out the proposed program, the Gaussian probability density function will prove very useful. This function, here called $G[(x - x_0), \eta]$, is given by:

$$\varrho(x) = G[(x - x_0), \eta] = \frac{1}{(2\pi)^{\frac{1}{2}} \eta} \exp \left[-\frac{1}{2}(x - x_0)^2/\eta^2 \right] \qquad (10.041)$$

and is illustrated in Fig. 10.02. It is symmetric about x_0 and it follows that $\langle x \rangle = x_0$. The expectation value of the squared deviation from x_0

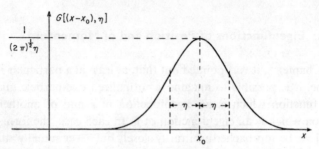

FIG. 10.02. The normalized one-dimensional Gaussian function.

is called the *variance*; it can be shown to be equal to the parameter η^2:

$$\langle (x-x_0)^2 \rangle = \int_{-\infty}^{\infty} (x-x_0)^2 \, \varrho(x) \, \mathrm{d}x = \eta^2. \qquad (10.042)$$

The denominator factors in (10.041) have been chosen so that the function remains normalized regardless of the value of η^2. The square root of the variance, namely η itself, has the same dimensions as x and is called the *standard deviation*. This quantity is in general a measure of the "width" of a probability density function and, in the case of the Gaussian, it has the particularly simple property of being equal to the distance from x_0 to either of the curve's two points of inflection which are symmetrically situated with respect to x_0. The relationship between η and the width is therefore obvious; a small value of η indicates that the curve has small width and rises to a very high value at $x = x_0$ whereas a large value of η has the opposite implications. It is this very property that makes the Gaussian useful in the present context. Thus, as $\eta \to 0$, $G[(x-x_0), \eta]$ becomes like the one-dimensional Dirac delta function $\delta(x-x_0)$. On the other hand, as $\eta \to \infty$, $G[(x-x_0), \eta]$ becomes the very antithesis of a delta function in that it is reduced to vanishing amplitude over the whole x continuum.

With the concepts developed in the preceding paragraph, it is easy to construct a wave function which, at time t_0, can approach either an eigenfunction of $\hat{x}$ or an eigenfunction of $\hat{p}_x$ according to the limit approached by η. The function in question involves the square root of (10.041) and is given by:

$$\Psi(x, t_0) = \frac{\exp ik_{0x}x}{(2\pi)^{\frac{1}{4}}\eta^{\frac{1}{2}}} \exp\left[-\tfrac{1}{4}(x-x_0)^2/\eta^2\right]; \qquad (10.043)$$

its properties may be summarized as follows:

$$\Psi^*\Psi(x, t_0) = G[(x-x_0), \eta]; \qquad (10.044)$$

$$\text{As } \eta \to 0, \quad \Psi(x, t_0) \sim \delta^{\frac{1}{2}}(x-x_0)\exp ik_{0x}x_0; \quad (10.045)$$

$$\text{As } \eta \to \infty, \quad \Psi(x, t_0) \sim B_1\exp ik_{0x}x. \qquad (10.046)$$

The distribution $\Phi(k_x, t_0)$ on momentum space, the dual of $\Psi(x, t_0)$, is very interesting. To obtain it, one applies (10.023):

$$\Phi(k_x, t_0) = \frac{1}{(2\pi)^{\frac{3}{4}}\eta^{\frac{1}{2}}} \int_{-\infty}^{\infty} \exp\left[-\tfrac{1}{4}(x-x_0)^2/\eta^2\right]\exp i(k_{0x}-k_x)x\, dx.$$

$$(10.047)$$

It is advantageous to make temporary use of the variables $x' = x-x_0$ and $k_x' = k_x-k_{0x}$. Then:

$$\Phi(k_x', t_0) = \frac{\exp-ik_x'x_0}{(2\pi)^{\frac{3}{4}}\eta^{\frac{1}{2}}} \int_{-\infty}^{\infty} \exp\left[-\tfrac{1}{4}x'^2/\eta^2\right]\exp-ik_x'x'\, dx'. \quad (10.048)$$

The exponent of the integrand may be rewritten as follows:

$$-\left\{\frac{x'^2}{4\eta^2} + ik_x'x'\right\} = -\left\{\left[\frac{x'}{2\eta} + i\eta k_x'\right]^2 + \eta^2 k_x'^2\right\}. \qquad (10.049)$$

If the quantity in square brackets above is regarded as a new variable u, equation (10.048) becomes:

$$\Phi(k_x', t_0) = \frac{\exp-ik_x'x_0}{(2\pi)^{\frac{3}{4}}\eta^{\frac{1}{2}}} 2\eta \int_{-\infty}^{\infty} \exp-u^2\, du \exp-\eta^2 k_x'^2. \quad (10.050)$$

The definite integral in this expression is equal to $\pi^{\frac{1}{2}}$ and the constant factors can be collected as follows:

$$\frac{2\eta\pi^{\frac{1}{2}}}{(2\pi)^{\frac{3}{4}}\eta^{\frac{1}{2}}} = \frac{(2\eta)^{\frac{1}{2}}}{(2\pi)^{\frac{1}{4}}}. \tag{10.051}$$

From this, one can recognize that $1/2\eta$ will become the standard deviation in momentum space; this quantity will be designated μ:

$$\mu = \frac{1}{2\eta}. \tag{10.052}$$

In terms of μ, $\Phi(k'_x, t_0)$ can be written:

$$\Phi(k'_x, t_0) = \frac{\exp - ik'_x x_0}{(2\pi)^{\frac{1}{4}}\mu^{\frac{1}{2}}} \exp\left[-\tfrac{1}{4}k'^2_x/\mu^2\right]. \tag{10.053}$$

If one now replaces k'_x by $k_x - k_{0x}$ in order to restore the variable k_x, the distribution on momentum space becomes:[†]

$$\Phi(k_x, t_0) = \frac{\exp - i(k_x - k_{0x})x_0}{(2\pi)^{\frac{1}{4}}\mu^{\frac{1}{2}}} \exp\left[-\tfrac{1}{4}(k_x - k_{0x})^2/\mu^2\right]. \tag{10.054}$$

The probability density on momentum space associated with $\Phi(k_x, t_0)$ is Gaussian; this and other interesting properties may be summarized as follows:

$$\Phi^*\Phi(k_x, t_0) = G[(k_x - k_{0x}), \mu]; \tag{10.055}$$

$$\text{As} \quad \eta \to 0 \quad \text{and} \quad \mu \to \infty, \quad \Phi(k_x, t_0) \sim C_1 \exp - ik_x x_0; \tag{10.056}$$

$$\text{As} \quad \eta \to \infty \quad \text{and} \quad \mu \to 0, \quad \Phi(k_x, t_0) \sim \delta^{\frac{1}{2}}(k_x - k_{0x}). \tag{10.057}$$

The results of this development show that if, at a particular time t_0, the function Ψ has a form approaching an exponential of $ik_{0x}x$ on

---

[†] The constant phase factor, $\exp ik_{0x}x_0$, which spoils the otherwise perfect symmetry between (10.043) and (10.054), is of no consequence. The symmetry in question could have been achieved by arbitrarily inserting the factor $\exp(-ik_{0x}x_0/2)$ into (10.043) when the latter was first defined.

configuration space, then the function Φ must have a form approaching the square root of a delta function on momentum space; both limiting forms indicate the presence of only one value of k_x, namely k_{0x}, and only one value of momentum, namely $\hbar k_{0x}$. Conversely, if the function Ψ has, at t_0, a form approaching the square root of a delta function of $(x - x_0)$ on configuration space, then the function Φ is automatically thrown into a form approaching an exponential of $-ik_x x_0$ on momentum space and this time both limiting forms indicate the presence of only one position, namely x_0. Regardless of the degree to which the limiting forms are approached, the standard deviations of Ψ and of Φ are inversely proportional so that as one of these functions becomes broad the other becomes narrow and vice versa. This phenomenon is discussed in more fundamental terms in Chapter 11.

It is a relatively simple matter to extend the results just obtained to three dimensions. To consider a very general situation, one can conceive of a wave function which is simply the product of three one-dimensional Gaussian-related functions of the form of (10.043), each based upon a different coordinate and each having a different value of η. A more specific possibility is that in which a three-dimensional system can be considered one-dimensional because two of the three factors have very large standard deviations (say η_y and η_z) whereas the third factor, which becomes the nominal wave function, is a function of x and t and assumes whatever form is dictated by the presence of x-directed forces and initial conditions. The simplest case of all is that in which all three factors have the same standard deviation η. The wave function then becomes:

$$\Psi(r, t_0) = \frac{\exp i k_0 \cdot r}{(2\pi)^{\frac{3}{4}} \eta^{\frac{3}{2}}} \exp \left[-\tfrac{1}{4} |r - r_0|^2 / \eta^2 \right]. \tag{10.058}$$

The dual function on momentum space is:

$$\Phi(k, t_0) = \frac{\exp i(k - k_0) \cdot r_0}{(2\pi)^{\frac{3}{4}} \mu^{\frac{3}{2}}} \exp \left[-\tfrac{1}{4} |k - k_0|^2 / \mu^2 \right], \tag{10.059}$$

where, as before, $\mu = 1/2\eta$. As $\eta \to 0$ and $\mu \to \infty$, both of these functions tend to become eigenfunctions of $\hat{r}$; as $\eta \to \infty$ and $\mu \to 0$, both tend to become eigenfunctions of $\hat{p}$. The forms of these eigenfunctions have already been discussed in Chapter 8.

10.05. The Unforced Particle in the Momentum Representation

Up to this point, the momentum representation has been formulated without reference to any specific physical application, i.e. there has been no attempt to require that the wave function $\Psi(r, t)$ should be a solution of the Schrödinger equation for any particular potential energy function. It is now time to consider how the momentum representation can actually be applied and, for this purpose, the constant potential energy $V = V_0$ suggests itself as an example that is at once simple and useful. It corresponds, of course, to the case in which the particle in question is not acted upon by forces. The treatment will be limited to the one-dimensional case to avoid undue prolixity; even in this form, it can sometimes be applied to three-dimensional situations in a way that has been recently indicated.

One begins by writing $\Psi(x, t)$ as a superposition of waves traveling in the x direction:

$$\Psi(x, t) = (2\pi)^{-\frac{1}{2}} \int\limits_{-\infty}^{\infty} \varphi(k_x) \exp i[k_x x - \omega(k_x)t] \, dk_x. \quad (10.060)$$

This, as may be readily verified, is a solution of the Schrödinger equation with constant potential energy provided that $\omega(k_x)$ is:

$$\omega(k_x) = \frac{\hbar k_x^2}{2m} + \frac{V_0}{\hbar}. \quad (10.061)$$

By comparison with (10.022), the complete distribution function on momentum space is seen to be:

$$\Phi(k_x, t) = \varphi(k_x) \exp -i\omega(k_x)t \quad (10.062)$$

and the corresponding probability density, namely $\Phi^*\Phi = \varphi^*\varphi$, is time-independent as would be expected in an unforced situation. The function $\varphi(k_x)$ is determined by the known form of Ψ at some initial time, say t_0. Thus one may apply (10.023) and, say:

$$\varphi(k_x)\exp-i\omega(k_x)t_0 = (2\pi)^{-\frac{1}{2}}\int_{-\infty}^{\infty}\Psi(x, t_0)\exp-ik_xx\,\mathrm{d}x. \quad (10.063)$$

This becomes:

$$\varphi(k_x) = (2\pi)^{-\frac{1}{2}}\exp i\omega(k_x)t_0 \int_{-\infty}^{\infty}\Psi(x, t_0)\exp-ik_xx\,\mathrm{d}x, \quad (10.064)$$

and the one-dimensional formulation for an unforced particle is thereby completed.

The theory just developed is nicely illustrated by supposing that $\Psi(x, t)$ is a moving wave packet with $\langle p_x \rangle = \hbar k_{0x}$, i.e. that it is a function of the form:

$$\Psi(x, t) = \exp i(k_{0x}x - \omega_0 t)f(x, t) \quad (10.065)$$

where $f(x, t)$ is, initially at least, a relatively smooth real envelope function and $\omega_0 = \omega(k_{0x})$. For simplicity, V_0 will be set equal to zero whereupon $\omega_0 = \hbar k_{0x}^2/2m$. It is also convenient to take the initial time t_0 as equal to zero. To provide a specific example that is relatively easy to analyze, let the packet envelope be initially a Gaussian-related function of standard deviation η centered about the origin.[†] One may therefore say:

$$\Psi(x, 0) = \frac{\exp ik_{0x}x}{(2\pi)^{\frac{1}{4}}\eta^{\frac{1}{2}}}\exp[-\tfrac{1}{4}x^2/\eta^2]. \quad (10.066)$$

This can be recognized as (10.043) with $x_0 = 0$ and it follows that $\varphi(k_x)$, which is equal to $\Phi(k_x, 0)$, can be obtained directly from (10.054) merely by setting x_0 equal to zero. Thus the labor of re-performing the operations indicated in (10.064) is avoided. The reasonable result, which shows that $\varphi(k_x)$ is also a Gaussian-related distribution centered

[†] This means that $\Psi^*\Psi(x, 0) = G[(x-0), \eta]$.

about k_{0x}, is as follows:

$$\varphi(k_x) = \frac{1}{(2\pi)^{\frac{1}{4}}\mu^{\frac{1}{2}}} \exp\left[-\tfrac{1}{4}(k_x - k_{0x})^2/\mu^2\right]. \qquad (10.067)$$

As before, $\mu = 1/2\eta$.

The future development of the wave function, given that (10.066) is its initial form, is now to be derived; to do this, one applies (10.060) to the known distribution function $\varphi(k_x)$:

$$\Psi(x, t) = \frac{1}{(2\pi)^{\frac{3}{4}}\mu^{\frac{1}{2}}} \int_{-\infty}^{\infty} \exp\left[-\tfrac{1}{4}(k_x - k_{0x})^2/\mu^2\right]$$

$$\times \exp\left[ik_x x - i\hbar k_x^2 t/2m\right] dk_x. \qquad (10.068)$$

The substitution of $k_x' = k_x - k_{0x}$ permits the exponent of the integrand to be written as follows:

$$-\tfrac{1}{4}(k_x'^2/\mu^2) + i(k_x' + k_{0x})x - (i\hbar t/2m)(k_x'^2 + 2k_x'k_{0x} + k_{0x}^2)$$

$$= i(k_{0x}x - \omega_0 t) - \left(\frac{1}{4\mu^2} + \frac{i\hbar t}{2m}\right)k_x'^2 + i\left(x - \frac{\hbar k_{0x}}{m}t\right)k_x'. \qquad (10.069)$$

Here $(\hbar k_{x0}/m)$, which is equal to $\langle p_x \rangle/m$, is the group velocity, v. Replacing $1/2\mu$ by η, one now has for $\Psi(x, t)$:

$$\Psi(x, t) = \frac{(2\eta)^{\frac{1}{2}} \exp i(k_{0x}x - \omega_0 t)}{(2\pi)^{\frac{3}{4}}}$$

$$\times \int_{-\infty}^{\infty} \exp\left[-(\eta^2 + i\hbar t/2m)k_x'^2 + i(x - vt)k_x'\right] dk_x'. \qquad (10.070)$$

The exponent of this simplified integrand becomes:

$$-\left[(\eta^2 + i\hbar t/2m)k_x'^2 - i(x - vt)k_x'\right] = -[Ak_x' - B]^2 + B^2 \qquad (10.071)$$

where:

$$\left. \begin{array}{l} A = (\eta^2 + i\hbar t/2m)^{\frac{1}{2}}; \\ B = i(x - vt)/2A. \end{array} \right\} \qquad (10.072)$$

The wave function can therefore be written:

$$\Psi(x, t) = \frac{(2\eta)^{\frac{1}{2}} \exp i(k_{0x}x - \omega_0 t)}{(2\pi)^{\frac{3}{4}} A} \exp B^2 \int_{-\infty}^{\infty} \exp -u^2 \, du. \quad (10.073)$$

Recognizing the definite integral as $\pi^{\frac{1}{2}}$, one obtains:

$$\Psi(x, r) = \frac{\exp i(k_{0x}x - \omega_0 t)}{(2\pi)^{\frac{1}{4}} (\eta + i\hbar t/2m\eta)^{\frac{1}{2}}} \exp \left[\frac{-(x - vt)^2}{4(\eta^2 + i\hbar t/2m)} \right]. \quad (10.074)$$

The expression for Ψ derived above clearly shows that, as time passes, the envelope function ceases to have the uniform phase which, by hypothesis, it possessed at $t = 0$. The probability density, which is given by:

$$\Psi^*\Psi(x, t) = \frac{1}{(2\pi)^{\frac{1}{2}} [\eta^2 + (\hbar t/2m\eta)^2]^{\frac{1}{2}}} \exp \left\{ \frac{-(x - vt)^2}{2[\eta^2 + (\hbar t/2m\eta)^2]} \right\},$$

$$(10.075)$$

retains Gaussian character but has a standard deviation η' which increases with time:

$$\eta' = [\eta^2 + (\hbar t/2m\eta)^2]^{\frac{1}{2}}. \quad (10.076)$$

This increasing of η' is a phenomenon often described as "the spreading of the wave packet" and is characteristic of an unforced particle. If the mass and the initial η are large, the spreading is slow and, in many situations, is negligible. If the mass and the initial η are small and μ is therefore large, however, the packet possesses a large intrinsic kinetic energy and the spreading becomes explosively rapid.

10.06.   The Stationary State in the Momentum Representation

In a stationary state, as has been frequently mentioned, the wave function is of the form:

$$\Psi(r, t) = \psi(r) \exp - i\omega t, \tag{10.077}$$

where ω is an absolute constant, i.e. not dependent upon k as it was in the previous section. By application of (10.032):

$$\Phi(k, t) = (2\pi)^{-\frac{3}{2}} \exp - i\omega t \int\limits_{\substack{\text{all} \\ \text{space}}} \psi(r) \exp - ik \cdot r \, d\tau. \tag{10.078}$$

It follows that $\Phi(k, t)$ is also of the form:

$$\Phi(k, t) = \varphi(k) \exp - i\omega t. \tag{10.079}$$

Thus the same factor, namely $\exp - i\omega t$, occurs in both Ψ and Φ; the functions ψ and φ are Fourier transforms of one another, and one may say:

$$\psi(r) = (2\pi)^{-\frac{3}{2}} \int\limits_{\substack{\text{all } k \\ \text{space}}} \varphi(k) \exp ik \cdot r \, d\tau_k; \tag{10.080}$$

$$\varphi(k) = (2\pi)^{-\frac{3}{2}} \int\limits_{\substack{\text{all} \\ \text{space}}} \psi(r) \exp - ik \cdot r \, d\tau; \tag{10.081}$$

$$\int\limits_{\substack{\text{all } k \\ \text{space}}} \varphi^*\varphi(k) \, d\tau_k = \int\limits_{\substack{\text{all} \\ \text{space}}} \psi^*\psi(r) \, d\tau = 1. \tag{10.082}$$

It is clearly seen that, in a stationary state, neither $\Psi^*\Psi$ which is equal to $\psi^*\psi$ nor $\Phi^*\Phi$ which is equal to $\varphi^*\varphi$ is a function of time.

CHAPTER 11

THE CONCEPT OF MEASUREMENT IN QUANTUM MECHANICS

11.01. Measurements: Classical and Quantum

As emphasized by Ehrenfest's theorem, the classical aspects of a given system are to be found in the behavior of the expectation values of its observables. In terms of these expectation values, it is easy to contemplate the classical process of measurement. If, for instance, one were to make a classical measurement of the energy of a harmonic oscillator with known elastic constant K, one would undertake to determine the amplitude A_0. The resultant value of energy would be the expectation value $\langle E \rangle = \frac{1}{2}KA_0^2$ as elaborated in Section 9.02. The making of this determination would inevitably involve some small interference with the motion of the oscillator even if such interference were no more than the temporary use of sufficient room illumination to enable the observer to see the mass at the moment of its maximum displacement. At the quantum level, this interference would be evidenced by permanent changes in the spectrum of the α_n coefficients illustrated in Fig. 9.01. It is likely that most of these coefficients would be affected in some way even if only slightly and that the value being determined, namely $\langle E \rangle$, would also be affected but only to a negligible degree if the measurement were performed with reasonable skill. This interference by the measuring apparatus is accepted as an unwelcome but tolerable effect which the classical experimenter

275

seeks to reduce or at least assess; beyond that he does not regard it as having fundamental importance. There is considerable contrast between this and the corresponding effect in the quantum measurement process which will now be described.

A quantum measurement of an observable A is an attempt to fix the value of A itself rather than an attempt to determine the value of $\langle A \rangle$; from the outset it is therefore conceptually different from a classical measurement even though both are inextricably connected with the statistical interpretation of quantum mechanics. At this point, it is well to recall that the expectation value of an observable has been defined as the average value which would be obtained if the distribution over which the average is taken were structured strictly in accord with *a priori* probability. Thus the very existence of an expectation value, say $\langle A \rangle$, indicates in general the existence of a distribution of values obeying a calculus of probabilities. Specifically, $\langle A \rangle$ can be written:

$$\langle A \rangle = \sum_n P_n A_n, \tag{11.001}$$

where A_n is a particular value which A may take and P_n is the corresponding probability that this value will be taken. It is clear, of course, that:

$$\sum_n P_n = 1. \tag{11.002}$$

One may compare (11.001) with (6.039) in order to recall the context in which the term "expectation value" was first introduced.

It is of the greatest relevance that an expression like (11.001) is produced whenever a given quantum state, here symbolized by Ψ, is represented in the eigenfunctions Ψ_n of an Hermitian operator such as $\hat{A}$. Thus:

$$\Psi = \sum_n \alpha_n \Psi_n, \tag{11.003}$$

where

$$\hat{A}\Psi_n = a_n \Psi_n \tag{11.004}$$

and, following (8.087), α_n is given by:

$$\alpha_n = \int\limits_{\substack{\text{all}\\\text{space}}} \Psi_n^* \Psi \, d\tau. \tag{11.005}$$

According to (8.109), $\langle A \rangle$ in the above situation can be written as:

$$\langle A \rangle = \sum_n \alpha_n^* \alpha_n a_n \tag{11.006}$$

where, by (8.089),

$$\sum_n \alpha_n^* \alpha_n = 1. \tag{11.007}$$

Each of these formulas can be adapted to the case in which a continuum of values rather than a set of discrete values is available to A. The important point here, however, is the similarity between (11.001) and (11.006) on the one hand and (11.002) and (11.007) on the other. It is seen that one must associate with $\langle A \rangle$ a distribution of values of A and a corresponding set of probabilities; it is as if there were a set of hypothetical events at each of which a determination of A is made and made with a certain probability. The conclusion suggests itself that the only value which A can take in one of these determinations is one of the eigenvalues a_n and that the probability that $A = a_n$, supposing the system to be originally in quantum state Ψ, is simply $\alpha_n^* \alpha_n$, where α_n is as given in (11.005). Such a determination is known as a *quantum measurement*. In some cases, it is possible to design an actual physical apparatus which will perform a given quantum measurement whereas in others it is difficult to construct anything except a "thought experiment". In any event, the idea of a quantum measurement is commonplace in the literature and has come to be recognized as one of the basic concepts of quantum mechanics.[†]

In general, both classical measurements and quantum measurements change the wave function, i.e. change the quantum state, of the system upon which they are performed. A classical measurement is almost invariably performed upon a system which is already in a superposition of states with respect to the observable whose measurement is contemplated. After the classical measurement, the system

[†] A Stern–Gerlach apparatus is an excellent example of a realizable quantum measuring device. Feynman, who has employed the ideas of superposition of states and of quantum measurements to introduce the subject of quantum mechanics, makes very effective use of a modified Stern–Gerlach apparatus as a model to illustrate these concepts. See ref. 35.

is still in a superposition of states with respect to the measured observable; the small difference between the pre-measurement and the post-measurement superpositions is not regarded as conceptually important and may be made practically negligible. By contrast, a quantum measurement of an observable, say A, must *by its very definition* leave the system in one of the eigenstates of $\hat{A}$ with A equal to the corresponding eigenvalue. A quantum measurement may therefore constitute a relatively major interaction between the system measured and the measuring apparatus; the nature of this interaction is such that in general the initial state is obliterated and a new final state is substituted.

To make the idea of a quantum measurement perfectly clear, suppose that the four eigenstates Ψ_1, Ψ_2, Ψ_3, and Ψ_4 of the operator $\hat{A}$ form a complete set of states for the description of the given system. This means that if the observable A is to have a definite value, the latter must be one of the four corresponding eigenvalues a_1, a_2, a_3, or a_4. The initial state of the system will in general be a superposition of the Ψ_n as follows:

$$\Psi = \sum_{n=1}^{4} \alpha_n \Psi_n. \tag{11.008}$$

A typical plot of the squared magnitude of the α_n coefficients versus A for this initial state is presented in Fig. 11.01. This plot shows probabilities of 0.4, 0.1, 0.3, and 0.2 that a quantum measurement will yield $A = a_1, a_2, a_3$, or a_4, respectively. The value of $\langle A \rangle$ for the super-

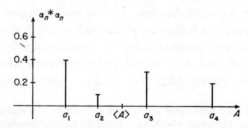

FIG. 11.01. A typical distribution of probability coefficients for the quantum measurement of an observable A. Eigenvalues a_n and expectation value $\langle A \rangle$ are shown.

position in question, which is readily calculable from the eigenvalues and their probabilities, is also shown.

In general, quantum mechanics is powerless to predict the outcome of any single given quantum measurement; it can predict only the probability of the outcome. This is the sense in which quantum mechanics is said to be a non-deterministic theory. Suppose, now, that the measurement is performed and that it happens to yield $A = a_3$. The final state Ψ' of the system will then be simply Ψ_3; another way of expressing this is to say that in the final state $|\alpha_3| = 1$ and $\alpha_1 = \alpha_2 = \alpha_4 = 0$. The latter situation is illustrated in Fig. 11.02. The system

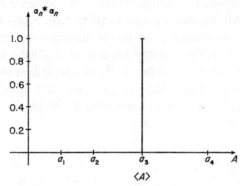

Fig. 11.02. The distribution of probability coefficients immediately after a quantum measurement has yielded the value $A = a_3$.

is now no longer in a superposition of states with respect to $\hat{A}$ but is in an eigenstate of $\hat{A}$; the expectation value $\langle A \rangle$ has changed significantly and has become identically equal to a_3. If another quantum measurement of A is performed immediately[†] upon this system, the result will inevitably be that $A = A_3$. This is an example of the unusual case in which a quantum measurement (i) does not change the wave function of the system and (ii) has a predictable outcome.

In view of the preceding discussion on quantum measurements, it is easy to visualize a quantum measuring apparatus as a type of

---

[†] If the final state Ψ' is a stationary state, then the word "immediately" is not needed in this statement.

"sorting" or "filtering" device. Thus, in the context of the same example, one may suppose that the apparatus can receive the system
in its initial state Ψ and indicate the presence of the final state $\Psi' = \Psi_3$
by ejecting the system through a "channel" designated by the quan-
tum number 3. Other channels, representing other possible outcomes
of the experiment, would be labeled 1, 2, or 4. If, as in Fig. 11.03, 1000
identical systems, each in initial state Ψ, were fed into the apparatus,
one would expect that about 400 of these would be ejected through
channel 1, 100 through channel 2, 300 through channel 3, and 200
through channel 4. It must be emphasized that the term "sorting
device" is used only as a simile. There is no predestination implied
in Fig. 11.03. All the entering systems are in state Ψ and are therefore
identical and indistinguishable in the quantum-mechanical sense;
it is strictly a matter of probability that a number of these will be placed
in final state Ψ_1, another number in Ψ_2, etc. According to the conventionally accepted interpretation of quantum mechanics, namely the
interpretation of the "Copenhagen school" which traces its origin

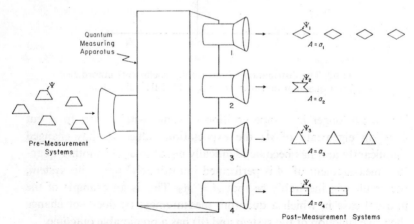

Fig. 11.03. Symbolic diagram of a quantum measuring apparatus.
The pre-measurement systems are all in an identical state Ψ; the post-
measurement systems have been placed in one of four possible
eigenstates Ψ_1, Ψ_2, Ψ_3, or Ψ_4 of operator $\hat{A}$ where A is the observable
being measured. In each instance, the latter has assumed the value
of one of the four corresponding eigenvalues a_1, a_2, a_3, or a_4.

to Bohr, a quantum "measuring" apparatus is really misnamed; such an apparatus does not detect that a particular system is in a certain final state, rather it places the system in its final state and does so with a probability that depends upon the degree to which the final state was involved in the composition of the initial state! The basic paradox of quantum mechanics exhibits itself here with unusual clarity; a distribution of measurement results is generated obeying a known calculus of probabilities without any apparent internal mechanism to explain how such a distribution comes into being. Many physicists accept this at face value, reasoning that the ultimate theory of the universe will probably contain elements which are incomprehensible in terms abstracted from macroscopic experience; hence, if quantum theory is the ultimate theory, it is not surprising that a paradox of the type just described should be incorporated in its makeup. Others, not satisfied with such a state of affairs, incline toward "hidden variable" theories which have been mentioned earlier in this book. On this viewpoint, the pre-measurement systems of Fig. 11.03, although quantum mechanically indistinguishable, are actually distinguishable in some yet more fundamental way. This question, which still remains controversial, has inspired a considerable volume of research[36] to which the interested reader is referred.

11.02. The Uncertainty Principle

In the previous section, it was emphasized that when a state Ψ of a given system is represented in the eigenstates of a particular operator, say $\hat{A}$, the coefficients in this representation are the elements which determine the probabilities of the various outcomes which a quantum measurement of the value of A (hereafter called simply a "quantum A measurement") can have. It is not difficult to imagine another observable, say B, with its operator $\hat{B}$, its eigenstates Ψ'_m, and their corresponding eigenvalues b_m. It can easily happen that the states Ψ'_m also constitute a complete set in which the initial state, or any other state, of the given system can be represented. Thus a state Ψ can simultaneously have two representations, one in the eigenstates of $\hat{A}$ and the other in the eigenstates of $\hat{B}$:

$$\Psi = \sum_n \alpha_n \Psi_n = \sum_m \beta_m \Psi'_m. \tag{11.009}$$

With each of these two representations there is associated a probability distribution with probability coefficients $\alpha_n^* \alpha_n$ and expectation value $\langle A \rangle$ in the A case and with probability coefficients $\beta_m^* \beta_m$ and expectation value $\langle B \rangle$ in the B case.

The possible coexistence of two (or even more) representations of a given state Ψ raises some interesting questions. One may find, for instance, that the operators $\hat{A}$ and $\hat{B}$ have no common eigenstates. This means that whenever Ψ is an eigenstate of $\hat{B}$ it is necessarily a superposition of eigenstates of $\hat{A}$ and conversely. If this is true, then a quantum A measurement which, by its very nature, places the system in an eigenstate of $\hat{A}$, will *ipso facto* destroy a pre-existing eigenstate of $\hat{B}$. In this particular case, then, the two quantum measurements are said to be *incompatible*. On the other hand, one may find that the two operators have a finite or an infinite number of common eigenstates in which case the two will be, to some degree at least, *compatible*. It becomes a matter of importance to ascertain under what conditions a given state Ψ may be a common eigenstate and, to attack this problem, the standard deviations, here called ΔA and ΔB respectively, of the two probability distributions should be considered. The standard deviation, introduced in Chapter 10, is defined as the positive square root of the expectation of the squared deviation from the expectation value. Thus ΔA is given by:

$$\Delta A = + \langle (A - \langle A \rangle)^2 \rangle^{\frac{1}{2}}. \tag{11.010}$$

This quantity measures the "spread" in the values of A or, in some sense, the "uncertainty" to be associated with a determination of A. Thus in Fig. 11.01 there is some uncertainty in the outcome of a determination of A and this uncertainty is reflected in the fact that the squared deviations of A from $\langle A \rangle$ are appreciable and impart to ΔA a value greater tham zero. In Fig. 11.02, on the other hand, there is no uncertainty; there is but one value of A and this occurs with unit probability. For this latter case, $A = \langle A \rangle$ and $\Delta A = 0$. To put the matter in still another way, the value of ΔA indicates unequivocally whether or not Ψ is an eigenstate of $\hat{A}$; if $\Delta A = 0$, then Ψ is an eigenstate and if $\Delta A > 0$, it is not.

The square of expression (11.010), i.e. the variance, can be readily expanded as follows:

$$(\Delta A)^2 = \langle A^2 - 2A\langle A \rangle + \langle A \rangle^2 \rangle. \tag{11.011}$$

The expectation value $\langle A \rangle$ is a single quantity derived from the distribution of values of A and therefore bears a fixed relationship to these values; in the context of the operation indicated in (11.011) it is to be treated as a numerical constant would be treated. Thus $\langle 2A\langle A \rangle \rangle = 2\langle A \rangle^2$ and one may write the following convenient formulas for the calculation of $(\Delta A)^2$:

$$
\begin{aligned}
(\Delta A)^2 &= \langle A^2 \rangle - \langle A \rangle^2 \\
&= \int_{\substack{\text{all} \\ \text{space}}} \Psi^* \hat{A}^2 \Psi \, d\tau - \left[\int_{\substack{\text{all} \\ \text{space}}} \Psi^* \hat{A} \Psi \, d\tau \right]^2 \\
&= \sum_n \alpha_n^* \alpha_n a_n^2 - \left[\sum_n \alpha_n^* \alpha_n a_n \right]^2.
\end{aligned} \tag{11.012}
$$

Similar expressions can be written for ΔB and for $(\Delta B)^2$.

To elucidate the relationship between ΔA and ΔB for a given quantum state Ψ, the following theorem is proposed:

THEOREM 11.01:

$$\boxed{\; \Delta A \, \Delta B \geqslant \tfrac{1}{2} \left| \int_{\substack{\text{all} \\ \text{space}}} \Psi^*(-i[\hat{A}, \hat{B}])\Psi \, d\tau \right| \;}. \tag{11.013}$$

This theorem expresses the celebrated HEISENBERG UNCERTAINTY PRINCIPLE[†] and will now be proved. For brevity, let the operator

[†] Heisenberg's original presentation of this principle was given in less general terms; see ref. 37. The form used here was first proved by H. P. Robertson in ref. 38 at a slightly later date.

$-i[\hat{A}, \hat{B}]$ which, by Theorem 8.04, is Hermitian, be denoted $\hat{M}$ and let the integral in (11.013), which is therefore real, be represented by the symbol[†] $\langle M \rangle$. Consider a real variable u with dimensions $[B/A]$; it is readily seen that the function $F(u)$, defined as follows, is real and greater than or equal to zero for all values of u since it is the integral of a function times its conjugate:

$$F(u) = \int_{\substack{\text{all} \\ \text{space}}} \{[(\hat{B} - \langle B \rangle \hat{1}) + iu(\hat{A} - \langle A \rangle \hat{1})]\Psi\}^*$$

$$\times [(\hat{B} - \langle B \rangle \hat{1}) + iu(\hat{A} - \langle A \rangle \hat{1})]\Psi \, d\tau. \qquad (11.014)$$

Letting $\hat{A} - \langle A \rangle \hat{1} = \hat{A}'$ and $\hat{B} - \langle B \rangle \hat{1} = \hat{B}'$ for convenience, one has:

$$F(u) = \int_{\substack{\text{all} \\ \text{space}}} \{[\hat{B}' + iu\hat{A}']\Psi\}^* [\hat{B}' + iu\hat{A}']\Psi \, d\tau$$

$$= \int_{\substack{\text{all} \\ \text{space}}} (\hat{B}'\Psi)^* \hat{B}'\Psi \, d\tau - iu \int_{\substack{\text{all} \\ \text{space}}} (\hat{A}'\Psi)^* \hat{B}'\Psi \, d\tau$$

$$+ iu \int_{\substack{\text{all} \\ \text{space}}} (\hat{B}'\Psi)^* \hat{A}'\Psi \, d\tau + u^2 \int_{\substack{\text{all} \\ \text{space}}} (\hat{A}'\Psi)^* \hat{A}'\Psi \, d\tau. \qquad (11.015)$$

Notice that, in taking this step, use was made of the fact that u is real. By Theorem 8.01, the above becomes:

$$F(u) = \int_{\substack{\text{all} \\ \text{space}}} \Psi^* \hat{B}'^2 \Psi \, d\tau - iu \int_{\substack{\text{all} \\ \text{space}}} \Psi^* \hat{A}' \hat{B}' \Psi \, d\tau$$

$$+ iu \int_{\substack{\text{all} \\ \text{space}}} \Psi^* \hat{B}' \hat{A}' \Psi \, d\tau + u^2 \int_{\substack{\text{all} \\ \text{space}}} \Psi^* \hat{A}'^2 \Psi \, d\tau. \qquad (11.016)$$

This is equivalent to:

$$F(u) = u^2 (\Delta A)^2 + u \int_{\substack{\text{all} \\ \text{space}}} \Psi^* (-i[\hat{A}', \hat{B}'])\Psi \, d\tau + (\Delta B)^2. \qquad (11.017)$$

[†] The notation employed here suggests that there is actually a physical observable M with which the Hermitian operator $-i[\hat{A}, \hat{B}]$ is associated. This is true; M is equal to $\hbar$ times a classical quantity known as the *Poisson bracket* of A and B. See Appendix D for details.

Since $\langle A \rangle \hat{1}$ and $\langle B \rangle \hat{1}$ freely commute with any other operator, it is readily seen that $[\hat{A}', \hat{B}'] = [\hat{A}, \hat{B}]$. $F(u)$ may therefore be written as the following expression in which every coefficient is seen to be real:

$$F(u) = u^2(\Delta A)^2 + u\langle M \rangle + (\Delta B)^2. \tag{11.018}$$

This is a quadratic function of the real variable u and, since it is greater than or equal to zero for all u, it must appear (when plotted) as a parabola which either lies entirely above the horizontal axis or touches that axis at most at one point. It follows that the roots of $F(u) = 0$ must be either complex or real and equal. These roots are given by the quadratic formula:

$$u = \frac{-\langle M \rangle \pm [\langle M \rangle^2 - 4(\Delta A)^2 (\Delta B)^2]^{\frac{1}{2}}}{2(\Delta A)^2} \tag{11.019}$$

and the following expresses the condition that the roots are as described:

$$4(\Delta A)^2 (\Delta B)^2 \geqslant \langle M \rangle^2. \tag{11.020}$$

This reduces to:

$$\Delta A \, \Delta B \geqslant \tfrac{1}{2} |\langle M \rangle|, \tag{11.021}$$

which is the expression to be proved. The magnitude signs are needed here because $\langle M \rangle$, although real, may be negative.

The basic question to which the uncertainty principle addresses itself is this: can Ψ simultaneously be an eigenstate of both $\hat{A}$ and $\hat{B}$, i.e. can the uncertainties in a proposed quantum A measurement and a proposed quantum B measurement both be zero? The answer is seen to depend primarily upon the commutator of $\hat{A}$ and $\hat{B}$. If $-i[\hat{A}, \hat{B}] = 0$, then $\langle M \rangle = 0$ for all Ψ and the answer is unequivocally affirmative; ΔA and ΔB can both be zero although they need not be so in every instance. On the other hand, if $-i[\hat{A}, \hat{B}] = \hbar\hat{1}$, as would be the case if A and B are canonically conjugate observables, then $\langle M \rangle = \hbar$ for all Ψ and the answer is unequivocally negative; if (by changing Ψ) one makes ΔA go to zero, then ΔB must go to infinity and vice versa. This important case is discussed in greater detail in the subsequent section.

The observables A and B may be such that $-i[\hat{A}, \hat{B}]$ is neither zero nor $\hbar\hat{1}$ but some more sophisticated operator. In such a situation, $\langle M \rangle$ is "state dependent" and may happen to vanish even though $\hat{A}$ and $\hat{B}$ do not commute. Then (11.013), although literally true, does not give the information it was expected to give. This case is of less fundamental importance than the two cases cited previously.

11.03. Realization of the Minimum Uncertainty Product

A Cartesian coordinate and its conjugate momentum, such as x and p_x, constitute the prime example of a pair of non-commuting observables. As pointed out in the previous section, $\langle M \rangle$ for this pair is equal to the constant $\hbar$ for all Ψ and the uncertainty principle yields:

$$\Delta x \, \Delta p_x \geqslant \tfrac{1}{2}\hbar. \qquad (11.022)$$

A quantum x measurement[†] performed at a particular time t_0 upon a system having x and p_x as two of its canonical variables would leave the wave function in a form approaching the square root of a Dirac delta function as far as its x dependence was concerned; Δp_x for this function would simultaneously approach infinity indicating extreme uncertainty in the definition of p_x or, in the context of representations, that an infinite continuum of eigenfunctions of p_x, all of essentially equal magnitude, would have to be superposed to yield the wave function in question. Corresponding statements could be made with respect to a quantum p_x measurement. It could also happen, of course, that the quantum state of the system is prepared by some means other than an x or a p_x measurement, in which case neither Δx nor Δp_x would approach an extreme value. Regardless of the method of preparation, it is interesting to consider the form that Ψ should have in order to minimize the product $\Delta x \Delta p_x$, i.e. in order to invoke the

[†] It is understood that a quantum x measurement can only be approximated, never literally achieved, and that the corresponding eigenstate of $\hat{x}$ is not realizable. Similar remarks apply to a quantum p_x measurement.

equality sign in (11.022) rather than the inequality. When the equality obtains, it is said that the *minimum uncertainty product* has been realized; the quantum state is then one in which x and p_x are defined as precisely as is mutually possible. Although the form of Ψ for this state may be guessed from material presented in Chapter 10, it is interesting to see a proof that that form is determined by the requirement of minimum uncertainty. Any other form will yield an uncertainty product greater than the minimum. This discussion, incidentally, applies not only to a one-dimensional system with instantaneous wave function $\Psi(x, t_0)$ but also to a system with more dimensions having a wave function that includes $\Psi(x, t_0)$ as a factor.

With reference to Section 11.02, it is seen that the equality $\Delta A \Delta B = \frac{1}{2}|\langle M \rangle|$ implies that the roots of $F(u) = 0$ are real and equal. The value of the double root, here called u_0, is then given by:

$$u_0 = \frac{-\langle M \rangle}{2(\Delta A)^2}. \tag{11.023}$$

According to (11.014), $F(u)$ is an integral over configuration space of the squared magnitude of a function; if $F(u) = 0$, the function in question must therefore vanish everywhere. Thus:

$$[\hat{B} - \langle B \rangle \hat{1} + iu_0(\hat{A} - \langle A \rangle \hat{1})]\Psi = 0. \tag{11.024}$$

The following substitutions are in order:

$$\left.\begin{array}{l} \hat{A} = \hat{x} = x; \\ \langle A \rangle = x_0; \\ \hat{B} = \hat{p}_x = -i\hbar\partial/\partial x; \\ \langle B \rangle = p_{0x}; \\ u_0 = -\hbar/2(\Delta x)^2. \end{array}\right\} \tag{11.025}$$

Since t_0 is a constant, $\Psi(x, t_0)$ is actually a function of x only and $\partial/\partial x$ can be changed to d/dx. Equation (11.024) becomes:

$$\left[-i\hbar\frac{d}{dx} - p_{0x} - \frac{i\hbar(x - x_0)}{2(\Delta x)^2}\right]\Psi(x, t_0) = 0. \tag{11.026}$$

It is convenient to work in the modified variable $x' = x - x_0$; substitution of this and subsequent division by $-i\hbar$ puts the above into the standard form for a first-order linear differential equation:

$$\frac{d\Psi}{dx'} + \left[-i\frac{p_{0x}}{\hbar} + \frac{x'}{2(\Delta x)^2} \right] \Psi = 0. \qquad (11.027)$$

Solution of this equation is routine. Only one arbitrary constant of integration is produced and the magnitude of this will be adjusted later to achieve normalization; it follows that the final result is determined to within an arbitrary phase factor. The solution in question is given by:

$$\Psi(x', t_0) = C \exp \left[\frac{ip_{0x}x'}{\hbar} - \frac{x'^2}{4(\Delta x)^2} \right]. \qquad (11.028)$$

Reinstatement of $x - x_0$ and evaluation of C yield:

$$\Psi(x, t_0) = \frac{\exp\left[ip_{0x}(x-x_0)/\hbar\right]}{(2\pi)^{\frac{1}{4}}(\Delta x)^{\frac{1}{2}}} \exp\left[-\tfrac{1}{4}(x-x_0)^2/(\Delta x)^2\right]. \qquad (11.029)$$

This is obviously a Gaussian-related function leading to a probability density with standard deviation Δx and expectation value x_0. Comparison with (10.043) is facilitated by noticing that $p_{0x} = \hbar k_{0x}$ and that $\Delta x = \eta$. It is seen that (11.029) and (10.043) differ by only the constant phase factor, $\exp\left[-ip_{0x}x_0/\hbar\right]$, which is of no consequence quantum mechanically; the family of functions studied in Section 10.04 therefore fulfills the criterion of minimum uncertainty. This is also evident from the fact that, by (10.052), $\eta\mu = \tfrac{1}{2}$ where $\mu = \Delta k_x = \Delta p_x/\hbar$.

Up to this point, the discussion has been concerned only with the form which may be impressed upon the wave function as of a particular time t_0; no reference has been made to the subsequent development of this function. To find the future form of a wave function which begins as the minimum uncertainty function (11.029), one must solve the Schrödinger equation incorporating the known potential energy and then require that Ψ equal (11.029) as an initial condition.

If the potential energy is that of a harmonic oscillator, then a minimum uncertainty function of proper Δx, multiplied by the factor $\exp -i\omega_0 t$, can endure indefinitely as the ground state of the system and the minimum uncertainty criterion is fulfilled at all times. If, however, there are no forces, then $\Psi(x, t)$ spreads as predicted by the derivation in Section 10.05, i.e. Δx increases with time whereas Δp_x, which is related to the corresponding probability density $\Phi^*\Phi(k_x)$ on momentum space, does not grow smaller. In this case, and generally, the condition of minimum uncertainty obtains for only an instant of time.

CHAPTER 12

THE HYDROGENIC ATOM

12.01. Separation of Center-of-mass Motion from Relative Motion

The term "hydrogenic atom" refers to an atom in which one electron (of charge $-e$) moves under the influence of a nucleus of charge Ze which, to a very good approximation, can be thought of as a point. The integer Z is the atomic number. On this definition, a doubly ionized lithium atom or a singly ionized helium atom, as well as a neutral hydrogen atom, is considered hydrogenic. The configuration, with an intertial frame of reference, is illustrated in Fig. 12.01. The mass of the electron will be called m_1 and that of the nucleus, m_2;

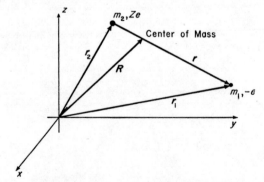

FIG. 12.01. Components of a hydrogenic atom (with finite nuclear mass) and an inertial Cartesian reference frame.

there are six degrees of freedom and one may begin by using the Cartesian coordinates x_1, y_1, z_1 of the electron and x_2, y_2, z_2 of the nucleus. For convenience in subsequent steps, these are combined into the vectors r_1 and r_2, also shown in the illustration.

The six coordinates mentioned in the preceding paragraph are not the most suitable for the contemplated analysis and, as a first step toward the acquisition of a more satisfactory set, consider the displacement R of the center of mass and the relative displacement r of the electron from the nucleus. These vectors, also depicted in Fig. 12.01, appear in the right-hand sides of the following equations which constitute their definitions:

$$m_1 r_1 + m_2 r_2 = (m_1 + m_2)R;\qquad(12.001)$$

$$r_1 - r_2 = r.\qquad(12.002)$$

Solving simultaneously:

$$r_1 = R + \frac{m_2}{m_1 + m_2} r;\qquad(12.003)$$

$$r_2 = R - \frac{m_1}{m_1 + m_2} r.\qquad(12.004)$$

The vectors R and r can be decomposed into components X, Y, Z, and x, y, z, respectively; these constitute a new set of six coordinates.

The kinetic energy of the system is given by:

$$T = \tfrac{1}{2}m_1\left|\dot{R} + \frac{m_2}{m_1+m_2}\dot{r}\right|^2 + \tfrac{1}{2}m_2\left|\dot{R} - \frac{m_1}{m_1+m_2}\dot{r}\right|^2.\quad(12.005)$$

This becomes:

$$T = \tfrac{1}{2}m_1\left[|\dot{R}|^2 + 2\frac{m_2}{m_1+m_2}\dot{R}\cdot\dot{r} + \left(\frac{m_2}{m_1+m_2}\right)^2|\dot{r}|^2\right]$$
$$+ \tfrac{1}{2}m_2\left[|\dot{R}|^2 - 2\frac{m_1}{m_1+m_2}\dot{R}\cdot\dot{r} + \left(\frac{m_1}{m_1+m_2}\right)^2|\dot{r}|^2\right].\quad(12.006)$$

Simplification produces:

$$T = \tfrac{1}{2}\left[(m_1+m_2)\left|\dot{\boldsymbol{R}}\right|^2 + \frac{m_1 m_2}{m_1+m_2}\left|\dot{\boldsymbol{r}}\right|^2\right]. \tag{12.007}$$

The quantity $m_1 m_2/(m_1+m_2)$ has the dimensions of mass and is smaller than the smaller of the two individual masses. It is the result that would be obtained if the two were added like resistors in parallel and is called the *reduced mass* μ. The expression for kinetic energy shows that the system may be regarded as a system of two pseudo-particles, one possessing the total mass $M = m_1+m_2$ at displacement $\boldsymbol{R}$ from the origin, the other having a mass equal to the reduced mass at displacement $\boldsymbol{r}$ from the origin. Thus (12.007) can be written:

$$T = \tfrac{1}{2}M\left|\dot{\boldsymbol{R}}\right|^2 + \tfrac{1}{2}\mu\left|\dot{\boldsymbol{r}}\right|^2. \tag{12.008}$$

The potential energy depends only upon the relative coordinates and is given by:

$$V = \frac{-Ze^2}{4\pi\varepsilon_0 r} = -\frac{\gamma}{r}, \tag{12.009}$$

where γ is a convenient symbol for $Ze^2/4\pi\varepsilon_0$. The Lagrangian in terms of the six new coordinates is therefore:

$$\mathscr{L} = \tfrac{1}{2}M(\dot{X}^2+\dot{Y}^2+\dot{Z}^2)+\tfrac{1}{2}\mu(\dot{x}^2+\dot{y}^2+\dot{z}^2)+\gamma(x^2+y^2+z^2)^{-\frac{1}{2}}. \tag{12.010}$$

Since $p_X = M\dot{X}$, $p_x = m\dot{x}$, etc., the Hamiltonian can be easily derived. It is:

$$\mathscr{H} = \frac{1}{2M}(p_X^2+p_Y^2+p_Z^2)+\frac{1}{2\mu}(p_x^2+p_y^2+p_z^2)-\gamma(x^2+y^2+z^2)^{-\frac{1}{2}}. \tag{12.011}$$

This function is clearly equal to the total energy.

Quantization is accomplished by converting every term in the Hamiltonian to the corresponding operator. Let:

$$\nabla_R = e_x\frac{\partial}{\partial X}+e_y\frac{\partial}{\partial Y}+e_z\frac{\partial}{\partial Z}; \tag{12.012}$$

$$\nabla = e_x\frac{\partial}{\partial x}+e_y\frac{\partial}{\partial y}+e_z\frac{\partial}{\partial z}. \tag{12.013}$$

Using these, the Hamiltonian operator becomes:

$$\hat{\mathcal{H}} = -\frac{\hbar^2}{2M}\,\nabla^2_{\boldsymbol{R}} - \frac{\hbar^2}{2\mu}\,\nabla^2 - \frac{\gamma}{r}\,. \tag{12.014}$$

Since the Hamiltonian does not contain the time, it is reasonable to look for stationary states. The wave function for a typical stationary state is called Ψ' for future convenience and may be written:

$$\Psi'(X, Y, Z, x, y, z, t) = \psi'(X, Y, Z, x, y, z)\exp(-iE't/\hbar), \tag{12.015}$$

where E' is the combined total energy of the center-of-mass motion and the relative motion. The time-independent Schrödinger equation becomes:

$$-\frac{\hbar^2}{2M}\,\nabla^2_{\boldsymbol{R}}\psi' - \frac{\hbar^2}{2\mu}\,\nabla^2\psi' - \frac{\gamma}{r}\,\psi' = E'\psi'. \tag{12.016}$$

The function ψ' can be further separated as follows:

$$\psi'(X, Y, Z, x, y, z) = \psi_{\boldsymbol{R}}(X, Y, Z)\,\psi(x, y, z). \tag{12.017}$$

Substitution and division by ψ' yield:

$$-\frac{\hbar^2}{2M}\,\frac{\nabla^2_{\boldsymbol{R}}\psi_{\boldsymbol{R}}}{\psi_{\boldsymbol{R}}} - \frac{\hbar^2}{2\mu}\,\frac{\nabla^2\psi}{\psi} - \frac{\gamma}{r} = E'. \tag{12.018}$$

Here the first term is a function of only X, Y, and Z; the second and third terms together constitute a function of only x, y, and z. Thus the respective sets of terms can be separately set equal to constants known as *separation constants* which, in this case, are the energies of the two distinct motions:

$$-\frac{\hbar^2}{2M}\,\frac{\nabla^2_{\boldsymbol{R}}\psi_{\boldsymbol{R}}}{\psi_{\boldsymbol{R}}} = E_{\boldsymbol{R}}; \tag{12.019}$$

$$-\frac{\hbar^2}{2\mu}\,\frac{\nabla^2\psi}{\psi} - \frac{\gamma}{r} = E. \tag{12.020}$$

Notice that $E_R + E = E'$. Equation (12.019) is the Schrödinger equation for the unforced motion of the center of mass and has plane wave solutions of which the following is typical:[†]

$$\psi_R = A_{\downarrow} \exp i\boldsymbol{K}\cdot\boldsymbol{R}; \qquad E_R = \hbar^2K^2/2M. \qquad (12.021)$$

As explained in an earlier chapter, a single plane wave solution like the above is an unattainable limiting case; a superposition of infinitely many such solutions, each with its appropriate time factor, is what would usually be found in a practical situation. This would involve a distribution of values of $\boldsymbol{K}$, E_R, and E' and would constitute a wave packet representing the center of mass of the atom. Since the wave mechanics of an unforced particle is already familiar, the center-of-mass motion will not be considered further. The more interesting part of the analysis is embodied in equation (12.020) for the relative motion and the study of this equation will occupy the remainder of the chapter.

12.02. Use of Spherical Polar Coordinates in the Analysis of the Relative Motion

The relative motion is the motion of the electron with respect to the nucleus; it involves the three coordinates x, y, and z which are the Cartesian components of the relative displacement vector $\boldsymbol{r}$. As outlined in the previous section, this motion is equivalent to that of a pseudoparticle of mass μ displaced from the origin by the same vector $\boldsymbol{r}$. In the present section, the spherical polar coordinates r, θ, and φ will be substituted for x, y, and z; thus $\psi = \psi(r, \theta, \varphi)$ and, if ∇^2 is written in the spherical polar system, equation (12.020) becomes:

$$-\frac{\hbar^2}{2\mu\psi}\left\{\frac{1}{r^2}\frac{\partial}{\partial r}\left(r^2\frac{\partial\psi}{\partial r}\right)\right.$$
$$\left.+\frac{1}{r^2}\left[\frac{1}{\sin\theta}\frac{\partial}{\partial\theta}\left(\sin\theta\frac{\partial\psi}{\partial\theta}\right)+\frac{1}{\sin^2\theta}\frac{\partial^2\psi}{\partial\varphi^2}\right]\right\}-\frac{\gamma}{r}=E. \qquad (12.022)$$

[†] For an explanation of the notation $A_{\downarrow}$, see Section 8.03.

In many of the steps undertaken in this analysis, it would be very time-consuming to explain *a priori* why certain forms are proposed as solutions or why certain substitutions or changes of variable are employed. Such explanations will therefore be abridged or omitted. It is, after all, sufficient if a proposed form can, upon substitution, be shown to satisfy the differential equation in question and to have the generality needed. In some cases, where standard forms are encountered, even these steps will be taken for granted since a reasonable acquaintance on the part of the reader in this area of mathematical physics can be presumed. Only in this way can tedium be avoided and maximum time devoted to the development of physical insights.

The first step in the solution of (12.022) is the assumption of the product solution $\psi = r^{-1}F(r)\,Y(\theta, \varphi)$; substitution of this followed by a multiplication by r^2 yields:

$$-\frac{\hbar^2 r^2}{2\mu F}\frac{d^2 F}{dr^2} - r^2\left(\frac{\gamma}{r} + E\right)$$
$$-\frac{\hbar^2}{2\mu Y}\left[\frac{1}{\sin\theta}\frac{\partial}{\partial\theta}\left(\sin\theta\frac{\partial Y}{\partial\theta}\right) + \frac{1}{\sin^2\theta}\frac{\partial^2 Y}{\partial\varphi^2}\right] = 0. \quad (12.023)$$

This says that a function of r only plus a function of θ and φ only is equal to zero. Accordingly, let the function of r be set equal to the separation constant $-\hbar^2 C/2\mu$, that of θ and φ, to $\hbar^2 C/2\mu$ where C is to be determined later. The r equation, recorded here for future reference, becomes:

$$-\frac{\hbar^2 r^2}{2\mu F}\frac{d^2 F}{dr^2} - r^2\left(\frac{\gamma}{r} + E\right) + \frac{\hbar^2 C}{2\mu} = 0. \quad (12.024)$$

Passing on to the θ, φ equation, one finds that $\hbar^2/2\mu$ can be factored out and the result written:

$$\frac{1}{Y\sin\theta}\frac{\partial}{\partial\theta}\left(\sin\theta\frac{\partial Y}{\partial\theta}\right) + \frac{1}{Y\sin^2\theta}\frac{\partial^2 Y}{\partial\varphi^2} + C = 0. \quad (12.025)$$

It is important to notice that this equation must have precisely this form whenever the potential energy depends upon r only and not

20*

upon θ or φ. Such a potential energy is called a *central potential* because it gives rise to a central force; with it is always associated the conservation of angular momentum. It follows that the solutions of equation (12.025) which are developed below are universally applicable to all cases of central force motion and, as may be suspected, are intimately connected with the concept of angular momentum.

To solve the θ, φ equation, it is desirable to let $Y(\theta, \varphi) = P(\theta)\Phi(\varphi)$; substitution of this into (12.025) followed by multiplication by $\sin^2 \theta$ gives:

$$\frac{\sin \theta}{P} \frac{d}{d\theta} \left(\sin \theta \frac{dP}{d\theta} \right) + C \sin^2 \theta + \frac{1}{\Phi} \frac{d^2\Phi}{d\varphi^2} = 0. \quad (12.026)$$

Once more the situation is a standard one; a function of θ only plus a function of φ only is equal to zero. For reasons of future convenience, let the former of these be equal to the separation constant m^2; the latter to $-m^2$. Thus:

$$\sin \theta \frac{d}{d\theta} \left(\sin \theta \frac{dP}{d\theta} \right) + (C \sin^2 \theta - m^2)P = 0; \quad (12.027)$$

$$\frac{d^2\Phi}{d\varphi^2} + m^2\Phi = 0. \quad (12.028)$$

The general solution of the second of these equations is:

$$\Phi = A \exp im\varphi + B \exp -im\varphi, \quad (12.029)$$

and, in order that the wave function may be single valued, only integer values of m can be considered. The number m, then, assumes the role of a quantum number and is the first of three such numbers to make its appearance. Actually, m will be allowed to range over both positive and negative integer values and only the first term in (12.029) will be retained. This is permissible because m enters the θ equation (12.027) as m^2 and, with a given functional form of the θ solution, there is associated a Φ function with positive m in one instance and one with negative m in the other.

To solve the θ equation, it is convenient to change the independent variable from θ to u where $u = \cos \theta$. Then:

$$\frac{d}{d\theta} = -\sin \theta \frac{d}{du}.$$
(12.030)

Substitution of this and subsequent division by $\sin^2 \theta$ yield:

$$\frac{d}{du} \left[(1-u^2) \frac{dP}{du} \right] + \left(C - \frac{m^2}{1-u^2} \right) P = 0.$$
(12.031)

It is advisable to study first the case in which $m = 0$; equation (12.031) is then readily recognized as Legendre's equation. The process for solving this equation in a quantum mechanical context is very similar to that employed in Section 7.07 for the harmonic oscillator, i.e. a series solution beginning with the zeroth power is substituted and a recurrence relation for the coefficients is obtained. In the present instance, the latter is:

$$a_{k+2} = \frac{k(k+1)-C}{(k+2)(k+1)} a_k.$$
(12.032)

If the series is infinite, it constitutes a function that cannot be normalized on the interval $-1 \leqslant u \leqslant 1$ and leads to a non-normalizable ψ function. Such a series is unacceptable. To make the series terminate, the constant C must take the special value $l(l+1)$ where l is a non-negative integer; this l is therefore the second of the quantum numbers to be encountered. The polynomials thus formed are the familiar *Legendre polynomials*, $P_l(u)$. The degree and the parity of each polynomial is indicated by l; a few examples are given in Table 12.01. The Legendre polynomials may also be generated by the well-known Rodrigues' formula:

$$P_l(u) = \frac{(-1)^l}{2^l l!} \frac{d^l}{du^l} (1-u^2)^l.$$
(12.033)

If $m \neq 0$, (12.031) becomes Legendre's associated equation. In somewhat expanded form, with C replaced by $l(l+1)$ and d/du re-

TABLE 12.01. *Legendre Polynomials*

Even parity polynomials	Odd parity polynomials
$P_0 = 1$	$P_1 = u$
$P_2 = \frac{1}{2}(-1 + 3u^2)$	$P_3 = \frac{1}{2}(-3u + 5u^3)$
$P_4 = \frac{1}{8}(3 - 30u^2 + 35u^4)$	$P_5 = \frac{1}{8}(15u - 70u^3 + 63u^5)$

placed by D, it is:

$$(1-u^2)D^2P - 2uDP + [l(l+1) - m^2/(1-u^2)]P = 0. \quad (12.034)$$

At this point, it is helpful to assume a solution structured as follows:

$$P(u) = (1-u^2)^{m/2}f(u). \quad (12.035)$$

Substitution of this into (12.034) yields:

$$(1-u^2)D^2f - 2(m+1)uDf + [l(l+1) - m(m+1)]f = 0. \quad (12.036)$$

Since $P_l(u)$ is a solution of (12.034) with $m = 0$, i.e. a solution of Legendre's equation, one may write:

$$(1-u^2)D^2P_l - 2uDP_l + l(l+1)P_l = 0. \quad (12.037)$$

If this is differentiated m times, one obtains after some effort:

$$(1-u^2)D^{m+2}P_l - 2(m+1)uD^{m+1}P_l$$
$$+ [l(l+1) - m(m+1)]D^mP_l = 0. \quad (12.038)$$

Comparison of (12.036) with (12.038) makes it possible to identify $f(u)$ as the mth derivative of $P_l(u)$:

$$f(u) = D^mP_l(u). \quad (12.039)$$

From this it is seen that $f(u)$ is a polynomial in u of degree $l-m$; if $m = l$, $f(u)$ is a constant and if $m > l$, $f(u)$ vanishes. Since $(1-u^2)^{m/2}$ is transcendental for odd m, the solution (12.035) is not always a polynomial. It is conventionally called an *associated Legendre function* and is designated by the two-index symbol $P_l^m(u)$:

$$P_l^m(u) = (1-u^2)^{m/2} D^m P_l(u). \tag{12.040}$$

With the aid of Rodrigues' formula, this becomes:

$$P_l^m(u) = \frac{(-1)^l}{2^l l!} (1-u^2)^{m/2} \frac{d^{l+m}}{du^{l+m}} (1-u^2)^l. \tag{12.041}$$

It may be noticed that this expression continues the definition of the associated Legendre function into the domain of negative m. A technique similar to that already used yields an ostensibly different but actually equivalent formula for $P_l^m(u)$, namely:

$$P_l^m(u) = \frac{(-1)^l (-1)^m}{2^l l!} \frac{(l+m)!}{(l-m)!} (1-u^2)^{-m/2} \frac{d^{l-m}}{du^{l-m}} (1-u^2)^l. \tag{12.042}$$

For positive m, this formula is more convenient than (12.041) since it involves a derivative of lower order. When $m = 0$, of course, the associated Legendre function becomes simply a Legendre polynomial; in other words, $P_l^0(u) = P_l(u)$. The reader is cautioned that some treatments, particularly those of a more theoretical mathematical nature, include an additional factor of $(-1)^m$ in each of the three preceding formulas, thus changing slightly the definition of $P_l^m(u)$. Tabulations and graphs of associated Legendre functions, using the definition given here, have appeared in the literature.[39]

Substitution of $-q$ for m in (12.042), where q is understood to be a positive integer, shows that:

$$P_l^{-q}(u) = (-1)^q \frac{(l-q)!}{(l+q)!} P_l^q(u). \tag{12.043}$$

This demonstrates that $P_l^{-q}(u)$ has the same functional form as $P_l^q(u)$ but differs therefrom by a multiplicative constant. This statement is true for the positive integer q provided $q \leqslant l$. If $q > l$, then $P_l^q(u)$ vanishes in accordance with earlier comments. When this occurs, however, $(l-q)!$ becomes infinite, the right-hand side of (12.043) becomes indeterminate, and it cannot be concluded that $P_l^{-q}(u) = 0$. Actually, $P_l^{-q}(u)$ with $q > l$ exists[†] as a function but is not normalizable and therefore not applicable. One thus reaches the important conclusion that for all associated Legendre functions useful in the quantum mechanics of central force motion, the quantum number m is restricted to the $2l+1$ integral values in the range:

$$-l \leqslant m \leqslant l. \tag{12.044}$$

The general solution of the θ, φ equation may now be written by combining $P_l^m(u)$ with $\exp im\varphi$:

$$Y_l^m(\theta, \varphi) = A_l^m P_l^m(\cos \theta) \exp im\varphi. \tag{12.045}$$

Here A_l^m is an arbitrary constant which will be evaluated in the subsequent section. It should now be obvious that a solution of the θ equation times any desired mixture of $\exp iq\varphi$ and $\exp -iq\varphi$ can be achieved by superposing, with suitably chosen coefficients, two functions like (12.045), one with a positive value of m and the other with a negative.

12.03. Spherical Harmonics

The normalizable solutions $Y_l^m(\theta, \varphi)$ of equation (12.025) are known as the *spherical harmonics*. Before evaluating the normalization constant A_l^m in (12.045), it is necessary to take notice of the orthogonality relation for the associated Legendre functions:[41]

$$\int_0^\pi P_{l'}^m(\cos \theta) P_{l'}^m(\cos \theta) \sin \theta \, d\theta = \int_{-1}^1 P_{l'}^m(u) P_{l'}^m(u) \, du$$
$$= \frac{2}{2l+1} \frac{(l+m)!}{(l-m)!} \delta_{l'l}. \tag{12.046}$$

[†] Mathematicians have studied the associated Legendre functions extensively and analytic continuations into unusual domains of both argument and indices are available. Reference 40 is particularly recommended in this context.

At the same time, the following orthogonality relation should be recalled:

$$\int_0^{2\pi} \exp - im'\varphi \exp im\varphi \, d\varphi = 2\pi\delta_{m'm}. \tag{12.047}$$

From these expressions, it can be seen that an integration of a product of two spherical harmonics over all solid angle yields:

$$\oint Y_{l'}^{m'*}(\theta, \varphi) Y_l^m(\theta, \varphi) \, d\Omega$$

$$= \int_0^{2\pi} \int_{-1}^1 A_{l'}^{m'*} P_{l'}^{m'}(u) \exp - im'\varphi \, A_l^m P_l^m(u) \exp im\varphi \, du \, d\varphi$$

$$= \left[|A_l^m|^2 \frac{4\pi}{2l+1} \frac{(l+m)!}{(l-m)!} \right] \delta_{l'l} \delta_{m'm}. \tag{12.048}$$

Normalization is achieved if the quantity in square brackets above is made equal to unity. This is conventionally done by giving A_l^m the following value:

$$A_l^m = + \left[\frac{2l+1}{4\pi} \frac{(l-m)!}{(l+m)!} \right]^{\frac{1}{2}} (-1)^m. \tag{12.049}$$

Since the spherical harmonics are already orthogonal, they become orthonormal with the adoption of (12.049). Their complete definition can then be given in two equivalent ways (in which the positive value of the square root is always taken):

$$Y_l^m = \frac{(-1)^{l+m}}{2^l l!} \left[\frac{2l+1}{4\pi} \frac{(l-m)!}{(l+m)!} \right]^{\frac{1}{2}}$$

$$(1-u^2)^{m/2} \left[\frac{d^{l+m}}{du^{l+m}} (1-u^2)^l \right] \exp im\varphi; \tag{12.050}$$

$$Y_l^m = \frac{(-1)^l}{2^l l!} \left[\frac{2l+1}{4\pi} \frac{(l+m)!}{(l-m)!} \right]^{\frac{1}{2}}$$

$$(1-u^2)^{-m/2} \left[\frac{d^{l-m}}{du^{l-m}} (1-u^2)^l \right] \exp im\varphi. \tag{12.051}$$

It follows from these that:

$$Y_l^{-q} = (-1)^q Y_l^{q*}. \tag{12.052}$$

Only the magnitude of A_l^m is determined by the normalization requirement; the phase angle of this constant is not determined and may, theoretically, be selected arbitrarily for every different pair of values of l and m. It is preferable, of course, if this angle is established by an analytic function of l or m or both and in such a way that it is either zero or π for integral values of these indices. The factor $(-1)^m$ in (12.049) fulfills these expectations and, in addition, puts the definition of Y_l^m into agreement with that used by Condon and Shortley[42] which has become practically standard in physics.[†] In spite of its wide acceptance, this definition has the peculiar and somewhat annoying effect of giving a plus sign to all the spherical harmonics of negative m and alternating signs to those of positive m. This can be noticed in Table 12.02 in which several of the associated Legendre functions and the corresponding spherical harmonics are displayed.

As has been emphasized, the spherical harmonics form an orthonormal set, i.e. they obey the simple relationship:

$$\oint Y_{l*}^{m*}(\theta, \varphi) Y_l^m(\theta, \varphi)\, \mathrm{d}\Omega = \delta_{l'l}\delta_{m'm}. \tag{12.053}$$

This set is also complete which means that any single-valued function $f(\theta, \varphi)$ satisfying reasonable conditions of continuity can be represented in spherical harmonics:

$$f(\theta, \varphi) = \sum_{l=0}^{\infty} \sum_{m=-l}^{l} \alpha_l^m Y_l^m(\theta, \varphi), \tag{12.054}$$

where:

$$\alpha_l^m = \oint f(\theta, \varphi)\, Y_l^{m*}(\theta, \varphi)\, \mathrm{d}\Omega. \tag{12.055}$$

It is easy to see in qualitative fashion how the Y_l^m functions can be superposed to represent a given $f(\theta, \varphi)$. In the angle φ, this superposition is simply a Fourier series. In the angle θ, the component func-

---

[†] In some treatments, $(-1)^m$ is included in the definition of $P_l^m(u)$ and omitted from A_l^m; this has no effect upon the final expressions for $Y_l^m(\theta, \varphi)$.

TABLE 12.02. *Associated Legendre Functions and Spherical Harmonics*

l	m	P_l^m	Y_l^m
0	0	1	$\left[\dfrac{1}{4\pi}\right]^{\frac{1}{2}}$
1	1	$\sin\theta$	$-\left[\dfrac{3}{4\pi}\cdot\dfrac{1}{2}\right]^{\frac{1}{2}}\sin\theta\exp i\varphi$
	0	$\cos\theta$	$\left[\dfrac{3}{4\pi}\right]^{\frac{1}{2}}\cos\theta$
	-1	$-\frac{1}{2}\sin\theta$	$\left[\dfrac{3}{4\pi}\cdot\dfrac{1}{2}\right]^{\frac{1}{2}}\sin\theta\exp -i\varphi$
	2	$3\sin^2\theta$	$\left[\dfrac{5}{4\pi}\cdot\dfrac{3}{8}\right]^{\frac{1}{2}}\sin^2\theta\exp 2i\varphi$
	1	$3\sin\theta\cos\theta$	$-\left[\dfrac{5}{4\pi}\cdot\dfrac{3}{2}\right]^{\frac{1}{2}}\sin\theta\cos\theta\exp i\varphi$
2	0	$\frac{3}{2}\cos^2\theta-\frac{1}{2}$	$\left[\dfrac{5}{4\pi}\right]^{\frac{1}{2}}(\frac{3}{2}\cos^2\theta-\frac{1}{2})$
	-1	$-\frac{1}{2}\sin\theta\cos\theta$	$\left[\dfrac{5}{4\pi}\cdot\dfrac{3}{2}\right]^{\frac{1}{2}}\sin\theta\cos\theta\exp -i\varphi$
	-2	$\frac{1}{8}\sin^2\theta$	$\left[\dfrac{5}{4\pi}\cdot\dfrac{3}{8}\right]^{\frac{1}{2}}\sin^2\theta\exp -2i\varphi$

tions exhibit (for given m) increasingly more structure with increasing l and, if sufficiently high values of this index are utilized, the θ dependence of any well-behaved function can be reproduced.

12.04. Orbital Angular Momentum Operators

In Section 12.01 it was shown that, for the atom under consideration, the total kinetic energy of the nucleus and the electron can be
viewed as the kinetic energy of a pseudoparticle of mass M which
is unforced plus that of a pseudoparticle of mass μ which is attracted
to the origin. It is not difficult to show that the total angular momentum of the system can be similarly partitioned. Thus:

$$r_1 \times m_1 \dot{r}_1 + r_2 \times m_2 \dot{r}_2 = R \times M\dot{R} + r \times \mu\dot{r}. \qquad (12.056)$$

The two terms on the right-hand side of this equation are separately
conserved since the pseudoparticle of mass M experiences no force
and that of mass μ experiences a central force. The second of these
terms is the angular momentum of the relative motion and only it is
of interest here. Henceforth it will be denoted by the vector L and
will be called the "orbital angular momentum":

$$L = r \times \mu\dot{r} = r \times p. \qquad (12.057)$$

It is a simple matter to express the Cartesian components of L; in
quantum mechanics, these become operators and may be written as
follows:

$$\left. \begin{array}{l} \hat{L}_x = \hat{y}\hat{p}_z - \hat{z}\hat{p}_y; \\ \hat{L}_y = \hat{z}\hat{p}_x - \hat{x}\hat{p}_z; \\ \hat{L}_z = \hat{x}\hat{p}_y - \hat{y}\hat{p}_x. \end{array} \right\} \qquad (12.058)$$

The commutation relations among the three operators of the preceding paragraph are interesting and important. One may begin by
evaluating the commutator of $\hat{L}_x$ and $\hat{L}_y$:

$$[\hat{L}_x, \hat{L}_y] = (\hat{y}\hat{p}_z - \hat{z}\hat{p}_y)(\hat{z}\hat{p}_x - \hat{x}\hat{p}_z) - (\hat{z}\hat{p}_x - \hat{x}\hat{p}_z)(\hat{y}\hat{p}_z - \hat{z}\hat{p}_y). \qquad (12.059)$$

Some terms, such as $\hat{z}\hat{p}_y\hat{z}\hat{p}_x$ and $\hat{z}\hat{p}_x\hat{z}\hat{p}_y$, contain commuting operators
and may be cancelled. The terms which remain are:

$$[\hat{L}_x, \hat{L}_y] = \hat{y}\hat{p}_z\hat{z}\hat{p}_x + \hat{z}\hat{p}_y\hat{x}\hat{p}_z - \hat{z}\hat{p}_x\hat{y}\hat{p}_z - \hat{x}\hat{p}_z\hat{z}\hat{p}_y. \qquad (12.060)$$

Only $\hat{z}$ and $\hat{p}_z$ fail to commute; one therefore has:

$$[\hat{L}_x, \hat{L}_y] = \hat{y}\hat{p}_x(\hat{p}_z\hat{z} - \hat{z}\hat{p}_z) + \hat{p}_y\hat{x}(\hat{z}\hat{p}_z - \hat{p}_z\hat{z}) = \hat{L}_z[\hat{z}, \hat{p}_z]. \quad (12.061)$$

Since $[\hat{z}, \hat{p}] = i\hbar\hat{1}$, this and the corresponding results for the other two pairs of operators may be summarized as follows:

$$\left. \begin{array}{l} [\hat{L}_x, \hat{L}_y] = i\hbar\hat{L}_z; \\ [\hat{L}_y, \hat{L}_z] = i\hbar\hat{L}_x; \\ [\hat{L}_z, \hat{L}_x] = i\hbar\hat{L}_y. \end{array} \right\} \quad (12.062)$$

This is often condensed into the elegant expression:

$$\hat{L} \times \hat{L} = i\hbar\hat{L}. \quad (12.063)$$

Another observable of significance is the magnitude squared of the total orbital angular momentum. The corresponding operator is:

$$\hat{L}^2 = \hat{L}\cdot\hat{L} = \hat{L}_x^2 + \hat{L}_y^2 + \hat{L}_z^2. \quad (12.064)$$

Once more, commutation relations are important. Consider:

$$[\hat{L}^2, \hat{L}_x] = [\hat{L}_x^2, \hat{L}_x] + [\hat{L}_y^2, \hat{L}_x] + [\hat{L}_z^2, \hat{L}_x]. \quad (12.065)$$

The first term on the right-hand side vanishes and the second may be written:

$$[\hat{L}_y^2, \hat{L}_x] = \hat{L}_y\hat{L}_y\hat{L}_x - \hat{L}_x\hat{L}_y\hat{L}_y. \quad (12.066)$$

Since $\hat{L}_x\hat{L}_y = \hat{L}_y\hat{L}_x + i\hbar\hat{L}_z$, the above becomes:

$$\begin{aligned} [\hat{L}_y^2, \hat{L}_x] &= \hat{L}_y\hat{L}_y\hat{L}_x - \hat{L}_y\hat{L}_x\hat{L}_y - i\hbar\hat{L}_z\hat{L}_y \\ &= \hat{L}_y\hat{L}_y\hat{L}_x - \hat{L}_y\hat{L}_y\hat{L}_x - i\hbar\hat{L}_z\hat{L}_y - i\hbar\hat{L}_y\hat{L}_z \\ &= -i\hbar[\hat{L}_y\hat{L}_z + \hat{L}_z\hat{L}_y]. \end{aligned} \quad (12.067)$$

Similarly:

$$[\hat{L}_z^2, \hat{L}_x] = i\hbar[\hat{L}_y\hat{L}_z + \hat{L}_z\hat{L}_y]. \quad (12.068)$$

Substitution into (12.065) shows that $\hat{L}^2$ and $\hat{L}_x$ do indeed commute. By symmetry, $\hat{L}^2$ mut also commute with $\hat{L}_y$ and $\hat{L}_z$ and one may say:

$$\left. \begin{array}{l} [\hat{L}^2, \hat{L}_x] = 0; \\ [\hat{L}^2, \hat{L}_y] = 0; \\ [\hat{L}^2, \hat{L}_z] = 0. \end{array} \right\} \quad (12.069)$$

The next step in the present study is to write each of the orbital angular momentum operators in terms of spherical polar coordinates. This is done by employing transformations such as the following:

$$\left.\begin{aligned}\hat{x} &= x = r \sin\theta \cos\varphi; \\ \hat{p}_x &= -i\hbar\frac{\partial}{\partial x} = -i\hbar\left(\frac{\partial r}{\partial x}\frac{\partial}{\partial r} + \frac{\partial\theta}{\partial x}\frac{\partial}{\partial\theta} + \frac{\partial\varphi}{\partial x}\frac{\partial}{\partial\varphi}\right).\end{aligned}\right\} \quad (12.070)$$

Similar formulas can be written for $\hat{y}$, $\hat{p}_y$, $\hat{z}$, and $\hat{p}_z$. The nine partial derivatives needed have been given in (1.018). When the necessary substitutions into (12.058) are made, r and $\partial/\partial r$ cancel out and the operators for the Cartesian components of orbital angular momentum are found to involve only θ and φ:

$$\left.\begin{aligned}\hat{L}_x &= i\hbar\left(\sin\varphi\frac{\partial}{\partial\theta} + \cos\varphi\cot\theta\frac{\partial}{\partial\varphi}\right); \\ \hat{L}_y &= i\hbar\left(-\cos\varphi\frac{\partial}{\partial\theta} + \sin\varphi\cot\theta\frac{\partial}{\partial\varphi}\right); \\ \hat{L}_z &= -i\hbar\frac{\partial}{\partial\varphi}.\end{aligned}\right\} \quad (12.071)$$

The last of these expressions could have been obtained by recalling that $L_z = p_\varphi$, the canonical momentum conjugate to φ. The operator $\hat{L}_x^2 = \hat{L}_x\hat{L}_x$ can now be derived in spherical polar form if care is taken to perform all the necessary differentiations. Adding this to the spherical polar forms of $\hat{L}_y^2$ and $\hat{L}_z^2$, one finds (after considerable effort) that:

$$\hat{L}_x^2 + \hat{L}_y^2 + \hat{L}_z^2 = -\hbar^2\left[\frac{\partial^2}{\partial\theta^2} + \cot\theta\frac{\partial}{\partial\theta} + \csc^2\theta\frac{\partial^2}{\partial\varphi^2}\right]. \quad (12.072)$$

It may be observed that:

$$\begin{aligned}\frac{1}{\sin\theta}\frac{\partial}{\partial\theta}\left(\sin\theta\frac{\partial}{\partial\theta}\right) &= \frac{1}{\sin\theta}\left(\sin\theta\frac{\partial^2}{\partial\theta^2} + \cos\theta\frac{\partial}{\partial\theta}\right) \\ &= \frac{\partial^2}{\partial\theta^2} + \cot\theta\frac{\partial}{\partial\theta},\end{aligned} \quad (12.073)$$

where the two terms on the extreme right can be recognized as terms in (12.072). The operator for the magnitude squared of the orbital angular momentum therefore becomes:

$$\hat{L}^2 = -\hbar^2\left[\frac{1}{\sin\theta}\frac{\partial}{\partial\theta}\left(\sin\theta\frac{\partial}{\partial\theta}\right) + \frac{1}{\sin^2\theta}\frac{\partial^2}{\partial\varphi^2}\right]. \quad (12.074)$$

It is easy to show that the spherical harmonics $Y_l^m(\theta, \varphi)$ are eigenfunctions of $\hat{L}_z$ and of $\hat{L}^2$, a situation which is clearly possible because these two operators commute. Since every spherical harmonic is a function of θ times exp $im\varphi$, by (12.071) it must be an eigenfunction of $\hat{L}_z$ with eigenvalue $m\hbar$. Thus:

$$L_z Y_l^m = m\hbar Y_l^m. \quad (12.075)$$

The operator $\hat{L}^2$ can be recognized in equation (12.025) which was formed after separation of the θ, φ factor from the r factor. This equation is satisfied by the Y_l^m and can be written:

$$\frac{1}{Y_l^m}\left[\frac{-1}{\hbar^2}\hat{L}^2 Y_l^m\right] + C = 0. \quad (12.076)$$

Since $C = l(l+1)$, this becomes:

$$\hat{L}^2 Y_l^m = l(l+1)\hbar^2 Y_l^m, \quad (12.077)$$

and Y_l^m is seen to be an eigenfunction of $\hat{L}^2$ with eigenvalue $l(l+1)\hbar^2$.

If the wave function of an atomic state is given by:

$$\Psi = r^{-1}F(r)Y_l^m(\theta, \varphi)\exp\left(-iEt/\hbar\right), \quad (12.078)$$

it follows that ψ itself is an eigenfunction of $\hat{L}^2$ and of $\hat{L}_z$ with eigenvalues as given in the preceding paragraph. Such a state is said to have the definite value $l(l+1)\hbar^2$ for the magnitude squared of its orbital angular momentum and the definite value $m\hbar$ for the component of this angular momentum along the z-axis. For given l there are, by (12.044), $2l+1$ states like (12.078). One of these has an m value of $-l$; others have $-l+1$, $-l+2$, etc., up to and including l. Thus the minimum and maximum z components are $-l\hbar$ and $l\hbar$, respectively.

The magnitude of these components is, appropriately, always less than $[l(l+1)]^{\frac{1}{2}}\hbar$.

By the very manner in which the spherical polar coordinate system is defined, the z-axis has a unique role to play and the spherical harmonics are eminently well suited to express quantum states which have a definite component of orbital angular momentum along this axis. Space, however, is isotropic and what is possible for the z-axis must also be possible for, say, the x-axis. Thus one must be able to construct a sequence of $2l+1$ quantum states which are eigenfunctions of $\hat{L}^2$ with quantum number l and, at the same time, are eigenfunctions of $\hat{L}_x$ with definite components of orbital angular momentum along the x-axis. To make this construction, one should assume the most general possible function of θ and φ for given l and write:

$$\Psi = r^{-1}F(r)\left[\sum_{m=-l}^{l}\alpha_l^m Y_l^m(\theta,\varphi)\right]\exp\left(-iEt/\hbar\right). \qquad (12.079)$$

This is a superposition involving $2l+1$ different spherical harmonics, each with its own coefficient α_l^m. To make it an eigenfunction of $\hat{L}_x$, it is only necessary to choose these $2l+1$ coefficients correctly. This is not an extremely difficult process but it does involve the rather sophisticated numerical technique of matrix diagonalization discussed in Chapter 13. From what was said earlier, it must be possible to choose these $2l+1$ coefficients in $2l+1$ different ways so that a whole sequence of eigenfunctions of $\hat{L}_x$ are generated, each with its distinct eigenvalue $m_x\hbar$ where m_x ranges by integers from $-l$ to l.

Again, by isotropy, what is true for the z- and the x-axes must be true for the y-axis and, for that matter, for any directed line in space. If such a line is described by the unit vector e, one may construct an operator $\hat{L}_e$ for the e component of orbital angular momentum. Then, for every given value of the quantum number l, there exists a sequence of $2l+1$ eigenfunctions of $\hat{L}_e$. Each member of this sequence will have the general form of (12.079), a unique set of expansion coefficients α_l^m, and a definite component of orbital angular momentum in the e direction. Thus there are potentially an infinite number of orbital angular momentum operators, one for each direction in space. It should be clear by now that isotropy is in no way compromised by the adoption of the z-based spherical polar coordinate system and the orthonormal set of spherical harmonics associated therewith. Only the ease or difficulty of expressing various eigenfunctions is affected by this adoption.

If a given function of θ and φ is an eigenfunction of one of the operators $\hat{L}_x$, $\hat{L}_y$, or $\hat{L}_z$, it cannot in general be an eigenfunction of either of the other two; the three sets of eigenfunctions of these operators are disjoint.[†] In view of the commutation relations (12.062) of these operators, this state of affairs is to be expected but to demonstrate its existence quantitatively requires more than a perfunctory appeal to the uncertainty principle. Suppose that a given quantum state has as its angular factor the single spherical harmonic Y_l^m; for this state, L_z has the definite value $m\hbar$ and the standard deviation (uncertainty) in the distribution over L_z, namely ΔL_z, vanishes. The following three applications of the uncertainty principle may now be made:

$$\Delta L_y\, \Delta L_z \geqslant \tfrac{1}{2}\hbar\, |\langle L_x\rangle|\,; \tag{12.080}$$

$$\Delta L_z\, \Delta L_x \geqslant \tfrac{1}{2}\hbar\, |\langle L_y\rangle|\,; \tag{12.081}$$

$$\Delta L_x\, \Delta L_y \geqslant \tfrac{1}{2}\hbar\, |\langle L_z\rangle|$$

$$\geqslant \tfrac{1}{2}\hbar^2\, |m|\,. \tag{12.082}$$

From the last of these expressions it might be supposed that ΔL_x, ΔL_y, or both could be zero if $m = 0$ but, as is shown below, this is not true. In (12.080) and (12.081), the left-hand sides vanish because $\Delta L_z = 0$. The right-hand sides, which cannot be negative, must also vanish and one may conclude:

$$\left.\begin{array}{l} \langle L_x\rangle = 0; \\ \langle L_y\rangle = 0. \end{array}\right\} \tag{12.083}$$

[†] The one exception occurs if there is no angular momentum. Then $l = 0$ and the only spherical harmonic which can exist is Y_0^0, a constant and a common eigenfunction of all three of the operators named above.

It follows that:

$$(\Delta L_x)^2 = \langle L_x^2 \rangle;$$
$$(\Delta L_y)^2 = \langle L_y^2 \rangle. \tag{12.084}$$

Therefore:

$$(\Delta L_x)^2 + (\Delta L_y)^2 = \langle L^2 \rangle - \langle L_z^2 \rangle = l(l+1)\hbar^2 - m^2\hbar^2. \tag{12.085}$$

By symmetry, an eigenfunction of $\hat{L}_z$ must have equal uncertainties n L_x and in L_y, i.e. $\Delta L_x = \Delta L_y$. This implies that:

$$(\Delta L_x)^2 = \tfrac{1}{2}[l(l+1) - m^2]\hbar^2 \tag{12.086}$$

and the final result is:

$$\Delta L_x = \Delta L_y = \{\tfrac{1}{2}[l(l+1) - m^2]\}^{\frac{1}{2}}\hbar. \tag{12.087}$$

This confirms the earlier statement that an eigenfunction of $\hat{L}_z$ cannot also be an eigenfunction of $\hat{L}_x$ or of $\hat{L}_y$ if $l > 0$. It is seen that the minimum value of ΔL_x is attained when $|m| = l$ and that the minimum product $\Delta L_x \Delta L_y$, which is equal to $(\Delta L_x)^2$, is given by $\tfrac{1}{2}\hbar^2 l$. This result is much more informative than (12.082); at the same time, it does not contradict (12.082).

By isotropy, the result of the preceding paragraph can be applied to an eigenfunction of $\hat{L}_x$ or of $\hat{L}_y$. Thus an eigenfunction of $\hat{L}^2$ and of $\hat{L}_x$ with a definite value of $\hat{L}_x$ equal to $m_x\hbar$ has uncertainties in L_y and L_z given by $\{\tfrac{1}{2}[l(l+1) - m^2]\}^{\frac{1}{2}}\hbar$. A corresponding statement can, of course, be made with respect to an eigenfunction of $\hat{L}_y$.

The foregoing discussion focuses upon special cases of a phenomenon which is really much more general. Assuming that $e' \neq \pm e$, it is found that the two operators $\hat{L}_e$ and $\hat{L}_{e'}$ do not commute and do not, for $l > 0$, have common eigenfunctions. Angular momentum can be quantized with respect to any direction. Thus if a quantum measurement of L_e is made, the result will be that $L_e = m_e\hbar$ where m_e is the familiar quantum number for the component of orbital angular momentum along e; this number takes an integral value. If the quantum L_e measurement is followed by a quantum $L_{e'}$ measurement, the result will be that $L_{e'} = m_{e'}\hbar$ with a similar remark about the significance of $m_{e'}$. The probabilities of occurrence of the various values of $m_{e'}$ are, of course, established by the nature of the former state of definite L_e. Once the $L_{e'}$ measurement is made, however, the former state of definite L_e is obliterated.

12.05. Solutions of the Radial Equation; Energy Levels

Let the r equation (12.024) be multiplied by F/r^2 and rearranged; also let $l(l+1)$ be substituted for C. The result is:

$$\underbrace{-\frac{\hbar^2}{2\mu}\frac{d^2F}{dr^2}}_{\hat{T}'F} + \underbrace{\left[\frac{l(l+1)\hbar^2}{2\mu r^2} - \frac{\gamma}{r}\right]F}_{V'(r)F} = EF. \tag{12.088}$$

This is a perfect example of an effective one-dimensional Schrödinger equation. The effective kinetic energy operator is easy to identify and the effective potential energy $V'(r)$ is seen to contain the "centrifugal potential energy", $l(l+1)\hbar^2/2\mu r^2$. Equation (12.088) should recall equation (2.028) for the Kepler problem in classical mechanics.

It is usual, in treating the hydrogenic atom, to orient the coordinate frame so that the z-axis coincides with the direction along which angular momentum is to be quantized. If such a direction does not exist *a priori*, the axes may be oriented arbitrarily but angular momentum should still be quantized along the z-axis for maximum simplification. The wave function then assumes the form of (12.078). The normalization condition may now be examined; since the volume element may be written as $d\Omega\, r^2\, dr$, one has:

$$\int_0^\infty \oint \Psi^*\Psi\, d\Omega\, r^2\, dr = \int_0^\infty F^*F\, dr \oint Y_l^{m*}Y_l^m\, d\Omega = 1. \tag{12.089}$$

By virtue of the separate normalization of the spherical harmonics, the integral over solid angle is equal to unity and it follows that:

$$\int_0^\infty F^*F\, dr = 1. \tag{12.090}$$

The task at hand is now easily described; it is to solve (12.088) subject to condition (12.090). A representative effective potential energy $V'(r)$ is illustrated in Fig. 12.02. The energy E of a typical bound stationary state is a negative quantity denoted by the short horizontal line; it must lie between the minimum of $V'(r)$ as a lower bound and zero as an upper bound. The reader should be able to deduce this from the background acquired in Chapter 7.

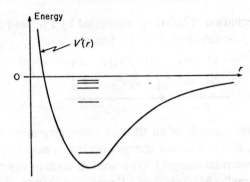

FIG. 12.02. $V'(r)$ for the effective one-dimensional Schrödinger equation (12.088). Positions of the expected energy levels of a few of the stationary states are shown qualitatively.

At very large r, where $V' \to 0$, equation (12.088) reduces to:

$$-\frac{\hbar^2}{2\mu}\frac{\mathrm{d}^2 F}{\mathrm{d}r^2} = EF. \qquad (12.091)$$

This has the following normalizable solution in which β is positive real:

$$F = A \exp(-\beta r). \qquad (12.092)$$

Through (12.091), the quantity β is directly related to the energy:

$$E = -\hbar^2\beta^2/2\mu; \qquad \beta = [-2\mu E]^{\frac{1}{2}}/\hbar. \qquad (12.093)$$

These results suggest the following substitution:

$$F(r) = G(r)\exp(-\beta r). \qquad (12.094)$$

Using this, the differential equation for $G(r)$ becomes:

$$-\frac{\hbar^2}{2\mu}\left[\frac{d^2G}{dr^2} - 2\beta\frac{dG}{dr}\right] + \left[\frac{l(l+1)\hbar^2}{2\mu r^2} - \frac{\gamma}{r}\right]G = 0. \quad (12.095)$$

On multiplication by $-2\mu/\hbar^2$, one has:

$$\frac{d^2G}{dr^2} - 2\beta\frac{dG}{dr} - \frac{l(l+1)G}{r^2} + \frac{2\mu\gamma G}{\hbar^2 r} = 0. \quad (12.096)$$

A change to a dimensionless independent variable is now in order. Let:

$$\xi = 2\beta r; \qquad r = \xi/2\beta. \quad (12.097)$$

If equation (12.096) is divided by $4\beta^2$, this substitution is readily performed. The result is:

$$\frac{d^2G}{d\xi^2} - \frac{l(l+1)G}{\xi^2} - \frac{dG}{d\xi} + \frac{\eta G}{\xi} = 0, \quad (12.098)$$

where η is a convenient shorthand defined by:

$$\eta = \frac{\mu\gamma}{\hbar^2\beta} = \frac{Ze^2}{4\pi\varepsilon_0\hbar}\left[\frac{\mu}{-2E}\right]^{\frac{1}{2}} = \frac{Ze^2}{2\varepsilon_0\hbar}\left[\frac{\mu}{-2E}\right]^{\frac{1}{2}}. \quad (12.099)$$

A series solution of equation (12.098) will now be assumed. Unlike previous examples, this series must begin with a power other than the zeroth in order to achieve needed flexibility. Thus:

$$G = \sum_{j=0}^{\infty} b_j\xi^{j+\sigma}; \quad (12.100)$$

$$\frac{dG}{d\xi} = \sum_{j=0}^{\infty} (j+\sigma)b_j\xi^{j+\sigma-1}; \quad (12.101)$$

$$\frac{d^2G}{d\xi^2} = \sum_{j=0}^{\infty} (j+\sigma)(j+\sigma-1)b_j\xi^{j+\sigma-2}. \quad (12.102)$$

In the first two terms of (12.098), let j be replaced by $k+1$; in the second two, let j be replaced by k. The result is:

$$\sum_{k=-1}^{\infty} (k+1+\sigma)(k+\sigma)b_{k+1}\xi^{k+\sigma-1} - l(l+1) \sum_{k=-1}^{\infty} b_{k+1}\xi^{k+\sigma-1}$$

$$- \sum_{k=0}^{\infty} (k+\sigma)b_k\xi^{k+\sigma-1} + \eta \sum_{k=0}^{\infty} b_k\xi^{k+\sigma-1} = 0. \qquad (12.103)$$

This becomes:

$$[\sigma(\sigma-1) - l(l+1)]b_0\xi^{\sigma-2}$$

$$+ \sum_{k=0}^{\infty} \{[(k+1+\sigma)(k+\sigma) - l(l+1)]b_{k+1} - [(k+\sigma) - \eta]b_k\}\xi^{k+\sigma-1} = 0.$$

$$(12.104)$$

The cofficient of each power of ξ must vanish. When this requirement is applied to $\xi^{\sigma-2}$, which is separate from the sum, it produces the *indicial equation*:

$$\sigma(\sigma-1) = l(l+1). \qquad (12.105)$$

This has the two roots, $\sigma = -l$ and $\sigma = l+1$. When the same requirement is applied to the higher powers of ξ, it gives the recurrence relation:

$$b_{k+1} = \frac{k+\sigma-\eta}{(k+1+\sigma)(k+\sigma) - l(l+1)} b_k. \qquad (12.106)$$

If $\sigma = -l$, one finds that the denominator of the above vanishes when k reaches the value of $2l$ and b_{2l+1} is therefore infinite. This is unacceptable and the root in question must be discarded. Upon substitution of the other root, $\sigma = l+1$, the recurrence relation becomes:

$$b_{k+1} = \frac{k+l+1-\eta}{(k+l+2)(k+l+1) - l(l+1)} b_k. \qquad (12.107)$$

As in earlier examples, it can be shown that if $G(\xi)$ becomes an infinite series, a non-normalizable solution results and it is concluded that η must take a value which causes this series to terminate. Thus η must be equal to an integer n where:

$$n \geqslant l+1. \qquad (12.108)$$

This n is the third quantum number to make its appearance and is known as the *principal quantum number*. Since $0 < l < \infty$, n has unity as a lower bound. It is usual to assign first a positive integral value to n and then let l be governed by (12.108). Using this and (12.044), it is a simple matter to summarize the ranges of the three quantum numbers:

$$\left. \begin{array}{l} n = 1, 2, 3, 4, \ldots \infty; \\ l = 0, 1, 2, 3, \ldots (n-1); \\ m = -l, -l+1, -l+2, \ldots l. \end{array} \right\} \tag{12.109}$$

The quantization of η is tantamount to a quantization of the energy. Thus if n is substituted for η in (12.099) and the latter solved for E, the result is a formula (which depends only on the quantum number n) for the energy levels of the stationary states of the hydrogenic atom:

$$E = E_n = \frac{-\mu e^4 Z^2}{8\varepsilon_0^2 h^2 n^2}. \tag{12.110}$$

This formula, which is in good agreement with spectroscopic data,[†] duplicates the Bohr formula (5.060) for the energy levels of neutral hydrogen ($Z = 1$) provided that the reduced mass μ is substituted for the electron mass m in the latter. The quantization of the energy produces, in turn, a quantization of β. Thus:

$$\beta = \beta_n = \frac{2\pi}{h} \left[\frac{2\mu^2 e^4 Z^2}{8\varepsilon_0^2 h^2 n^2} \right]^{\frac{1}{2}} = \frac{\pi \mu e^2 Z}{\varepsilon_0 h^2 n}. \tag{12.111}$$

With reference to equation (5.061) and with μ again substituted for m, the combination of constants $\varepsilon_0 h^2 / \pi \mu e^2$ can be recognized as the

---

[†] Although the wave mechanical solution given here is rigorous, the model of an electron attracted to a point nucleus by an inverse square force in a non-relativistic context ignores many subtle features of the actual atom. It is not surprising, therefore, that a number of very small discrepancies exist between the predictions of (12.110) and the spectroscopically determined energy levels, most of which exhibit fine structure. Explanations for the exact locations and for the splitting of these levels require a consideration of electron spin and, frequently, of other interesting discoveries and sophisticated developments of modern physics.

radius of the first Bohr orbit in hydrogen. This radius, which remains an important length in the wave mechanical development of atomic theory, is usually denoted by the letter a; β_n then becomes:[†]

$$\beta_n = \frac{Z}{na}.$$ (12.112)

Details of the stationary-state wave functions depend upon the structure of $G(\xi)$. Since the highest power of ξ in the latter is $n-l-1$ and since $\sigma = l+1$, it is seen that $G(\xi)$ depends upon both of the quantum numbers n and l and may be written:

$$G(\xi) = G_{nl}(\xi) = \xi^{l+1} \sum_{k=0}^{n-l-1} b_k \xi^k.$$ (12.113)

The coefficients b_k are determined up to a multiplicative constant by recurrence relation (12.107). The numerator of this relation should be rewritten with η replaced by n, the denominator can be factored, and the whole becomes:

$$\frac{b_{k+1}}{b_k} = \frac{-(n-l-1-k)}{(k+2l+2)(k+1)}.$$ (12.114)

Notice that the quantities in parentheses are non-negative and that the signs of the b_k therefore alternate. It will be shown that a standard set of polynomials, known as the *associated Laguerre polynomials*, have cofficients that obey (12.114) and that a member of this set may therefore represent the sum in (12.113). The reader is cautioned that there is considerabe disparity among authors in the definition of the associated Laguerre polynomials. The definition presented here is in accord with widespread usage in physics.

[†] The first Bohr radius for a hypothetical hydrogen atom with infinite nuclear mass is frequently denoted a_0, where:

$$a_0 = \frac{\varepsilon_0 h^2}{\pi m e^2} = \frac{4\pi \varepsilon_0 \hbar^2}{m e^2} = 0.52917715 \text{ Å}.$$

The numerical value is from [11]. Since a_0 involves the mass of only the electron rather than the masses of both electron and nucleus, it is considered more basic than a and its value is the one quoted in tables of fundamental constants.

A Laguerre polynomial[43] of integral index q is given by the following series:

$$L_q(\xi) = q! \sum_{j=0}^{q} \frac{(-1)^j q!}{j!(q-j)!} \frac{\xi^j}{j!}. \qquad (12.115)$$

The binomial coefficient $\binom{q}{j} = q!/j!(q-j)!$ may be recognized here and the terms of this series are very easy to write out. If (12.115) is now differentiated p times, $p \leqslant q$, the result is the associated Laguerre polynomial:

$$L_q^p(\xi) = q! \sum_{j=0}^{q} \frac{(-1)^j q!}{j!(q-j)} \frac{\xi^{j-p}}{(j-p)!}. \qquad (12.116)$$

It is seen that all terms from $j = 0$ to $j = p-1$, inclusive, vanish; one may therefore change the index of summation from j to $k = j-p$ and write:

$$L_q^p(\xi) = q! \sum_{k=0}^{q-p} \frac{(-1)^{k+p} q!}{(k+p)!(q-p-k)!} \frac{\xi^k}{k!}. \qquad (12.117)$$

This is to be identified with the sum of $b_k \xi^k$ in (12.113). Using the coefficients in (12.117), one finds that:

$$\frac{b_{k+1}}{b_k} = \frac{-(k+p)!(q-p-k)!\,k!}{(k+1+p)!(q-p-k-1)!(k+1)!} = \frac{-(q-p-k)}{(k+1+p)(k+1)}. $$
$$(12.118)$$

Agreement with recurrence relation (12.114) is achieved if:

$$\left. \begin{array}{l} p = 2l+1; \\ q = n+l. \end{array} \right\} \qquad (12.119)$$

It follows from these results that:

$$G_{nl}(\xi) = B_{nl}\xi^{l+1} L_{n+1}^{2l+1}(\xi), \qquad (12.120)$$

where B_{nl} is a constant to be evaluated by normalization. A few of the Laguerre polynomials and associated Laguerre polynomials are displayed in Table 12.03.

Since $G_{nl}(\xi)$ depends upon the quantum numbers n and l, the function $F(\xi)$ does likewise. By (12.094), one has:

$$F(\xi) = F_{nl}(\xi) = B_{nl} \exp\left(-\tfrac{1}{2}\xi\right) \xi^{l+1} L_{n+l}^{2l+1}(\xi). \qquad (12.121)$$

The normalization condition (12.090) requires the following:

$$\frac{|B_{nl}|^2}{2\beta_n} \int_0^\infty \exp\left(-\xi\right) \xi^{2l+2} \left[L_{n+l}^{2l+1}(\xi)\right]^2 \, d\xi = 1. \qquad (12.122)$$

The applicable integral is given by:[44]

$$\int_0^\infty \exp\left(-\xi\right) \xi^{p+1} \left[L_q^p(\xi)\right]^2 \, d\xi = \frac{(q!)^3 (2q-p+1)}{(q-p)!}. \qquad (12.123)$$

When appropriate substitutions for the indices are made, (12.122) becomes:

$$\frac{na|B_{nl}|^2}{2Z} \frac{[(n+l)!]^3 2n}{(n-l-1)!} = 1. \qquad (12.124)$$

All the associated Laguerre polynomials actually used in (12.121) have an odd upper index and therefore a negative leading term whereas a positive leading term would be more convenient. The situation is easily remedied by choosing a negative sign for B_{nl}; the latter is then given by:

$$B_{nl} = \frac{-1}{n} \left[\frac{Z(n-l-1)!}{a[(n+l)!]^3}\right]^{\frac{1}{2}}. \qquad (12.125)$$

It is well to remember the relationship between ξ and r:

$$\xi = \frac{2Zr}{na}. \qquad (12.126)$$

The solution of the effective one-dimensional problem is now complete and the effective wave function is simply $F_{nl}(2Zr/na) \exp(-iE_n t/\hbar)$ The magnitude squared of this function is the probability density on

TABLE 12.03. Laguerre Polynomials and Associated Laguerre Polynomials

$$L_0 = 1$$

$$L_1 = 1 - \xi$$
$$L_1^1 = -1$$

$$L_2 = 2 - 4\xi + \xi^2$$
$$L_2^1 = -4 + 2\xi$$
$$L_2^2 = 2$$

$$L_3 = 6 - 18\xi + 9\xi^2 - \xi^3$$
$$L_3^1 = -18 + 18\xi - 3\xi^2$$
$$L_3^2 = 18 - 6\xi$$
$$L_3^3 = -6$$

$$L_4 = 24 - 96\xi + 72\xi^2 - 16\xi^3 + \xi^4$$
$$L_4^1 = -96 + 144\xi - 48\xi^2 + 4\xi^3$$
$$L_4^2 = 144 - 96\xi + 12\xi^2$$
$$L_4^3 = -96 + 24\xi$$
$$L_4^4 = 24$$

$$L_5 = 120 - 600\xi + 600\xi^2 - 200\xi^3 + 25\xi^4 - \xi^5$$
$$L_5^1 = -600 + 1200\xi - 600\xi^2 + 100\xi^3 - 5\xi^4$$
$$L_5^2 = 1200 - 1200\xi + 300\xi^2 - 20\xi^3$$
$$L_5^3 = -1200 + 600\xi - 60\xi^2$$
$$L_5^4 = 600 - 120\xi$$
$$L_5^5 = -120$$

the r continuum which means that $F_{nl}^* F_{nl} \, dr$ is the probability that a quantum measurement of r will yield a result lying in the infinitesimal interval dr. The states of maximum angular momentum for given energy, namely those for which $l = n-1$, are the ones most closely related to the circular orbit states of the old quantum theory. It is not surprising, then, to find that $F_{n,\,n-1}^* F_{n,\,n-1}$ has a single maximum which occurs at $r = n^2 a/Z$, i.e. at the Bohr radius for such a state. The probability densities with respect to r for other types of states have more than one maximum and are not so easily interpreted in these terms.

12.06. The Hydrogenic Wave Functions

The ψ functions for the bound stationary states of a hydrogenic atom can now be assembled. Denoted $\psi_{nlm}(r, \theta, \varphi)$, they are dependent upon all three quantum numbers and are given by:

$$\psi_{nlm} = r^{-1} F_{nl}(\xi) \, Y_l^m(\theta, \varphi) = (2Z/na)\xi^{-1} F_{nl}(\xi) \, Y_l^m(\theta, \varphi). \quad (12.127)$$

With obvious substitutions from the preceding section, one may now write the bound stationary state wave functions with time dependence in finished form as follows:

$$\Psi_{nlm} = \left(\frac{Z}{a}\right)^{\frac{3}{2}} \frac{2}{n^2[(n+l)!\,(n-l-1)!]^{\frac{1}{2}}} \left[\frac{-(n-l-1)!\,\xi^l}{(n+l)!} L_{n+l}^{2l+1}(\xi) \right]$$

$$\times \exp\left(-\tfrac{1}{2}\xi\right) Y_l^m(\theta, \varphi) \exp\left(-iE_n t/\hbar\right). \quad (12.128)$$

An attempt has been made here to group the various factors for easy evaluation; the polynomial in square brackets has a coefficient of magnitude unity for its highest power. The radial dependence of these functions, embodied in the part which contains ξ, is given for a few states in Table 12.04. This dependence is different from that of the corresponding F_{nl} functions by the factor ξ^{-1}.

The influence of the atomic number Z upon a stationary-state wave function is interesting. Suppose that ξ' is a particular value of the argument at which the radial dependence has some special feature, e.g. a maximum or a zero. The corresponding radius r' at which this special feature occurs is then given by $na\xi'/2Z$. If Z is now increased, r' must decrease; the wave function must therefore shrink, assuming a form that is geometrically similar to, but more compact than, its previous form. At the same time, the value of the wave function must increase everywhere, through the factor $Z^{\frac{3}{2}}$ in (12.128), thus maintaining normalization. This shrinking in response to an increase in Z exemplifies the tighter binding of an electron to a nucleus of greater electrostatic charge.

TABLE 12.04. *Radial Dependence of the Hydrogenic Bound Stationary States*

$$\xi = 2Zr/na$$

n	l	$\dfrac{2\left[\dfrac{-(n-l-1)!\,\xi^l}{(n+l)!}\,L_{n+l}^{2l+1}(\xi)\right]\exp\left(-\tfrac{1}{2}\xi\right)}{n^2[(n+l)!\,(n-l-1)!]^{\frac{1}{2}}}$
1	0	$2\exp\left(-\tfrac{1}{2}\xi\right)$
2	0	$\dfrac{1}{2(2)^{\frac{1}{2}}}[2-\xi]\exp\left(-\tfrac{1}{2}\xi\right)$
2	1	$\dfrac{1}{2(6)^{\frac{1}{2}}}[\xi]\exp\left(-\tfrac{1}{2}\xi\right)$
3	0	$\dfrac{1}{9(3)^{\frac{1}{2}}}[6-6\xi+\xi^2]\exp\left(-\tfrac{1}{2}\xi\right)$
3	1	$\dfrac{1}{9(6)^{\frac{1}{2}}}[4\xi-\xi^2]\exp\left(-\tfrac{1}{2}\xi\right)$
3	2	$\dfrac{1}{9(30)^{\frac{1}{2}}}[\xi^2]\exp\left(-\tfrac{1}{2}\xi\right)$

The integral of the product $\Psi^*_{n'l'm'}\Psi_{nlm}$ over all of space must vanish if $l' \neq l$ or if $m' \neq m$ by virtue of the orthogonality of the spherical harmonics. To prove a similar statement for $n' \neq n$ by appeal to the properties of the associated Laguerre polynomials is difficult. It is much easier to use the fact that the Ψ_{nlm} are eigenfunctions of an Hermitian operator (the Hamiltonian for the relative motion) with eigenvalues E_n which are different for every different value of n. Orthogonality with respect to n follows immediately and the situation may be summarized by saying:

$$\int_{\substack{\text{all} \\ \text{space}}} \Psi^*_{n'l'm'}\Psi_{nlm}\,d\tau = \delta_{n'n}\delta_{l'l}\delta_{m'm}. \tag{12.129}$$

Thus the stationary-state wave functions form an orthonormal set. This set is complete, however, only with respect to functions which vanish properly at infinity. These include all bound (but not necessarily stationary) states such as, for example, the quasi-classical orbiting states which can be formed at high quantum numbers. Excluded, i.e. not representable in the Ψ_{nlm} set of functions, are the continuum states. A continuum state in this context is similar to a continuum state of, say, the finite rectangular well in that it represents a situation in which the electron is projected from infinity (with a given kinetic energy) toward the nucleus. After interacting with the latter, the electron may leave the scene along any one of an infinite number of possible departure directions according to a calculable probability distribution. Since the potential energy is zero at infinity, a continuum state is a state of positive total energy and one cannot expect to be able to synthesize such a state by superposing states of negative total energy.

Except for the ground state, Ψ_{100}, all the stationary states of the hydrogenic atom are, according to the theoretical model used here, grouped into degenerate subsets. Thus the four states Ψ_{200}, Ψ_{21-1}, Ψ_{210}, and Ψ_{211}, which have principal quantum number equal to two, have a common energy E_2 and form a four-fold degenerate subset. With reference to (12.109), it is seen that for a given value of n

there are n possible values of l and that for a given value of l there are $2l+1$ possible values of m. Thus there is an n-fold degeneracy with respect to l and, less importantly for present purposes, a $(2l+1)$-fold degeneracy with respect to m. The latter occurs whenever the force is central and the potential energy is spherically symmetric. On the other hand, the l degeneracy occurs whenever the spatial dependence of the potential energy has the so-called Coulomb form of $-\gamma/r$ which corresponds to a force that is not only central but also inverse square in character. Any perturbing influence which, while maintaining spherical symmetry, changes the r dependence of the potential energy to a non-Coulombic form makes the energies of the stationary states dependent upon l as well as upon n. Such an influence will, in general, separate the energy of a state like Ψ_{200} from the energy of the subset of states Ψ_{21-1}, Ψ_{210}, and Ψ_{211} to form two distinct levels, i.e. it will remove the l degeneracy. As indicated in an earlier footnote (p. 315) minute separations of this type do occur in actual hydrogenic atoms but they are of insufficient magnitude to merit consideration in the present context.

When the hydrogenic spectral terms were plotted in Fig. 5.09, there was no hint of the multiple character of terms for $n > 1$ although this character is also a feature of the old quantum theory, as is evident when elliptical as well as circular orbits are considered. Some of this can be illustrated, without introducing too much complication, by arranging the terms in a number of vertical columns, each corresponding to a different value of l as in Fig. 12.03. It is not convenient and, fortunately, not necessary to illustrate the m degeneracy on a diagram of this type.

A diagram like Fig. 12.03 would acquire much more physical significance if the terms with the same value of n but different values of l could be given significantly different energies for then it would become easily possible to pinpoint the origin and the destination of each radiative transition in terms of both n and l. Nature has provided just such an opportunity in the spectra of the alkali metals. An atom of one of these metals can, for many purposes, be treated as a one-electron atom because the valence electron is relatively loosely bound whereas the other electrons, called the "core" electrons, are tightly bound and, collectively, form the equivalent of a spherically symmetric static charge distribution around the nucleus. The valence electron moves in this environment under the influence of the net electrostatic

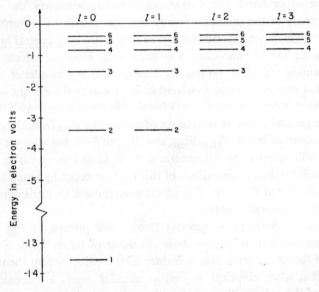

FIG. 12.03. Energy levels of hydrogen showing principal quantum number n adjacent to level and orbital angular momentum quantum number l at top of column. Fine structure (splitting of levels due to electron spin and other effects) is invisible on a diagram of this size. Interval between $n = 1$ and $n = 2$ levels is not to scale.

field which, because of the presence of the distributed charge, is obviously non-Coulombic. An alkali metal atom may be thought of as a spherically symmetric but non-hydrogenic one-electron atom; the same spherical harmonics are applicable to it but the F_{nl} functions, found by solving its effective one-dimensional Schrödinger equation, are different from the ones encountered in the previous section. It is through these functions that the energy levels become dependent upon

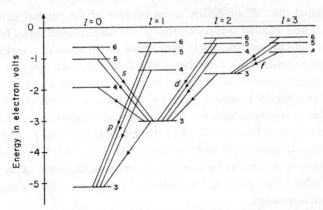

FIG. 12.04. Energy levels of sodium showing principal quantum number n adjacent to level and orbital angular momentum quantum number l at top of column. Some transitions belonging to the sharp (s), principal (p), diffuse (d), and fundamental (f) series are shown. Fine structure (splitting of levels due to electron spin and other effects) is invisible on a diagram of this size and precise identification of initial and terminal levels of transitions in the context of fine structure is not attempted.

l as well as upon n. Thus, when a term diagram for sodium is constructed[†] as in Fig. 12.04, a rich variety of distinct energy levels is found.

Before the formulation of either the old quantum theory or of wave mechanics, early spectroscopists had already discovered the term diagrams for the alkali metals in essentially the form shown in Fig. 12.04 simply by measuring the wavelengths of the various series of transitions, four of which are illustrated. One of these series, characterized by unusually well-defined lines, was designated the "s" or sharp series. The series which terminates on the lowest level (the ground state) was designated the "p" or principal series. Still another was

---

[†] It may be noticed that the lowest value of n in this diagram is three. This is a consequence of the Pauli exclusion principle; states with $n = 1$ and $n = 2$ have been pre-empted by the core electrons and are not available to the valence electron. The actual value of n is not a crucial element from the point of view of spectroscopic interpretation, however, since energy is already indicated by the vertical positions of the terms. Consequently, n is often de-emphasized or omitted altogether from diagrams of this type.

designated the "d" or diffuse series because of the general nature of its lines and another was called the "f" or fundamental series because its lines had an approximately hydrogenic distribution of wavelengths. Only at a considerably later date was it found that the terms at which the s series transitions originate have $l = 0$, those at which the p series transitions originate have $l = 1$, those at which the d series transitions originate have $l = 2$, and those at which the f series transitions originate have $l = 3$. Thus the letters s, p, d, and f came to be permanently associated with l values of 0, 1, 2, and 3, respectively. This system proved to be convenient and was extended as in Table 12.05 where the letters are alphabetical beginning with f except for j which is omitted.

TABLE 12.05. *Spectroscopic Letter Notation*

l value	0	1	2	3	4	5	6	7	8	9	...
Letter	s	p	d	f	g	h	i	k	l	m	...

Using the letter notation, it has become customary to designate a one-electron state like Ψ_{320}, for example, as a "$3d$" state where the "3" indicates that $n = 3$ and the "d" indicates that $l = 2$. This is done regardless of the nature of the atom involved so long as, in some sense, the electron in question can be conceived in isolation from the others. Thus the hydrogenic ground state Ψ_{100} is a "$1s$" state, the states Ψ_{21-1}, Ψ_{210}, and Ψ_{211} are all "$2p$" states, etc. The energy levels of m-degenerate states such as the latter three remain coincident even in an environment where the l degeneracy is removed but may be separated by the application of a magnetic field because such a field destroys the spherical symmetry. For this reason, the quantum number m is sometimes called the *magnetic quantum number*. In this general context, it should be stated that the capital letters S, P, D, F, etc., are substituted for the small letters when it is desired to indicate the total orbital angular momentum quantum number for an entire atom rather than the orbital angular momentum quantum number for a single electron.

With these remarks, this very brief introduction to the vast subject of atomic wave functions and the origin of spectra is somewhat reluctantly concluded. If the interest and curiosity of the reader have been stimulated, the treatment has been well justified.

CHAPTER 13

MATRIX MECHANICS

13.01. The Non-commutative Algebra of Matrices

A *matrix* is a rectangular array of numbers which, by the use of
certain algebraic rules, can be combined with another such array
to yield a third as a result. The numbers which form the array are
called the *elements* of the matrix and are in general complex. A vertical
set of elements within the array is called a *column* and a horizontal
set is called a *row*. Thus the typical matrix shown below has M
columns and N rows. Its elements are customarily written within a
pair of large brackets (or, at times, parentheses) and, strictly for convenience, the whole may be designated by a single symbol such as
(G). Thus:

$$(G) = \begin{bmatrix} g_{11} & g_{12} & \cdots & g_{1M} \\ g_{21} & g_{22} & \cdots & g_{2M} \\ \cdot & \cdot & \cdots & \cdot \\ g_{N1} & g_{N2} & \cdots & g_{NM} \end{bmatrix}. \tag{13.001}$$

Notice that (G) is not a value but a very abbreviated notation which
stands for the entire matrix regarded as a single mathematical entity.
The subscripts of the elements are called *indices* and, by convention,
the first index denotes the row to which the element belongs and the
second denotes the column. To be considered equal, two matrices
must be identical in all respects; every element of one must be equal

to the corresponding element of the other. Matrices have great utility in many areas of mathematical physics and are especially valuable in quantum mechanics; as may be recalled from Section 6.01, this discipline was first expressed rigorously in matrix form.

Two matrices (G) and (G') can be added (subtracted) only if $M' = M$ and $N' = N$. In that event, the sum (difference) is the matrix whose elements are the sums (differences) of the individual elements of (G) and (G'). For example:

$$\begin{pmatrix} a & b & c \\ d & e & f \end{pmatrix} \pm \begin{pmatrix} p & q & r \\ s & t & u \end{pmatrix} = \begin{pmatrix} a\pm p & b\pm q & c\pm r \\ d\pm s & e\pm t & f\pm u \end{pmatrix}. \quad (13.002)$$

Multiplication of a matrix by a single number, often called a *scalar* in this context, is equally simple. The elements of the product matrix are the products of the scalar with the elements of the original matrix:

$$s\begin{pmatrix} a & b & c \\ d & e & f \end{pmatrix} = \begin{pmatrix} sa & sb & sc \\ sd & se & sf \end{pmatrix} = \begin{pmatrix} a & b & c \\ d & e & f \end{pmatrix}s. \quad (13.003)$$

Such a multiplication, unlike another type to be considered shortly, is always commutative.

The matrices used in quantum mechanics almost invariably belong to one of the three following types:

$$\text{(i) a } \textit{column} \text{ matrix, } (X) = \begin{bmatrix} x_1 \\ x_2 \\ \vdots \\ x_N \end{bmatrix}, \quad (13.004)$$

$$\text{(ii) a } \textit{row} \text{ matrix, } \quad (Y) = (y_1 \, y_2 \, \ldots \, y_N), \quad (13.005)$$

$$\text{and} \quad \text{(iii) a } \textit{square} \text{ matrix,} \quad (A) = \begin{bmatrix} a_{11} & a_{12} & \ldots & a_{1N} \\ a_{21} & a_{22} & \ldots & a_{2N} \\ \vdots & \vdots & \cdots & \vdots \\ a_{N1} & a_{N2} & \ldots & a_{NN} \end{bmatrix}. \quad (13.006)$$

The processes described and the definitions given in the remainder of this chapter pertain to one or more of these types. The analogy whereby a single number is called a scalar can be extended to entities of higher rank. Thus a column matrix is often called a *column vector* and a row matrix, a *row vector*.

In matrix multiplication, a square matrix may be multiplied onto a column matrix (provided both have the same value of N) to yield another column matrix as follows:

$$\begin{bmatrix} a_{11} & a_{12} & \ldots & a_{1N} \\ a_{21} & a_{22} & \ldots & a_{2N} \\ \cdot & \cdot & \cdot & \cdot \\ a_{N1} & a_{N2} & \ldots & a_{NN} \end{bmatrix} \begin{bmatrix} x_1 \\ x_2 \\ \vdots \\ x_N \end{bmatrix} = \begin{bmatrix} a_{11}x_1 + a_{12}x_2 + \ldots + a_{1N}x_N \\ a_{21}x_1 + a_{22}x_2 + \ldots + a_{2N}x_N \\ \cdot \quad \cdot \quad \cdot \quad \cdot \\ a_{N1}x_1 + a_{N2}x_2 + \ldots + a_{NN}x_N \end{bmatrix}.$$

$$(13.007)$$

Notice that the elements of the nth row of the square are multiplied onto the elements of the column and the results summed to produce the nth element of the product. Similarly, a row matrix may be multiplied onto a square matrix to yield another row matrix as follows:

$$(y_1 \quad y_2 \ldots y_N) \begin{bmatrix} a_{11} & a_{12} & \ldots & a_{1N} \\ a_{21} & a_{22} & \ldots & a_{2N} \\ \cdot & \cdot & \cdot & \cdot \\ a_{N1} & a_{N2} & \ldots & a_{NN} \end{bmatrix}$$

$$= ([y_1 a_{11} + y_2 a_{21} + \ldots + y_N a_{N1}] [y_1 a_{12} + y_2 a_{22} + \ldots + y_N a_{N2}]$$
$$\ldots [y_1 a_{1N} + y_2 a_{2N} + \ldots + y_N a_{NN}]).$$

$$(13.008)$$

The same rule, row multiplied onto column, has been used here; the elements of the row are multiplied onto the elements of the nth column of the square and the results summed to produce the nth element of the product. Notice that a column matrix cannot be multiplied onto a square matrix nor a square matrix onto a row matrix, i.e. if the order of the factors in (13.007) or (13.008) is reversed, the multiplication is not defined.

A square matrix can be multiplied onto a square matrix to produce another square matrix. Thus:

$$
\begin{bmatrix}
a_{11} & a_{12} & \dots & a_{1N} \\
a_{21} & a_{22} & \dots & a_{2N} \\
\cdot & \cdot & \cdot & \cdot \\
a_{N1} & a_{N2} & \dots & a_{NN}
\end{bmatrix}
\begin{bmatrix}
b_{11} & b_{12} & \dots & b_{1N} \\
b_{21} & b_{22} & \dots & b_{2N} \\
\cdot & \cdot & \cdot & \cdot \\
b_{N1} & b_{N2} & \dots & b_{NN}
\end{bmatrix}
=
\begin{bmatrix}
c_{11} & c_{12} & \dots & c_{1N} \\
c_{21} & c_{22} & \dots & c_{2N} \\
\cdot & \cdot & \cdot & \cdot \\
c_{N1} & c_{N2} & \dots & c_{NN}
\end{bmatrix}.
$$

$$(13.009)$$

Once more the same rule applies; the element c_{jk} of the product square is formed by multiplying the elements of the jth row of the first square onto those of the kth column of the second square and summing. This may be expressed algebraically as follows:

$$c_{jk} = \sum_{l=1}^{N} a_{jl} b_{lk}. \qquad (13.010)$$

If the order of the factors in (13.009) is reversed, the multiplication is still possible and one has:

$$
\begin{bmatrix}
b_{11} & b_{12} & \dots & b_{1N} \\
b_{21} & b_{22} & \dots & b_{2N} \\
\cdot & \cdot & \cdot & \cdot \\
b_{N1} & b_{N2} & \dots & b_{NN}
\end{bmatrix}
\begin{bmatrix}
a_{11} & a_{12} & \dots & a_{1N} \\
a_{21} & a_{22} & \dots & a_{2N} \\
\cdot & \cdot & \cdot & \cdot \\
a_{N1} & a_{N2} & \dots & a_{NN}
\end{bmatrix}
=
\begin{bmatrix}
c'_{11} & c'_{12} & \dots & c'_{1N} \\
c'_{21} & c'_{22} & \dots & c'_{2N} \\
\cdot & \cdot & \cdot & \cdot \\
c'_{N1} & c'_{N2} & \dots & c'_{NN}
\end{bmatrix},
$$

$$(13.011)$$

where:

$$c'_{jk} = \sum_{l=1}^{N} b_{jl} a_{lk}. \qquad (13.012)$$

It is easily verified that c'_{jk} is in general not equal to c_{jk} and therefore the multiplication of square matrices is in general non-commutative. As with the operators treated in Chapter 8, one may define the commutator of two square matrices as:

$$[(A), (B)] = (A)(B) - (B)(A). \qquad (13.013)$$

Following the rule of row multiplied onto column, it is clear that a row matrix can be multiplied onto a column matrix to produce a scalar, i.e. a single number:

$$(y_1 \quad y_2 \ \ldots \ y_N) \begin{bmatrix} x_1 \\ x_2 \\ \vdots \\ x_N \end{bmatrix} = [y_1 x_1 + y_2 x_2 + \ \ldots \ + y_N x_N]. \quad (13.014)$$

A column matrix can also be multiplied onto a row matrix but the result is quite different; it is, as the reader may verify, an $N \times N$ square matrix. Moreover, it is a special type of square matrix for not every square matrix can be so formed.

The *main diagonal* of a square matrix is the set of diagonally positioned elements beginning at the upper left and ending at the lower right; in (13.006), for example, it consists of the elements a_{11}, a_{22}, ..., a_{NN}. If a matrix is modified by exchanging the positions of all the elements symmetrically with respect to the main diagonal, it is said to be *transposed* and the new matrix which is formed in this way is known as the *transpose* of the original matrix. The transpose of matrix (A) is symbolized by $(\tilde{A})$. Thus:

$$\overbrace{\begin{bmatrix} a_{11} & a_{12} & \ldots & a_{1N} \\ a_{21} & a_{22} & \ldots & a_{2N} \\ \cdot & \cdot & \cdot & \cdot \\ a_{N1} & a_{N2} & \ldots & a_{NN} \end{bmatrix}} = \begin{bmatrix} a_{11} & a_{21} & \ldots & a_{N1} \\ a_{12} & a_{22} & \ldots & a_{N2} \\ \cdot & \cdot & \cdot & \cdot \\ a_{1N} & a_{2N} & \ldots & a_{NN} \end{bmatrix}. \quad (13.015)$$

The concept of transposition also applies to a row or a column matrix; by this process, a row becomes a column and a column a row:

$$\overbrace{\begin{bmatrix} x_1 \\ x_2 \\ \vdots \\ x_N \end{bmatrix}} = (x_1 \quad x_2 \ \ldots \ x_N); \quad (13.016)$$

$$\overbrace{(y_1 \quad y_2 \ \ldots \ y_N)} = \begin{bmatrix} y_1 \\ y_2 \\ \vdots \\ y_N \end{bmatrix}. \quad (13.017)$$

Two important features of every square matrix are the *trace*, Tr (A), and the *determinant*, det (A). The trace is simply the sum of the elements on the main diagonal:

$$\text{Tr}\,(A) = a_{11} + a_{22} + \ldots + a_{NN}, \qquad (13.018)$$

and the determinant is the determinant of the array of elements defined in the usual way:

$$\det\,(A) = \begin{vmatrix} a_{11} & a_{12} & \ldots & a_{1N} \\ a_{21} & a_{22} & \ldots & a_{2N} \\ . & . & . & . \\ a_{N1} & a_{N2} & \ldots & a_{NN} \end{vmatrix}. \qquad (13.019)$$

It is interesting that, because of the nature of the process of multiplication of a matrix by a scalar, defined in (13.003), one has:

$$\det\,[s(A)] = s^N \det\,(A). \qquad (13.020)$$

The unit matrix, symbolized by (1), is defined as the matrix for which all elements on the main diagonal are unity and all others are zero:

$$(1) = \begin{bmatrix} 1 & 0 & \ldots & 0 \\ 0 & 1 & \ldots & 0 \\ . & . & . & . \\ 0 & 0 & \ldots & 1 \end{bmatrix}. \qquad (13.021)$$

Whenever this matrix is employed as a factor in a multiplication, the product is identically equal to the other factor, i.e. every matrix is unchanged after multiplication with the unit matrix. It is understood, of course, that a unit matrix with the proper value of N is used.

If det $(A) \neq 0$, the square matrix (A) has an inverse matrix, also square, written $(A)^{-1}$. The latter has the property that:

$$(A)^{-1}(A) = (A)(A)^{-1} = (1). \qquad (13.022)$$

An inverse matrix is intimately related to the process of solving a set of simultaneous linear equations. If (A) is a square matrix containing the coefficients, (X) a column matrix containing the unknowns, and

(R) another column matrix containing the known right-hand sides, the set can be written:

$$(A)(X) = (R). \tag{13.023}$$

If the inverse matrix $(A)^{-1}$ is available, it may be multiplied from the left onto both sides of (13.023) with the result that a universal solution, valid for any (R), is produced:

$$(X) = (A)^{-1}(R). \tag{13.024}$$

The matrix elements of $(A)^{-1}$ are automatically obtained when (13.023) is solved by the determinant method; the process for doing this is well known.

In quantum mechanics, the *Hermitian adjoint* of a given matrix is important. Denoted $(A)^{\dagger}$, it is defined by:

$$(A)^{\dagger} = (\widetilde{A})^*. \tag{13.025}$$

Thus, to find the Hermitian adjoint of a given matrix, one transposes the matrix and takes the complex conjugate of every element. Column matrices, row matrices, and square matrices have Hermitian adjoints, as the following examples show:

$$
\begin{bmatrix} x_1 \\ x_2 \\ \vdots \\ x_N \end{bmatrix}^{\dagger} = (x_1^* \quad x_2^* \ \ldots \ x_N^*);
$$

$$
(y_1 \quad y_2 \ \ldots \ y_N)^{\dagger} = \begin{bmatrix} y_1^* \\ y_2^* \\ \vdots \\ y_N^* \end{bmatrix};
$$

$$
\begin{bmatrix} a_{11} & a_{12} & \ldots & a_{1N} \\ a_{21} & a_{22} & \ldots & a_{2N} \\ \cdot & \cdot & \cdots & \cdot \\ a_{N1} & a_{N2} & \ldots & a_{NN} \end{bmatrix}^{\dagger} = \begin{bmatrix} a_{11}^* & a_{21}^* & \ldots & a_{N1}^* \\ a_{12}^* & a_{22}^* & \ldots & a_{N2}^* \\ \cdot & \cdot & \cdots & \cdot \\ a_{1N}^* & a_{2N}^* & \ldots & a_{NN}^* \end{bmatrix}. \tag{13.026}
$$

If a square matrix is equal to its own Hermitian adjoint, it is said to be *self-adjoint* or *Hermitian*. To be Hermitian, a matrix must have diagonal elements that are real and off-diagonal elements that form complex conjugate pairs, symmetrically disposed with respect to the main diagonal. The following is an example:

$$\begin{bmatrix} a & k+ip & l+iq \\ k-ip & b & m+ir \\ l-iq & m-ir & c \end{bmatrix},\tag{13.027}$$

where the quantities a, b, c, k, l, m, p, q, and r are real.

It is left as an exercise to show that if:

$$(G) = (A)(B)(C) \dots (K),\tag{13.028}$$

then:

$$(\widetilde{G}) = (\widetilde{K}) \dots (\widetilde{C})(\widetilde{B})(\widetilde{A}).\tag{13.029}$$

From this result it follows that:

$$(G)^{\dagger} = (K)^{\dagger} \dots (C)^{\dagger}(B)^{\dagger}(A)^{\dagger}.\tag{13.030}$$

13.02. Matrix Formulation of Quantum Mechanics

In the matrix formulation of quantum mechanics, the following concepts are especially important:

   (i) A class C of normalizable (but not necessarily normalized) functions. These functions may be literally functions of coordinates defined upon a physical space of interest as in wave mechanics or they may be more abstract objects which are considered "functions" because they can act as operands and as results in operational statements. The examples studied here illustrate the former possibility; examples illustrating the latter also occur frequently in quantum applications.

(ii) A set of operators, $\hat{A}$, $\hat{B}$, ... which can operate upon the members of C and which have the property that the results of their operation are also members of C.

(iii) One or more complete orthonormal sets of functions, chosen from C, in which any member of C may be represented. Each such set is generated by an Hermitian operator belonging to the set of operators mentioned in (ii) and is called a *basis*.

The class C of functions may be very restricted; e.g. it may be the class of all functions of θ and φ which are eigenfunctions of $\hat{L}^2$ with $l = 1$. In that event, $\hat{L}^2$, $\hat{L}_x$, $\hat{L}_y$, and $\hat{L}_z$ are typical members, but not the only members, of the set of operators mentioned in (ii) and the functions Y_1^1, Y_1^0, and Y_1^{-1} constitute a typical basis, but not the only basis, in which the members of C may be represented. In this particular example, every basis consists of three orthonormal functions and, in the associated matrix development, $N = 3$. In another example, an entirely different class of functions might be involved and N might have a different finite value or, perhaps, an infinite value. In some cases, members of C can serve as actual wave functions of quantum states, complete with time dependence. In other cases, they can serve as ψ-functions and, in still others, as factors of ψ-functions which are more conveniently studied in separation from the whole. The reader should be cognizant of the flexibility present here and, at the same time, be prepared to assemble the complete wave function of a quantum state with all the required factors when it is necessary to do so.

Let f be a member of C; the construction of a representation of f in the basis formed by the functions χ_k means that f is to be written as:

$$f = \sum_k \alpha_k \chi_k, \qquad (13.031)$$

where:

$$\alpha_k = \int \chi_k^* f \, d\tau. \qquad (13.032)$$

Here and elsewhere in this chapter, $\int \ldots d\tau$ denotes integration over the pertinent configuration space. The set of coefficients α_k constitutes the representation in question; the basis functions act as a sort of language and the set of coefficients is the expression for f in this language. In the matrix formulation of quantum mechanics, the α_k become the elements of a column matrix which, quite appropriately, will be designated by the symbol $(f)_\chi$. Thus:

$$(f)_\chi = \begin{bmatrix} \alpha_1 \\ \alpha_2 \\ \alpha_3 \\ \cdot \\ \cdot \end{bmatrix}. \qquad (13.033)$$

The subscript which indicates the basis may be omitted when it is clearly understood that an entire analysis is being carried out in a single well-known basis or when a basis-independent result is being stated. From (13.033) it follows that:

$$(f)_\chi^\dagger = (\alpha_1^* \quad \alpha_2^* \quad \alpha_3^* \ \ldots). \qquad (13.034)$$

If f is normalized, then in any basis $(f)^\dagger (f) = 1$.

At times, it is desirable to change from one basis to another. Thus if the functions φ_j constitute another basis suitable for the representation of f, one may say:

$$f = \sum_j \beta_j \varphi_j, \qquad (13.035)$$

where, as before,

$$\beta_j = \int \varphi_j^* f \, d\tau. \qquad (13.036)$$

The β_j coefficients also become the elements of a column matrix and one has:

$$(f)_\varphi = \begin{bmatrix} \beta_1 \\ \beta_2 \\ \beta_3 \\ \cdot \\ \cdot \end{bmatrix}. \qquad (13.037)$$

Substitution of (13.031) into (13.036) produces the formula which translates from one representation to the other:

$$\beta_j = \int \varphi_j^* \left(\sum_k \alpha_k \chi_k \right) d\tau$$

$$= \sum_k \left(\int \varphi_j^* \chi_k \, d\tau \right) \alpha_k. \tag{13.038}$$

This is tantamount to the multiplication of a square matrix onto the column $(f)_\chi$ to yield the column $(f)_\varphi$:

$$\begin{bmatrix} \beta_1 \\ \beta_2 \\ \beta_3 \\ \cdot \end{bmatrix} = \begin{bmatrix} U_{11} & U_{12} & U_{13} \ldots \\ U_{21} & U_{22} & U_{23} \ldots \\ U_{31} & U_{32} & U_{33} \ldots \\ \cdot & \cdot & \cdot \end{bmatrix} \begin{bmatrix} \alpha_1 \\ \alpha_2 \\ \alpha_3 \\ \cdot \end{bmatrix} \tag{13.039}$$

where the elements of the square matrix are given by:

$$U_{jk} = \int \varphi_j^* \chi_k \, d\tau. \tag{13.040}$$

Transformation from the φ basis back to the χ basis requires the substitution of (13.035) into (13.032). Thus one finds:

$$\alpha_k = \int \chi_k^* \left(\sum_j \beta_j \varphi_j \right) d\tau$$

$$= \sum_j \left(\int \chi_k^* \varphi_j \, d\tau \right) \beta_j. \tag{13.041}$$

This represents the multiplication of another matrix, say (U'), onto the column $(f)_\varphi$ to retrieve the column $(f)_\chi$. Evidently the matrix elements of (U') are:

$$U'_{kj} = \int \chi_k^* \varphi_j \, d\tau, \tag{13.042}$$

and, equally evidently, $(U') = (U)^{-1}$. Comparison of (13.040) with (13.042) shows that $U'_{kj} = U_{jk}^*$, i.e. that (U') is also equal to $(U)^\dagger$. Thus it is a simple matter to state the mathematical property which distinguishes the matrix (U):

$$(U)^{-1} = (U)^\dagger. \tag{13.043}$$

A matrix having this property is said to be *unitary*. It is easy to show that the quantity $(f)^{\dagger}(f)$ is invariant if (f) is transformed, as in (13.039), by a unitary matrix.

The transformations of the preceding paragraph were transformations of representation only in which the underlying function remained unchanged. It is now time to investigate transformations in which one function is actually changed into a different function and the whole operation is represented in a single basis. Consider, for example, an operator $\hat{B}$. The action of this operator upon f can be symbolized in two different ways:

$$\hat{B}f = f', \tag{13.044}$$

or:

$$\begin{bmatrix} B_{11} & B_{12} & B_{13} \ldots \\ B_{21} & B_{22} & B_{23} \ldots \\ B_{31} & B_{32} & B_{33} \ldots \\ \cdot & \cdot & \cdot \; \cdot \; \cdot \\ \cdot & \cdot & \cdot \; \cdot \; \cdot \end{bmatrix} \begin{bmatrix} \alpha_1 \\ \alpha_2 \\ \alpha_3 \\ \cdot \\ \cdot \end{bmatrix} = \begin{bmatrix} \alpha'_1 \\ \alpha'_2 \\ \alpha'_3 \\ \cdot \\ \cdot \end{bmatrix}. \tag{13.045}$$

Assuming that both f and f' are represented in the χ basis, one may say:

$$f = \sum_k \alpha_k \chi_k, \quad \hat{B}f = \sum_k \alpha_k \hat{B}\chi_k = f', \tag{13.046}$$

and

$$f' = \sum_j \alpha'_j \chi_j. \tag{13.047}$$

The coefficient α'_j is given by:

$$\alpha'_j = \int \chi_j^* f' \, d\tau = \int \chi_j^* \left(\sum_k \alpha_k \hat{B}\chi_k \right) d\tau. \tag{13.048}$$

This becomes:

$$\sum_k \left(\int \chi_j^* \hat{B}\chi_k \, d\tau \right) \alpha_k = \alpha'_j, \tag{13.049}$$

which is precisely equivalent to (13.045). Thus the matrix elements of $(\hat{B})$ in the χ basis are given by:

$$(B_{jk})_\chi = \int \chi_j^* \hat{B}\chi_k \, d\tau. \tag{13.050}$$

It may be noticed that whereas the representations of functions are column or row matrices, the representations of operators are always square matrices.

If the original operator $\hat{B}$ is Hermitian, (13.050) may be modified by (8.065) as follows:

$$(B_{jk})_\chi = \int \chi_k (\hat{B}\chi_j)^* \, d\tau. \tag{13.051}$$

However, by reversing the order of the indices and taking the conjugate of (13.050), one has:

$$(B_{kj})_\chi^* = \left(\int \chi_k^* \hat{B}\chi_j \, d\tau \right)^* = \int \chi_k (\hat{B}\chi_j)^* \, d\tau. \tag{13.052}$$

Evidently $(B_{jk})_\chi$ and $(B_{kj})_\chi^*$ are equal and would be so in any basis. Thus if operator $\hat{B}$ is Hermitian, the matrix $(\hat{B})$ is equal to its Hermitian adjoint and is therefore also Hermitian.

Once more the possibility of a transformation to a new basis should be considered. If the new basis is the φ basis, the representations of f and f' become:

$$\left. \begin{array}{l} (f)_\varphi = (U)(f)_\chi; \\ (f')_\varphi = (U)(f')_\chi. \end{array} \right\} \tag{13.053}$$

Here (U) is the unitary matrix of (13.039) and (13.040). One may now rewrite (13.045) in abbreviated form:

$$(\hat{B})_\chi (f)_\chi = (f')_\chi. \tag{13.054}$$

Both sides of this may be multiplied by (U) from the left; also $(U)^{-1}(U)$, which is equal to (1), may be inserted between $(\hat{B})_\chi$ and $(f)_\chi$. The result is:

$$(U)(\hat{B})_\chi (U)^{-1}(U)(f)_\chi = (U)(f')_\chi. \tag{13.055}$$

This says that:

$$(U)(\hat{B})_\chi (U)^{-1}(f)_\varphi = (f')_\psi, \tag{13.056}$$

and the transformed operator is evidently given by:

$$(\hat{B})_\varphi = (U)(\hat{B})_\chi (U)^{-1}. \tag{13.057}$$

Multiplication of $(U)^{-1}$ from the left and (U) from the right onto both sides of the above produces the corresponding transformation in the reverse direction:[†]

$$(\hat{B})_\chi = (U)^{-1} (\hat{B})_\varphi (U).$$ (13.058)

Transformations such as (13.057) and (13.058) are called *similarity transformations* and matrices such as $(\hat{B})_\chi$ and $(\hat{B})_\varphi$, which represent the same operator in different bases, are said to be *similar* even though their elements are in general quite different. The following properties of similarity transformations are stated as self-evident:

  (i) The unit matrix (1) and the null matrix[‡] (0) are invariant under similarity transformation.

 (ii) If it is true that $(\hat{B}) \neq (1)$ and $(\hat{B}) \neq (0)$ in one basis, then the same is true in every basis in which $\hat{B}$ could conceivably be represented. This follows from (i) for if $(\hat{B})$ were equal to (1) in one basis, it would be equal to (1) in every basis; if it were equal to (0) in one basis, it would be equal to (0) in every basis.

When two operators act in sequence on a given function, the result is equivalent to the action of a single operator regarded as the product of the original two. The same is true of two square matrices acting in sequence on a given column matrix. The strict parallelism between these cases is emphasized by the following theorem.

THEOREM 13.01. *Given two operators $\hat{A}$ and $\hat{B}$. In any basis suitable for the representation of both, the representation of the product $\hat{A}\hat{B}$ is equal to the matrix product of the representations of the individual operators taken in the same order, i.e. to $(\hat{A})(\hat{B})$.*

Taking χ as a mutually suitable but otherwise arbitrary basis, the statement to be proved is:

$$\int \chi_j^* \hat{A}\hat{B}\chi_k \, d\tau = \sum_l \int \chi_j^* \hat{A}\chi_l \, d\tau \int \chi_l^* \hat{B}\chi_k \, d\tau.$$ (13.059)

[†] Notice that the designations "(U)" and "$(U)^{-1}$" are interchangeable; the matrix which has been called (U) above could have been called $(U)^{-1}$ and vice versa. Thus there is no significance to the fact that the leading unitary matrix may carry the superscript "-1" in one similarity transformation whereas the trailing unitary matrix may carry it in another.

[‡] The null matrix is a matrix with all elements equal to zero.

Let the function $\hat{B}\chi_k$ be called Γ for convenience and let the representation of this function be the set of coefficients γ_l. Thus:

$$\hat{B}\chi_k = \Gamma = \sum_l \gamma_l \chi_l. \tag{13.060}$$

Then:

$$\int \chi_j^* \hat{A}\Gamma \, d\tau = \sum_l \int \chi_j^* \hat{A}\chi_l \, d\tau \int \chi_l^* \Gamma \, d\tau. \tag{13.061}$$

The final integral on the right-hand side is simply γ_l, hence:

$$\int \chi_j^* \hat{A}\Gamma \, d\tau = \int \chi_j^* \hat{A}\left(\sum_l \gamma_l \chi_l\right) d\tau. \tag{13.062}$$

This is an identity and the theorem is therefore true. It follows that the representation of the commutator of two operators is equal to the matrix commutator of the representations of the individual operators and, from the properties of similarity transformations enumerated earlier, that commutativity and non-commutativity are invariant under such transformations. This, of course, is an expected result; every operator has an independent existence which transcends the limitations of any particular representation. An intrinsic property like the commutativity of two operators should therefore be independent of the choice of basis in which the operators are represented.

A familiar feature of operators is the fact that they can have eigenfunctions and eigenvalues. Thus if f_n is an eigenfunction of $\hat{B}$ with eigenvalue B_n, $\hat{B}f_n = B_n f_n$. Written in matrix form, this becomes:

$$(\hat{B})(f_n) = B_n(f_n), \tag{13.063}$$

and the operation of square matrix $(\hat{B})$ on column matrix (f_n) is equivalent to a multiplication of (f_n) by the scalar eigenvalue B_n. This relationship is true in any basis and the column matrix (f_n) is often said to be an *eigenvector* of $(\hat{B})$. If one considers two eigenfunctions f_m and f_n of $\hat{B}$, it is a simple matter to show that in any basis:

$$\int f_m^* f_n \, d\tau = (f_m)^\dagger (f_n). \tag{13.064}$$

The product of the row and the column in this expression is called an

Hermitian product. If the eigenfunctions are orthogonal, this product will vanish and the two eigenvectors are also said to be orthogonal. It follows that if $(\hat{B})$ is Hermitian, any two of its eigenvectors are orthogonal if they belong to different eigenvalues. Even a set of degenerate eigenvectors can be orthogonalized and normalized by a process analogous to that given in Section 8.06 and if $(\hat{B})$ is an $N \times N$ Hermitian matrix, it is always possible to construct a set of N orthonormal eigenvectors (f_1), (f_2), ..., (f_N), i.e. eigenvectors which have the property that:

$$(f_m)^{\dagger} (f_n) = \delta_{nm}. \tag{13.065}$$

The matrix process for finding the expectation value of an observable, say B, may now be explored. In wave mechanics, this quantity is given by the familiar formula:

$$\langle B \rangle = \int \Psi^* \hat{B} \Psi \, d\tau. \tag{13.066}$$

If Ψ is represented in, say, the χ basis:[†]

$$\begin{aligned}
\langle B \rangle &= \int \left(\sum_j \zeta_j^* \chi_j^* \right) \hat{B} \left(\sum_k \zeta_k \chi_k \right) d\tau. \\
&= \sum_j \zeta_j^* \sum_k \left(\int \chi_j^* \hat{B} \chi_k \, d\tau \right) \zeta_k \\
&= \sum_j \zeta_j^* \sum_k (B_{jk})_{\chi} \zeta_k. \tag{13.067}
\end{aligned}$$

The matrix equivalent of this expression is easily constructed. It is:

$$\langle B \rangle = (\zeta_1^* \quad \zeta_2^* \quad \zeta_3^* \ldots) \begin{bmatrix} B_{11} & B_{12} & B_{13} & \ldots \\ B_{21} & B_{22} & B_{23} & \ldots \\ B_{31} & B_{32} & B_{33} & \ldots \\ \cdot & \cdot & \cdot & \cdot \\ \cdot & \cdot & \cdot & \cdot \end{bmatrix} \begin{bmatrix} \zeta_1 \\ \zeta_2 \\ \zeta_3 \\ \cdot \end{bmatrix} \tag{13.068}$$

which may be written more concisely as:

$$\langle B \rangle = (\Psi)^{\dagger} (\hat{B}) (\Psi). \tag{13.069}$$

[†] If the time dependence of Ψ is not vested in the basis functions χ_k, it must be vested in the coefficients ζ_k.

23*

Since this expression is true in any basis, the subscripts have been omitted. It is easy to show, strictly by matrix methods, that if $(\hat{B})$ is Hermitian, $\langle B \rangle$ must be real.

The Hamiltonian operator $\mathcal{H}$ has its matrix $(\mathcal{H})$. If the time dependence is in the representation rather than in the basis, one may write the Schrödinger equation as follows:

$$(\mathcal{H})(\Psi) = i\hbar \frac{\partial}{\partial t}(\Psi). \qquad (13.070)$$

If $(\Psi) = (\Psi_n)$, a stationary state, then:

$$(\Psi_n) = (\psi_n)\exp(-i\omega_n t), \qquad (13.071)$$

and (ψ_n) is an eigenvector of $(\mathcal{H})$ with eigenvalue $E_n = \hbar\omega_n$:

$$(\mathcal{H})(\psi_n) = E_n(\psi_n). \qquad (13.072)$$

By now it should be evident that each one of the operator expressions of wave mechanics has its analog in matrix mechanics; all that is needed is a suitable basis for the required representations. Thus, as stated in Section 6.01, the two branches of quantum mechanics are theoretically equivalent. As intimated at the beginning of this section, however, matrix mechanics is also able to deal with certain abstract operators and eigenfunctions which have no analog in wave mechanics. The operators and eigenfunctions for the intrinsic angular momentum (spin) of many atomic and subatomic particles are prime examples. In any event, an advantage is sometimes to be gained by working in the matrix branch because of the effectiveness with which numerical techniques can be brought to bear upon matrix expressions.

13.03. Eigenvalues and Eigenvectors; the Diagonalization of a Matrix

The central problem of matrix mechanics is that of finding the N eigenvalues and the N eigenvectors of a given $N \times N$ matrix. To illustrate the technique, a numerical example will be used; suppose that one is given the following 3×3 matrix which represents the Hermitian operator $\hat{B}$ in the χ basis:

$$(\hat{B}_\chi) = \begin{bmatrix} 2 & -2i & 2i \\ 2i & 1 & 0 \\ -2i & 0 & 3 \end{bmatrix}. \tag{13.073}$$

It is required to find the three eigenvalues B_1, B_2, and B_3 and the corresponding eigenvectors:

$$(f_1)_\chi = \begin{bmatrix} \alpha_{11} \\ \alpha_{21} \\ \alpha_{31} \end{bmatrix}; \quad (f_2)_\chi = \begin{bmatrix} \alpha_{12} \\ \alpha_{22} \\ \alpha_{32} \end{bmatrix}; \quad (f_3)_\chi = \begin{bmatrix} \alpha_{13} \\ \alpha_{23} \\ \alpha_{33} \end{bmatrix}. \tag{13.074}$$

Notice that double subscripts are needed for the elements of the eigenvectors. The set of equations to be satisfied is:

$$\begin{bmatrix} 2 & -2i & 2i \\ 2i & 1 & 0 \\ -2i & 0 & 3 \end{bmatrix} \begin{bmatrix} \alpha_{1n} \\ \alpha_{2n} \\ \alpha_{3n} \end{bmatrix} = B_n \begin{bmatrix} \alpha_{1n} \\ \alpha_{2n} \\ \alpha_{3n} \end{bmatrix}, \tag{13.075}$$

where $n = 1, 2, 3$. This is equivalent to $[(\hat{B})_\chi - B_n(1)] (f_n)_\chi = 0$ which may be expressed in expanded form as follows:

$$\begin{bmatrix} 2-B_n & -2i & 2i \\ 2i & 1-B_n & 0 \\ -2i & 0 & 3-B_n \end{bmatrix} \begin{bmatrix} \alpha_{1n} \\ \alpha_{2n} \\ \alpha_{3n} \end{bmatrix} = 0. \tag{13.076}$$

As discussed in Section 7.04, a non-trivial solution exists only if the determinant of the square matrix in the above vanishes. The equation expressing this requirement is called the *secular equation*, an algebraic equation of degree N in B_n as an unknown. In the present example, the secular equation is the cubic:

$$\begin{vmatrix} 2-B_n & -2i & 2i \\ 2i & 1-B_n & 0 \\ -2i & 0 & 3-B_n \end{vmatrix} = -10 - 3B_n + 6B_n^2 - B_n^3 = 0. \tag{13.077}$$

The roots of this equation are:

$$B_1 = -1; \quad B_2 = 2; \quad B_3 = 5. \quad (13.078)$$

These are the three eigenvalues; they must be real because the operator $\hat{B}$ is Hermitian.[†] It is usually best to list the B_n in order of increasing value, as has been done.

To find the eigenvector $(f_1)_\chi$ belonging to the eigenvalue B_1, one must go back to (13.076) and insert B_1 for B_n. The resulting equations now have definite coefficients:

$$\begin{bmatrix} 3 & -2i & 2i \\ 2i & 2 & 0 \\ -2i & 0 & 2 \end{bmatrix} \begin{bmatrix} \alpha_{11} \\ \alpha_{21} \\ \alpha_{31} \end{bmatrix} = 0. \quad (13.079)$$

These equations have, by design, a null determinant. They are therefore non-independent and, in general, any two of them can be used in the subsequent step. Regardless of which two are chosen, only the ratios among the α_{j1} can be found. Let the second and third of these equations be selected. Then:

$$\left.\begin{array}{r} 2i\alpha_{11} + 2\alpha_{21} \qquad = 0; \\ -2i\alpha_{11} \qquad + 4\alpha_{31} = 0. \end{array}\right\} \quad (13.080)$$

These yield the following results:

$$\left.\begin{array}{l} \alpha_{21} = -\ i\alpha_{11}; \\ \alpha_{31} = \ \tfrac{1}{2}i\alpha_{11}. \end{array}\right\} \quad (13.081)$$

It is now possible to find the actual values of the α_{j1} (up to an arbitrary common phase factor) by the requirement of normalization. Thus:

$$|\alpha_{11}|^2 + |\alpha_{21}|^2 + |\alpha_{31}|^2 = |\alpha_{11}|^2 + |\alpha_{11}|^2 + \tfrac{1}{4}|\alpha_{11}|^2$$
$$= \tfrac{9}{4}|\alpha_{11}|^2 = 1. \quad (13.082)$$

[†] If it is known that the roots are real, the difficulty of finding these quantities is greatly reduced. Approximate graphical location followed by numerical refinement to any desired degree of accuracy becomes a workable method. The example used here, contrived for ease of presentation, is exceptional; usually the roots are irrational numbers requiring lengthy decimal expressions.

If the phase angle of α_{11} is chosen equal to zero, one finds that $\alpha_{11} = \frac{2}{3}$, $\alpha_{21} = -i\frac{2}{3}$, and $\alpha_{31} = i\frac{1}{3}$. To obtain the second eigenvector, let B_2 be substituted into (13.076):

$$\begin{bmatrix} 0 & -2i & 2i \\ 2i & -1 & 0 \\ -2i & 0 & 1 \end{bmatrix} \begin{bmatrix} \alpha_{12} \\ \alpha_{22} \\ \alpha_{32} \end{bmatrix} = 0. \qquad (13.083)$$

Repetition of the previous procedure yields $\alpha_{12} = \frac{1}{3}$, $\alpha_{22} = i\frac{2}{3}$, and $\alpha_{32} = i\frac{2}{3}$. Finally, to find the third eigenvector, let B_3 be substituted into (13.076):

$$\begin{bmatrix} -3 & -2i & 2i \\ 2i & -4 & 0 \\ -2i & 0 & -2 \end{bmatrix} \begin{bmatrix} \alpha_{13} \\ \alpha_{23} \\ \alpha_{33} \end{bmatrix} = 0. \qquad (13.084)$$

The process now gives $\alpha_{13} = \frac{2}{3}$, $\alpha_{23} = i\frac{1}{3}$, and $\alpha_{33} = -i\frac{2}{3}$. To summarize:

$$(f_1)_\chi = \begin{bmatrix} \frac{2}{3} \\ -i\frac{2}{3} \\ i\frac{1}{3} \end{bmatrix}; \qquad (f_2)_\chi = \begin{bmatrix} \frac{1}{3} \\ i\frac{2}{3} \\ i\frac{2}{3} \end{bmatrix}; \qquad (f_3)_\chi = \begin{bmatrix} \frac{2}{3} \\ i\frac{1}{3} \\ -i\frac{2}{3} \end{bmatrix}. \qquad (13.085)$$

Since no two of the eigenvalues are equal, these eigenvectors constitute a non-degenerate set and are therefore orthogonal. Since they were deliberately normalized, they are orthonormal, i.e. they satisfy (13.065) as the reader may readily verify.

The finding of the eigenvalues and the eigenvectors does not exhaust the interesting features of an analysis of this type. Consider the matrix (R) formed by juxtaposing the three eigenvectors:

$$(R) = \begin{bmatrix} \alpha_{11} & \alpha_{12} & \alpha_{13} \\ \alpha_{21} & \alpha_{22} & \alpha_{23} \\ \alpha_{31} & \alpha_{32} & \alpha_{33} \end{bmatrix}. \qquad (13.086)$$

Since the columns of (R) are orthonormal, the Hermitian adjoint of (R), if multiplied onto (R), will produce the unit matrix. It follows that:

$$(R)^\dagger = \begin{bmatrix} \alpha_{11}^* & \alpha_{21}^* & \alpha_{31}^* \\ \alpha_{12}^* & \alpha_{22}^* & \alpha_{32}^* \\ \alpha_{13}^* & \alpha_{23}^* & \alpha_{33}^* \end{bmatrix} = (R)^{-1}, \qquad (13.087)$$

and that (R) is unitary. The matrix $(R)^{-1}$ has the power to transform from the χ basis to a new basis in which the eigenvectors have unusually simple forms. This new basis will be called the ξ basis. Thus:

$$\left. \begin{aligned} (R)^{-1}(f_1)_\chi = (f_1)_\xi = \begin{bmatrix} 1 \\ 0 \\ 0 \end{bmatrix}; \\[2mm] (R)^{-1}(f_2)_\chi = (f_2)_\xi = \begin{bmatrix} 0 \\ 1 \\ 0 \end{bmatrix}; \\[2mm] (R)^{-1}(f_3)_\chi = (f_3)_\xi = \begin{bmatrix} 0 \\ 0 \\ 1 \end{bmatrix}. \end{aligned} \right\} \qquad (13.088)$$

From these, it follows that:

$$f_1 = \xi_1, \quad f_2 = \xi_2, \quad f_3 = \xi_3, \qquad (13.089)$$

and the functions ξ_n, which form the ξ basis, are simply the eigenfunctions of $\hat{B}$. If (13.075) is transformed over to the ξ basis, one has:

$$(R)^{-1}(\hat{B})_\chi(R)(R)^{-1}(f_n)_\chi = B_n(R)^{-1}(f_n)_\chi. \qquad (13.090)$$

Here $(R)^{-1}(\hat{B})_\chi(R)$ can be recognized as $(\hat{B})_\xi$, the representation of $\hat{B}$ in the ξ basis. Since the columns of (R) are the eigenvectors of $(\hat{B})_\chi$, the operation $(\hat{B})_\chi(R)$ yields:

$$(\hat{B})_\chi(R) = \begin{bmatrix} B_1\alpha_{11} & B_2\alpha_{12} & B_3\alpha_{13} \\ B_1\alpha_{21} & B_2\alpha_{22} & B_3\alpha_{23} \\ B_1\alpha_{31} & B_2\alpha_{32} & B_3\alpha_{33} \end{bmatrix}. \qquad (13.091)$$

If $(R)^{-1}$ is now allowed to act upon the above, it will produce the unit column vectors of (13.088), each multiplied by its respective eigenvalue. The result is:

$$(R)^{-1}(B)_\chi(R) = \begin{bmatrix} B_1 & 0 & 0 \\ 0 & B_2 & 0 \\ 0 & 0 & B_3 \end{bmatrix} = \begin{bmatrix} -1 & 0 & 0 \\ 0 & 2 & 0 \\ 0 & 0 & 5 \end{bmatrix} = (B)_\xi. \quad (13.092)$$

Thus, in the basis formed by its eigenfunctions, an operator assumes the so-called *diagonal form* in which the diagonal elements are the eigenvalues and all other elements are zero. In this special basis, the set of equations (13.075) becomes very simple, e.g.:

$$\begin{bmatrix} B_1 & 0 & 0 \\ 0 & B_2 & 0 \\ 0 & 0 & B_3 \end{bmatrix} \begin{bmatrix} 1 \\ 0 \\ 0 \end{bmatrix} = B_1 \begin{bmatrix} 1 \\ 0 \\ 0 \end{bmatrix}, \text{ etc.} \quad (13.093)$$

In view of the preceding discussion and of other remarks, it is evident that an operator $\hat{B}$, its eigenvalues B_n, and its eigenfunctions f_n constitute a set of independently existing entities whose interrelationships are maintained regardless of the basis in which $\hat{B}$ and the f_n are represented. The eigenvalues, in particular, are precisely the same in every basis, i.e. they are invariant under similarity transformation, and in some sense the inner nature of an operator is portrayed by its set of eigenvalues. The eigenvalues are the roots of the secular equation, hence every coefficient of this equation is also an invariant. The coefficient of B_n^{N-1} and the coefficient of B_n^0 are two especially well-known examples equal to $(-1)^{N-1} \operatorname{Tr}(\hat{B})$ and to $\det(\hat{B})$, respectively. The other invariant coefficients can also be expressed in relatively simple ways.[45]

If the operator $\hat{B}$ has a degenerate subset of eigenvalues, the matrix $(\hat{B})$ will also have such a subset and the eigenvectors belonging to this subset will not necessarily be orthogonal. As mentioned earlier, however, they can be readily orthogonalized; an example requiring a procedure of this type is included as an exercise.

This section concludes with some remarks about the simultaneous diagonalization of two matrices.

THEOREM 13.02. *If $\hat{A}$ and $\hat{B}$ do not commute, then there is no basis in which $(\hat{A})$ and $(\hat{B})$ are both diagonal.*

To prove this, let the contrary be assumed; let χ be a basis in which $(\hat{A})$ and $(\hat{B})$ are both diagonal. If the two matrices are diagonal, however, they commute and if they commute in χ, they commute in every basis. But this can happen only if the corresponding operators $\hat{A}$ and $\hat{B}$ commute, which contradicts the hypothesis. The theorem is therefore true.

The converse of Theorem 13.02 is stated separately; it is more difficult to prove.

THEOREM 13.03. *If $\hat{A}$ and $\hat{B}$ commute, then there is at least one basis in which $(\hat{A})$ and $(\hat{B})$ are both diagonal.*

Obviously there is a basis in which $(\hat{A})$ is diagonal; let this be the φ basis. In φ, one therefore has:

$$(\hat{A})_\varphi = \begin{bmatrix} A_{11} & 0 & 0 & 0 & \dots \\ 0 & A_{22} & 0 & 0 & \dots \\ 0 & 0 & A_{33} & 0 & \dots \\ 0 & 0 & 0 & A_{44} & \dots \\ \cdot & \cdot & \cdot & \cdot & \cdot \end{bmatrix}. \tag{13.094}$$

One might also have:

$$(\hat{B})_\varphi = \begin{bmatrix} B_{11} & B_{12} & B_{13} & B_{14} & \dots \\ B_{21} & B_{22} & B_{23} & B_{24} & \dots \\ B_{31} & B_{32} & B_{33} & B_{34} & \dots \\ B_{41} & B_{42} & B_{43} & B_{44} & \dots \\ \cdot & \cdot & \cdot & \cdot & \cdot \end{bmatrix} \tag{13.095}$$

It is a simple matter to show that the commutator of these matrices is given by:

$$[(\hat{A}), (\hat{B})]_\varphi = \begin{bmatrix} 0 & [A_{11}-A_{22}]B_{12} & [A_{11}-A_{33}]B_{13} & \dots \\ [A_{22}-A_{11}]B_{21} & 0 & [A_{22}-A_{33}]B_{23} & \dots \\ [A_{33}-A_{11}]B_{31} & [A_{33}-A_{22}]B_{32} & 0 & \dots \\ \cdot & \cdot & \cdot & \cdot \end{bmatrix}$$

$$\tag{13.096}$$

This is zero by hypothesis, therefore:

$$[A_{jj} - A_{kk}]B_{jk} = 0 \quad \text{for all } j \text{ and } k. \tag{13.097}$$

If all the eigenvalues of $(\hat{A})$ are distinct, i.e. if there is no degeneracy in $(\hat{A})$, then it follows from (13.097) that all the off-diagonal elements of $(\hat{B})$ must vanish and that $(\hat{B})$ also must be diagonal in φ. This, of course, satisfies the theorem. Suppose, however, that two of the eigenvalues of $(\hat{A})$, say A_{11} and A_{22}, are degenerate and are equal to the same number A'. Then B_{12} and B_{21} need not be zero and the two matrices will in general assume the following forms in φ:

$$\begin{bmatrix} A' & 0 & 0 & 0 & \dots \\ 0 & A' & 0 & 0 & \dots \\ 0 & 0 & A_{33} & 0 & \dots \\ 0 & 0 & 0 & A_{44} & \dots \\ \cdot & \cdot & \cdot & \cdot & \cdot \end{bmatrix}; \quad \begin{bmatrix} B_{11} & B_{12} & 0 & 0 & \dots \\ B_{21} & B_{22} & 0 & 0 & \dots \\ 0 & 0 & B_{33} & 0 & \dots \\ 0 & 0 & 0 & B_{44} & \dots \\ \cdot & \cdot & \cdot & \cdot & \cdot \end{bmatrix}.$$

$$\tag{13.098}$$

It is now possible to make another similarity transformation to a basis, say μ, in which $(\hat{B})$ is entirely diagonal. Such a transformation will be mediated by a unitary matrix of the form:

$$(R) = \begin{bmatrix} \alpha_{11} & \alpha_{12} & 0 & 0 & \dots \\ \alpha_{21} & \alpha_{22} & 0 & 0 & \dots \\ 0 & 0 & 1 & 0 & \dots \\ 0 & 0 & 0 & 1 & \dots \\ \cdot & \cdot & \cdot & \cdot & \cdot \end{bmatrix}. \tag{13.099}$$

This transformation must leave $(\hat{A})_\varphi$ unchanged because the sub-matrix,

$$\begin{pmatrix} A' & 0 \\ 0 & A' \end{pmatrix} = A'(1), \tag{13.100}$$

is invariant under similarity transformation. In the μ basis, therefore, both matrices are diagonal and the theorem is true.

Theorems 13.02 and 13.03 together assert that if and only if two operators such as $\hat{A}$ and $\hat{B}$ commute is there a basis μ in which their matrix representations $(\hat{A})$ and $(\hat{B})$ are both diagonal. If they do commute, then, there is a common similarity transformation from the basis in which they were originally represented to the μ basis. Since the unitary matrix which effects this transformation is composed of juxtaposed eigenvectors, it follows that these eigenvectors must be common to both $(\hat{A})$ and $(\hat{B})$. This confirms what was already shown in wave mechanics, namely that commuting operators can have common eigenstates; in each such state, the corresponding observables simultaneously have definite values.

13.04. Solution of a Quantum Mechanical Problem by Matrix Methods

For this section, a simple but illustrative one-dimensional problem which demonstrates the power of matrix methods has been selected. Specifically, the problem is to find the ground-state energy and the

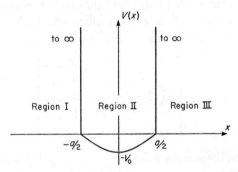

FIG. 13.01. An infinite well with non-constant potential energy in region II.

ground-state wave function of the infinite well shown in Fig. 13.01 in which the potential energy is given by:

$$V(x) = \begin{cases} -V_0 \cos(\pi x/a) & \text{in} \quad -(a/2) < x < (a/2); \\ \infty & \text{elsewhere.} \end{cases} \quad (13.101)$$

The procedure to be used can also serve to determine the energies and the wave functions of the excited states, as will be indicated. The class C of functions in this context consists of all the single-valued well-behaved functions of x which vanish outside the interval $-(a/2) < x < (a/2)$. As a basis, the ψ_n functions of the infinite rectangular well of Section 7.04 will be chosen and will be denoted χ_j. Thus:

$$
\left.
\begin{aligned}
\chi_0 &= (2/a)^{\frac{1}{2}} \cos (\pi x/a); \\
\chi_1 &= (2/a)^{\frac{1}{2}} \sin (2\pi x/a); \\
\chi_2 &= (2/a)^{\frac{1}{2}} \cos (3\pi x/a); \\
&\cdot \quad \cdot \quad \cdot \quad \cdot \quad \cdot \quad \cdot \quad \cdot \quad \cdot
\end{aligned}
\right\}
\qquad (13.102)
$$

These functions belong to C since they are understood to vanish outside the interval stated above. They are non-degenerate eigenfunctions of the Hamiltonian for the infinite rectangular well, $-(\hbar^2/2m) \partial^2/\partial x^2$, which consists of only the kinetic energy term and they form an orthonormal set eminently well suited to the purposes at hand.

The ground-state function ψ_0 for the well of Fig. 13.01 will theoretically require an infinite number of terms when represented in the χ basis:

$$
\psi_0 = \sum_{j=0}^{\infty} \alpha_{j0} \chi_j . \qquad (13.103)
$$

Analogously, $(\psi_0)_\chi$ will be an infinite column matrix with the α_{j0} as its elements. In spite of the infinite lengths of these expressions, it is intuitively evident that ψ_0 will not be a highly structured function; it will, in fact, resemble the function χ_0 and will require for its representation a relatively large coefficient α_{00} together with coefficients of higher order which are much smaller and which decrease rapidly with increasing j. The method to be used capitalizes upon this and assumes that ψ_0 can be adequately represented in a finite number of terms. For illustrative purposes three terms will be chosen although, for very accurate work, a larger number would be required. One therefore begins by saying:

$$
\psi_0 = \alpha_{00} \chi_0 + \alpha_{10} \chi_1 + \alpha_{20} \chi_2 . \qquad (13.104)
$$

This corresponds to the column matrix of three elements:

$$(\psi_0)_\chi = \begin{bmatrix} \alpha_{00} \\ \alpha_{10} \\ \alpha_{20} \end{bmatrix}. \tag{13.105}$$

The task at hand can now be outlined very simply; first one must represent the Hamiltonian of the well of Fig. 13.01 as a 3×3 square matrix in the χ basis, then find its lowest eigenvalue and the eigenvector belonging thereto. The Hamiltonian in question is given by:

$$\hat{\mathcal{H}} = -\frac{\hbar^2}{2m}\frac{\partial^2}{\partial x^2} + V(x), \tag{13.106}$$

and, to find its χ representation, it is convenient to deal with the kinetic energy and the potential energy separately. Thus:

$$(T_{jk})_\chi = \int_{-a/2}^{a/2} \chi_j^* \left(-\frac{\hbar^2}{2m}\frac{\partial^2}{\partial x^2}\right)\chi_k \, dx. \tag{13.107}$$

Since the χ_j are eigenfunctions of $\hat{T}$, the corresponding matrix is diagonal; it is easily found to be:

$$(\hat{T})_\chi = E_{0r}\begin{bmatrix} 1 & 0 & 0 \\ 0 & 4 & 0 \\ 0 & 0 & 9 \end{bmatrix}, \tag{13.108}$$

where E_{0r}, a convenient constant equal to the ground-state energy of the infinite rectangular well, is given by:

$$E_{0r} = \frac{\hbar^2\pi^2}{2ma^2}. \tag{13.109}$$

The matrix elements of the potential energy are:

$$(V_{jk})_\chi = \int_{-a/2}^{a/2} \chi_j^*[-V_0 \cos(\pi x/a)]\chi_k \, dx. \tag{13.110}$$

Therefore:

$$(V_{00})_\chi = -2V_0(2/a) \int_0^{a/2} \cos^3(\pi x/a)\, dx = \frac{-V_0}{\pi}\left(\frac{8}{3}\right). \quad (13.111)$$

It is easily seen that the elements $(V_{jk})_\chi$ with one even and one odd index must vanish and, after some effort, the complete matrix is found to be:

$$(\hat{V})_\chi = \frac{-V_0}{\pi}\begin{bmatrix} \frac{8}{3} & 0 & \frac{8}{15} \\ 0 & \frac{32}{15} & 0 \\ \frac{8}{15} & 0 & \frac{72}{35} \end{bmatrix}. \quad (13.112)$$

One cannot proceed further without knowing the precise value of V_0 in terms of, say, E_{0r}. Since this is strictly an illustrative example, a value of V_0 will be chosen which incorporates certain numerical conveniences and, at the same time, makes the potential energy of Fig. 13.01 reasonably different from that of the infinite rectangular well. This value is:

$$V_0 = 0.525\pi E_{0r} = 1.6493\, E_{0r}. \quad (13.113)$$

Thus the depression of the bottom of the well of Fig. 13.01 is slightly greater than one and one-half times the ground state energy of the infinite rectangular well. The complete Hamiltonian matrix may now be expressed as follows:

$$(\hat{\mathscr{H}})_\chi = E_{0r}\left\{ \begin{bmatrix} 1 & 0 & 0 \\ 0 & 4 & 0 \\ 0 & 0 & 9 \end{bmatrix} - \begin{bmatrix} 1.400 & 0 & 0.280 \\ 0 & 1.120 & 0 \\ 0.280 & 0 & 1.080 \end{bmatrix} \right\}$$

$$= E_{0r}\begin{bmatrix} -0.400 & 0 & -0.280 \\ 0 & 2.880 & 0 \\ -0.280 & 0 & 7.920 \end{bmatrix}. \quad (13.114)$$

The eigenvalues may now be found by the methods of Section 13.03. They are:

$$\left.\begin{aligned} E_0 &= -0.4094 E_{0r}; \\ E_1 &= 2.8800 E_{0r}; \\ E_2 &= 7.9294 E_{0r}. \end{aligned}\right\} \quad (13.115)$$

The lowest of these is of interest here and the corresponding normalized eigenvector is given by:

$$(\psi_0)_\chi = \begin{bmatrix} 0.9994 \\ 0.0000 \\ 0.0336 \end{bmatrix}. \tag{13.116}$$

As expected, the ground state wave function is composed of a relatively large amount of χ_0 and a much smaller amount of χ_2. The odd-parity χ_1 is missing entirely. The ground-state energy, $-0.4094E_{0r}$, lies above the very bottom of the well by $1.2399E_{0r}$ which means that, with respect to the bottom of the well as a datum, E_0 is higher than the corresponding energy for the infinite rectangular well. This is reasonable since the higher values of V on either side of the lowest point of the well of Fig. 13.01 tend to raise the energies of the stationary states.

The energy of the first excited state ψ_1 is not well represented by E_1 in (13.115) since the use of a matrix of such limited size is tantamount to the assumption that $\psi_1 = \chi_1$, i.e. that none of the higher order odd-parity basis functions are required in the representation of ψ_1. To remedy this, one should repeat the process with, say, the following assumed form for the representation of ψ_1:

$$\psi_1 = \alpha_{11}\chi_1 + \alpha_{21}\chi_2 + \alpha_{31}\chi_3, \tag{13.117}$$

and then construct and diagonalize another 3×3 Hamiltonian matrix appropriate to the above expression. The result will be a rather large value for α_{11}, a much smaller value for α_{31}, and a zero value for α_{21} for reasons of parity.

By now it can be appreciated that it is wasteful of effort to include terms of more than one parity in the assumed representation of any one of the stationary states. To find ψ_6, for example, one might well begin by saying:

$$\psi_6 = \alpha_{46}\chi_4 + \alpha_{66}\chi_6 + \alpha_{86}\chi_8. \tag{13.118}$$

The associated Hamiltonian matrix would contain no zeros and would be more difficult to diagonalize than those invoked previously in this section. It would give a good value of E_6, however, and a corresponding eigenvector $(\psi_6)_\chi$ in which α_{66} would be nearly unity and the other two elements, although much smaller, would not be zero.

It is hoped that this simple example, for which calculations can easily be envisaged at any desired level of accuracy, will demonstrate the very significant possibilities of matrix mechanics. To acquire the same information by wave mechanics, one would have to solve a rather sophisticated differential equation even though the potential energy function of Fig. 13.01 is relatively uncomplicated. The forms of the stationary-state solutions thus produced would probably be less easy to work with than the expansions (in a simple set of basis functions) which are obtained through the application of matrix methods.

APPENDIX A

ELECTROMAGNETIC INTERACTION ENERGIES IN TERMS OF LOCAL POTENTIALS

THE electric and magnetic interaction energies between a charged particle with fields $\mathbf{E}'$ and $\mathbf{B}'$ and a given electromagnetic field symbolized by $\mathbf{E}$ and $\mathbf{B}$ are:

$$\left.\begin{array}{l} U_{Ei} = \varepsilon_0 \int \mathbf{E}' \cdot \mathbf{E} \, d\tau; \\ U_{Bi} = \mu_0^{-1} \int \mathbf{B}' \cdot \mathbf{B} \, d\tau. \end{array}\right\} \tag{A.01}$$

Here $d\tau$ is a volume element and the integrals are taken over all of space. To transform the expression for U_{Ei}, one may employ $\mathbf{E} = -\nabla\varphi - \partial\mathbf{A}/\partial t$. Then:

$$U_{Ei} = -\varepsilon_0 \int \mathbf{E}' \cdot \nabla\varphi \, d\tau - \varepsilon_0 \int \mathbf{E}' \cdot (\partial\mathbf{A}/\partial t) \, d\tau. \tag{A.02}$$

Using the formula $\nabla \cdot (\varphi\mathbf{E}') = \varphi\nabla \cdot \mathbf{E}' + \mathbf{E}' \cdot \nabla\varphi$ and the Maxwell equation $\nabla \cdot \mathbf{E}' = \varepsilon_0^{-1} \varrho$, the above becomes:

$$U_{Ei} = -\varepsilon_0 \int \nabla \cdot (\varphi\mathbf{E}') \, d\tau + \int \varphi\varrho \, d\tau - \varepsilon_0 \int \mathbf{E}' \cdot (\partial\mathbf{A}/\partial t) \, d\tau. \tag{A.03}$$

By the divergence theorem, the first integral can be transformed to a surface integral of $\varphi\mathbf{E}' \cdot \mathbf{n}$ on the sphere at infinity where $\mathbf{n}$ is the unit outward normal; with reasonable assumptions about the asymptotic behavior of φ and $\mathbf{E}'$ at infinity, this integral vanishes. The charge density of the particle is ϱ and, as the latter is concentrated into a point, the second integral in (A.03) reduces to the total charge times the local value of the scalar potential. The electric interaction energy therefore becomes:

$$U_{Ei} = q\varphi - \varepsilon_0 \int \mathbf{E}' \cdot (\partial \mathbf{A}/\partial t) \, d\tau. \qquad (A.04)$$

The expression for U_{Bi} may be rewritten in terms of the vector potential as:

$$U_{Bi} = \mu_0^{-1} \int \mathbf{B}' \cdot (\nabla \times \mathbf{A}) d\tau. \qquad (A.05)$$

Using the formula $\nabla \cdot (\mathbf{A} \times \mathbf{B}') = \mathbf{B}' \cdot (\nabla \times \mathbf{A}) - \mathbf{A} \cdot (\nabla \times \mathbf{B}')$ and the Maxwell equation $\mu_0^{-1} \nabla \times \mathbf{B}' = \mathbf{J} + \varepsilon_0 \partial \mathbf{E}'/\partial t$, this becomes:

$$U_{Bi} = \mu_0^{-1} \int \nabla \cdot (\mathbf{A} \times \mathbf{B}') \, d\tau + \int \mathbf{A} \cdot \mathbf{J} \, d\tau + \varepsilon_0 \int \mathbf{A} \cdot (\partial \mathbf{E}'/\partial t) \, d\tau. \qquad (A.06)$$

Again the first integral vanishes by reasoning similar to that used earlier. The current density of the particle is $\mathbf{J}$ and, as the latter is concentrated into a point, the second integral reduces to the scalar product of $q\dot{\mathbf{r}}$ with the local value of the vector potential. The magnetic interaction energy therefore becomes:

$$U_{Bi} = q\dot{\mathbf{r}} \cdot \mathbf{A} + \varepsilon_0 \int \mathbf{A} \cdot (\partial \mathbf{E}'/\partial t) \, d\tau. \qquad (A.07)$$

APPENDIX B

CANONICITY OF THE TRANSFORMATION GENERATED BY $G_b(q_j, P_j, t)$

THE object of this appendix is to demonstrate that the transformation generated by $G_b(q_j, p_j, t)$ and described by the formulas (3.016), (3.017), and (3.018), is canonical, i.e. that if Hamilton's equations are fulfilled with respect to q_j, p_j, t, and $\mathcal{H}$, then — under this transformation — they are also fulfilled with respect to Q_j, P_j, t, and $\mathcal{K}$. The proof to be used is based upon a recent article by Ludford and Yannitell[46] and is remarkable in that no variational principle is invoked.

For simplicity, the demonstration will be carried out for a system with but one degree of freedom; extension to more degrees of freedom is not excessively difficult. Thus there are only two old variables q and p, and only two new variables Q and P; $G_b = G_b(q, P, t)$. Partial derivatives of G_b taken in a regime in which q and P are independent will be marked by primes; in most of the development, however, the partial derivatives are taken in a regime in which q and p are independent and these will have no special distinguishing mark. The transformation formulas are:

$$p = \left(\frac{\partial G_b}{\partial q}\right)'; \qquad Q = \left(\frac{\partial G_b}{\partial P}\right)', \qquad \mathcal{K} = \mathcal{H} + \mathcal{R}, \qquad \text{(B.01)}$$

where:

$$\mathcal{R} = {}^r\!\left(\frac{\partial G_b}{\partial t}\right)'. \qquad \text{(B.02)}$$

The differential of G_b may be written:

$$dG_b = \left(\frac{\partial G_b}{\partial q}\right)' dq + \left(\frac{\partial G_b}{\partial P}\right)' dP + \mathcal{R} dt. \qquad \text{(B.03)}$$

Substitution from the transformation and expansion of $dP(q_j, p_j, t)$ produces:

$$dG_b = p\, dq + Q\left[\frac{\partial P}{\partial q} dq + \frac{\partial P}{\partial p} dp + \frac{\partial P}{\partial t} dt\right] + \mathcal{R} dt. \qquad \text{(B.04)}$$

From this, one may identify:

$$\frac{\partial G_b}{\partial q} = p + Q\frac{\partial P}{\partial q}; \qquad \text{(B.05)}$$

$$\frac{\partial G_b}{\partial p} = Q\frac{\partial P}{\partial p}; \qquad \text{(B.06)}$$

$$\frac{\partial G_b}{\partial t} = \mathcal{R} + Q\frac{\partial P}{\partial t}. \qquad \text{(B.07)}$$

Equating the partial derivative of (B.05) with respect to p and the partial derivative of (B.06) with respect to q, one obtains:

$$1 + \frac{\partial Q}{\partial p}\frac{\partial P}{\partial q} + Q\frac{\partial^2 P}{\partial q\, \partial p} = \frac{\partial Q}{\partial q}\frac{\partial P}{\partial p} + Q\frac{\partial^2 P}{\partial p\, \partial q}. \qquad \text{(B.08)}$$

This becomes:

$$\frac{\partial Q}{\partial q}\frac{\partial P}{\partial p} - \frac{\partial P}{\partial q}\frac{\partial Q}{\partial p} = 1, \qquad \text{(B.09)}$$

where the expression on the left is known as the *Lagrange bracket*[†] of q and p with respect to Q and P. From (B.05), (B.06), and (B.07), the following additional Lagrange bracket expressions are readily derived:

$$\frac{\partial Q}{\partial t}\frac{\partial P}{\partial q} - \frac{\partial P}{\partial t}\frac{\partial Q}{\partial q} = \frac{\partial \mathcal{R}}{\partial q}; \qquad \text{(B.10)}$$

$$\frac{\partial Q}{\partial t}\frac{\partial P}{\partial p} - \frac{\partial P}{\partial t}\frac{\partial Q}{\partial p} = \frac{\partial \mathcal{R}}{\partial p}. \qquad \text{(B.11)}$$

[†] A Lagrange bracket involves a summation over all the degrees of freedom but since only one degree of freedom is considered here, no summation is required.

By (B.01), one has:

$$\frac{\partial \mathcal{K}}{\partial q} = \frac{\partial \mathcal{H}}{\partial q} + \frac{\partial \mathcal{R}}{\partial q}, \tag{B.12}$$

which may be written:

$$\frac{\partial \mathcal{K}}{\partial q} = \frac{\partial Q}{\partial t}\frac{\partial P}{\partial q} - \frac{\partial P}{\partial t}\frac{\partial Q}{\partial q} + \left(\frac{\partial Q}{\partial q}\frac{\partial P}{\partial p} - \frac{\partial P}{\partial q}\frac{\partial Q}{\partial p}\right)\frac{\partial \mathcal{H}}{\partial q}. \tag{B.13}$$

If the quantity $(\partial P/\partial q)\,(\partial Q/\partial q)\,(\partial \mathcal{H}/\partial p)$ is subtracted and added on the right, one obtains:

$$\frac{\partial \mathcal{K}}{\partial q} = -\frac{\partial P}{\partial q}\frac{\partial Q}{\partial q}\frac{\partial \mathcal{H}}{\partial p} + \frac{\partial Q}{\partial q}\frac{\partial P}{\partial p}\frac{\partial \mathcal{H}}{\partial q} - \frac{\partial P}{\partial t}\frac{\partial Q}{\partial q}$$

$$+ \frac{\partial P}{\partial q}\frac{\partial Q}{\partial q}\frac{\partial \mathcal{H}}{\partial p} - \frac{\partial P}{\partial q}\frac{\partial Q}{\partial p}\frac{\partial \mathcal{H}}{\partial q} + \frac{\partial Q}{\partial t}\frac{\partial P}{\partial q}. \tag{B.14}$$

This becomes:

$$\frac{\partial \mathcal{K}}{\partial q} = -\left[\left\{\frac{\partial P}{\partial q}\frac{\partial \mathcal{H}}{\partial p} - \frac{\partial \mathcal{H}}{\partial q}\frac{\partial P}{\partial p}\right\} + \frac{\partial P}{\partial t}\right]\frac{\partial Q}{\partial q}$$

$$+ \left[\left\{\frac{\partial Q}{\partial q}\frac{\partial \mathcal{H}}{\partial p} - \frac{\partial \mathcal{H}}{\partial q}\frac{\partial Q}{\partial p}\right\} + \frac{\partial Q}{\partial t}\right]\frac{\partial P}{\partial q}. \tag{B.15}$$

The quantities within the braces are *Poisson brackets* of P and $\mathcal{H}$ and of Q and $\mathcal{H}$, respectively, with respect to q and p; structures such as these are discussed at some length in Appendix D. By Appendix D or by direct application of Hamilton's equations in the old variables, one finds that:

$$\frac{\partial \mathcal{K}}{\partial q} = -\dot{P}\frac{\partial Q}{\partial q} + \dot{Q}\frac{\partial P}{\partial q}. \tag{B.16}$$

However, the derivative $\partial \mathcal{K}/\partial q$ can also be written:

$$\frac{\partial \mathcal{K}}{\partial q} = \frac{\partial \mathcal{K}}{\partial Q}\frac{\partial Q}{\partial q} + \frac{\partial \mathcal{K}}{\partial P}\frac{\partial P}{\partial q}. \tag{B.17}$$

Two expressions similar to (B.16) and (B.17) can also be obtained beginning with $\partial \mathcal{H}/\partial p$. In both cases, equating coefficients yields Hamilton's equations in the new variables:

$$\dot{Q} = \frac{\partial \mathcal{K}}{\partial P} \; ; \qquad \dot{P} = -\frac{\partial \mathcal{K}}{\partial Q} \; . \tag{B.18}$$

The expression $\dot{\mathcal{K}} = \partial \mathcal{K}/\partial t$ follows as a consequence of the above as in (2.013). Relations (B.18) demonstrate the canonicity of the transformation.

APPENDIX C

MOST PROBABLE DISTRIBUTION OF ENERGY AMONG CAVITY MODES

SUPPOSE that a fixed amount of energy U_0 is distributed among an infinite number of electromagnetic cavity modes by putting n_1 energy units of magnitude ε_1 into the first set of ΔN_1 modes, n_2 energy units of magnitude ε_2 into the second set of ΔN_2 modes, etc. The total number of ways of realizing this particular distribution is a function of the set of numbers $n_1, n_2, \ldots$ and is given by:

$$W(n_i) = \prod_i \frac{(n_i + \Delta N_i - 1)!}{n_i! \, (\Delta N_i - 1)!} \, . \tag{C.01}$$

The numbers $n_1, n_2, \ldots$ must, of course, be such that:

$$U(n_i) = \sum_i n_i \varepsilon_i = U_0. \tag{C.02}$$

The object of this derivation is to find the most probable distribution, i.e. that distribution (described by $n_1 = n_1', \ n_2 = n_2', \ \ldots$) for which W is maximum subject to the constraint that (C.02) must be obeyed.

The problem just stated may be compared to that of finding, in ordinary three-dimensional space, a point P on a given surface G where the scalar function $f(x, y, z)$ is maximum (or minimum) from the point of view of an observer constrained to remain upon G. Let G be described by the equation $g(x, y, z) = g_0$, a constant. In general, a surface of constant f will intersect G in a curve but, if the maximum

364

(minimum) exists, there will be a surface F of constant f which will touch G at but a single point. This is the point P sought and at this point ∇f and ∇g are colinear which is to say that F and G have, at P, a common normal. Thus the necessary condition for the existence of the maximum (minimum) is that there be a point at which:

$$\nabla f = \beta \, \nabla g, \tag{C.03}$$

where β is a positive or negative constant of proportionality known as a *Lagrange multiplier*.

To apply this theory to the problem of maximizing W, one must think in terms of an infinite-dimensional space coordinatized by the n_i. Also, since $\ln W$ is a monotonic function of W, and is easier to work with than W, it is both acceptable and preferable to maximize $\ln W$. The necessary condition, expressed by the proportionality of the two gradients, is:

$$\frac{\partial}{\partial n_i} \ln W = \beta \, \frac{\partial U}{\partial n_i} \, . \tag{C.04}$$

A discussion of the sufficient conditions for the occurrence of a maximum will not be undertaken here; physical reasoning is adequate to show that $\ln W$ has a unique maximum for given U_0 to which the distribution determined below must correspond. Equation (C.04) becomes:

$$\frac{\partial}{\partial n_i} \sum_j \left[\ln (n_j + \Delta N_j - 1)! - \ln n_j! - \ln (\Delta N_j - 1)! \right] = \beta \varepsilon_i. \tag{C.05}$$

In this analysis, ΔN_j is the number of modes whose frequencies lie within the interval Δv_j. Because of the granularity of the mode arrangement in M vector space (Fig. 5.06a), it becomes difficult to conceptualize the ΔN_j in the limit $\Delta v_j \to 0$ and it becomes correspondingly difficult to work with an equation like (C.05). Fortunately, the maximizing distribution sought cannot be dependent upon the size of the Δv_j so long as the latter are small. It is expedient, then, to take these quantities to be small but non-zero in which case the ΔN_j become large (except at the statistically insignificant frequencies of very low value) and the analysis becomes enormously easier. In these circumstances, one may invoke the Stirling approximation for large factorials:

$$\ln x! \approx x \ln x - x, \tag{C.06}$$

from which it follows that:

$$\frac{\partial}{\partial n_i} \ln x! \approx \frac{\partial x}{\partial n_i} \ln x. \tag{C.07}$$

When this result is applied to the term in (C.05) for which $j = i$, one finds:

$$\ln (n_i + \Delta N_i - 1) - \ln n_i = \beta \varepsilon_i. \tag{C.08}$$

This becomes:

$$\frac{n_i + \Delta N_i - 1}{n_i} = \exp \beta \varepsilon_i, \tag{C.09}$$

or:

$$\frac{\Delta N_i - 1}{n_i} = (\exp \beta \varepsilon_i) - 1. \tag{C.10}$$

In view of earlier approximations, unity can be neglected in comparison with ΔN_i. The set of quantities n_i, which are to be called n_i' since they constitute the maximizing set, are then given by:

$$n_i' = \frac{\Delta N_i}{(\exp \beta \varepsilon_i) - 1}. \tag{C.11}$$

Substitution of this back into (C.02) produces:

$$U_0 = \sum_i \frac{\varepsilon_i \Delta N_i}{(\exp \beta \varepsilon_i) - 1}. \tag{C.12}$$

The quantities ε_i and ΔN_i in the right-hand side are known and therefore the unknown multiplier β is determined by the amount of total energy U_0. Careful inspection of (C.12) shows that as $U_0 \to \infty$, $\beta \to 0$ and vice versa. Thus, although it might be a difficult task, it is theoretically possible to solve (C.12) for β thereby completing the explicit determination of the n_i'.

APPENDIX D

POISSON BRACKETS

GIVEN a system with D degrees of freedom and with coordinates q_j and momenta p_j. The total time derivative of the observable $A(q_j, p_j, t)$ may be written:

$$\frac{dA}{dt} = \frac{\partial A}{\partial t} + \sum_{j=1}^{D} \left(\frac{\partial A}{\partial q_j} \dot{q}_j + \frac{\partial A}{\partial p_j} \dot{p}_j \right). \tag{D.01}$$

By Hamilton's equations, this becomes:

$$\frac{dA}{dt} = \frac{\partial A}{\partial t} + \sum_{j=1}^{D} \left(\frac{\partial A}{\partial q_j} \frac{\partial \mathcal{H}}{\partial p_j} - \frac{\partial \mathcal{H}}{\partial q_j} \frac{\partial A}{\partial p_j} \right). \tag{D.02}$$

The summation in this expression is known as the *Poisson bracket* of A and $\mathcal{H}$ and will be denoted $\{A, \mathcal{H}\}$. Thus:

$$\frac{dA}{dt} = \frac{\partial A}{\partial t} + \{A, \mathcal{H}\}. \tag{D.03}$$

Since both A and $\mathcal{H}$ are functions of q_j, p_j, and t, it follows that the Poisson bracket $\{A, \mathcal{H}\}$ is a function of these same variables and is *also an observable*.

Since $\langle A \rangle$ is an average of many, perhaps infinitely many, values of A taken under specified conditions,[†] it follows that $\langle dA/dt \rangle =$

† See Section 6.05.

$= \mathrm{d}\langle A \rangle / \mathrm{d}t$. If the expectation value of both sides of (D.03) is taken, one therefore finds:

$$\frac{\mathrm{d}\langle A \rangle}{\mathrm{d}t} = \langle \partial A / \partial t \rangle + \langle \{A, \mathcal{H}\} \rangle. \tag{D.04}$$

Comparison of this with (9.008) shows that:

$$\langle \{A, \mathcal{H}\} \rangle = \hbar^{-1} \int_{\substack{\text{all} \\ \text{space}}} \Psi^*(-i[\hat{A}, \hat{\mathcal{H}}])\Psi \, \mathrm{d}\tau \tag{D.05}$$

and it is seen that $(i\hbar)^{-1}[\hat{A}, \hat{\mathcal{H}}]$, which is Hermitian, is the quantum mechanical operator for the real observable $\{A, \mathcal{H}\}$. Notice that the vanishing of $\{A, \mathcal{H}\}$ implies the commutativity of $\hat{A}$ and $\hat{\mathcal{H}}$ and vice versa and it is most interesting that a quantum mechanical commutator should have, in the classical domain, such a closely related analog as the Poisson bracket.

Poisson brackets do not necessarily have to involve the Hamiltonian; they may be defined with respect to any two classical observables. Thus:

$$\{A, B\} = \sum_{j=1}^{D} \left(\frac{\partial A}{\partial q_j} \frac{\partial B}{\partial p_j} - \frac{\partial B}{\partial q_j} \frac{\partial A}{\partial p_j} \right). \tag{D.06}$$

Using this, it is found that:

$$\langle \{A, B\} \rangle = \hbar^{-1} \int_{\substack{\text{all} \\ \text{space}}} \Psi^*(-i[\hat{A}, \hat{B}])\Psi \, \mathrm{d}\tau \tag{D.07}$$

and the operator for $\{A, B\}$ is $(i\hbar)^{-1} [\hat{A}, \hat{B}]$.

The study of Poisson brackets affords many interesting insights. Thus, by substitution into (D.06), it is seen that for any two coordinates or for any two momenta:

$$\{q_m, q_n\} = 0; \quad \{p_m, p_n\} = 0. \tag{D.08}$$

The Poisson bracket of a coordinate and a momentum also vanishes unless the two are canonical conjugates, in which case it is equal to unity. Thus:

$$\{q_m, p_n\} = \delta_{mn}. \tag{D.09}$$

It can be shown that Poisson brackets are invariant under canonical transformation, hence a set of $2D$ observables F_m and G_n cannot be a set of canonical variables unless, like the coordinates and momenta considered above, they obey:

$$\{F_m, F_n\} = 0; \quad \{G_m, G_n\} = 0; \quad \{F_m, G_n\} = \delta_{mn}. \quad \text{(D.10)}$$

The quantum mechanical commutation relations (12.062) for the Cartesian components of angular momentum are well known. Not so well known, perhaps, are the corresponding classical expressions in terms of Poisson brackets:

$$\begin{aligned}
\{L_x, L_y\} &= L_z; \\
\{L_y, L_z\} &= L_x; \\
\{L_z, L_x\} &= L_y.
\end{aligned} \quad \text{(D.11)}$$

It follows from this and from (D.10) that only one of L_x, L_y, and L_z can be a member of a set of canonical variables and that only one can be a *canonical* constant of the motion even though it frequently happens, from a classical point of view, that all three are constant.

REFERENCES

1. WHITTAKER, E. T., *Analytical Dynamics of Particles and Rigid Bodies*, Cambridge University Press, 1937, 4th ed., Chapter XIII.
2. FEYNMAN, R. P., LEIGHTON, R. B., and SANDS, M., *The Feynman Lectures on Physics*, Addison-Wesley, Reading, Mass., 1964, Vol. II, p. 19–8.
3. RUSSELL, A. M., *J. Appl. Phys.* **33**, 970 (1962).
4. TER HAAR, D., *Elements of Hamiltonian Mechanics*, Pergamon Press, Oxford, 1971, Chapter 5, Section 2.
5. SYMON, K. R., *Mechanics*, Addison-Wesley, Reading, Mass., 1960, 2nd ed., Section 8–10.
6. TER HAAR, D., *The Old Quantum Theory*, Pergamon Press, Oxford, 1967.
7. LORD RAYLEIGH, *Philos. Mag.* **49**, 539 (1900).
8. JEANS, J. H., *Philos. Mag.* **10**, 91 (1905).
9. REIF, F., *Fundamentals of Statistical and Thermal Physics*, McGraw-Hill, New York, 1965, Section 7.5.
10. PLANCK, M., *Verh. Dtsch. Phys. Ges. Berlin* **2**, 202 and 237 (1900).
11. TAYLOR, B. N., PARKER, W. H., and LANGENBERG, D. N., *Rev. Mod. Phys.* **41**, 375 (1969).
12. EINSTEIN, A., *Ann. Phys. (Leipz.)* **17**, 132 (1905).
13. HARNWELL, G. P. and LIVINGOOD, J. J., *Experimental Atomic Physics*, McGraw-Hill, New York, 1933, Section 6–8.
14. MILLIKAN, R. A., *Phys. Rev.* **7**, 335 (1916).
15. *Proceedings of the International Colloquium on Optical Properties and Electronic Structure of Metals and Alloys*, North-Holland, Amsterdam, 1966, Section V.
16. RUTHERFORD, E., *Philos. Mag.* **21**, 669 (1911).
17. BOHR, N., *Philos. Mag.* **26**, 1 (1913).
18. ISHIWARA, J., as referenced in [36], p. 92.
19. WILSON, W., *Philos. Mag.* **29**, 795 (1915).
20. SOMMERFELD, A., *Münchener Ber.* **1915**, 425 and 459; *Ann. Phys. (Leipz.)* **51**, 1 and 125 (1916).
21. COMPTON, A. H., *Phys. Rev.* **21**, 483 and 715 (1923).
22. DE BROGLIE, L., *Recherches sur la théorie des Quanta* (doctoral thesis), University of Paris, 1924. For English translation (by HASLETT, J. W.), see *Am. J. Phys.* **40**, 1315 (1972).
23. DAVISSON, C. J. and GERMER, L. H., *Phys. Rev.* **30**, 705 (1927).

24. FRIEDRICH, W., KNIPPING, P., and VON LAUE, M., *Münchener Ber.* **1912**, 303.
25. BRAGG, W. L., *Proc. Camb. Philos. Soc.* **17**, 43 (1912).
26. LUDWIG, G., *Wave Mechanics*, Pergamon Press, Oxford, 1968.
27. HEISENBERG, W., *Z. Phys.* **33**, 879 (1925).
28. BORN, M. and JORDAN, P., *Z. Phys.* **34**, 858 (1925).
29. BORN, M., HEISENBERG, W., and JORDAN, P., *Z. Phys.* **35**, 557 (1926).
30. DIRAC, P. A. M., *Proc. R. Soc. A* **109**, 642 (1925) and *Proc. Camb. Philos. Soc.* **23**, 412 (1926).
31. SCHRÖDINGER, E., *Ann. Phys. (Leipz.)* **79**, 489 (1926). See also **79**, 361; **79**, 734; **80**, 437; **81**, 109 (1926).
32. BORN, M., *Z. Phys.* **37**, 863 and **38**, 803 (1926); *Nature* **119**, 354 (1927).
33. JOHNSON, D. E. and JOHNSON, J. R., *Mathematical Methods in Engineering and Physics*, Ronald Press, New York, 1965, Chapter 9.
34. EHRENFEST, P., *Z. Phys.* **45**, 455 (1927).
35. FEYNMAN, R. P., LEIGHTON, R. B., and SANDS, M., *The Feynman Lectures on Physics*, Addison-Wesley, Reading, Mass., 1965, Vol. III, Chapter 5.
36. JAMMER, M., *The Conceptual Development of Quantum Mechanics*, McGraw-Hill, New York, 1966, Chapter 9.
37. HEISENBERG W., *Z. Phys.* **43**, 172 (1927).
38. ROBERTSON, H. P., *Phys. Rev.* **34**, 163 (1929).
39. JAHNKE, E., EMDE, F., and LÖSCH, F., *Tables of Higher Functions*, McGraw-Hill, New York, 1960, 6th ed., Chapter VIII.
40. ERDÉLYI, A. *et al.*, *Higher Transcendental Functions* (The Bateman Manuscript Project), McGraw-Hill, New York, 1953, Volume I, Chapter III.
41. See [33], Section 4.10.
42. CONDON, E. U. and SHORTLEY, G. H., *The Theory of Atomic Spectra*, Cambridge University Press, 1951, Chapter III, Section 4.
43. MARGENAU, H. and MURPHY, G. M., *The Mathematics of Physics and Chemistry*, D. Van Nostrand, New York, 1956, 2nd ed., Section 2.16.
44. See ref. 43, Section 3.11.,
45. FRANKLIN, J. N., *Matrix Theory*, Prentice-Hall, Englewood Cliffs, N.J., 1968, Section 4.3.
46. LUDFORD, G. S. S. and YANNITELL, D. W., *Am. J. Phys.* **36**, 231 (1968).

SELECTED SUPPLEMENTARY REFERENCES

BOHM, D., *Quantum Theory*, Prentice-Hall, New York, 1951.

BOROWITZ, S., *Fundamentals of Quantum Mechanics*, W. A. Benjamin, New York, 1967.

CORBEN, H. C. and STEHLE, P., *Classical Mechanics*, John Wiley, New York, 1960, 2nd ed.

CROPPER, W. H., *The Quantum Physicists*, Oxford University Press, London, 1970.

DICKE, R. H. and WITTKE, J. P., *Introduction to Quantum Mechanics*, Addison-Wesley, Reading, Mass., 1960.

GOLDSTEIN, H., *Classical Mechanics*, Addison-Wesley, Reading, Mass., 1950.

MERZBACHER, E., *Quantum Mechanics*, John Wiley, New York, 1970, 2nd ed.

MESSIAH, A., *Quantum Mechanics* (in two volumes), North-Holland, Amsterdam, 1958.

PAULING, L. and WILSON, E. B., *Introduction to Quantum Mechanics*, McGraw-Hill, New York, 1935.

POWELL, J. L. and CRASEMANN, B., *Quantum Mechanics*, Addison-Wesley, Reading, Mass., 1961.

SAXON, D. S., *Elementary Quantum Mechanics*, Holden-Day, San Francisco, 1968.

SCHIFF, L. I., *Quantum Mechanics*, McGraw-Hill, New York, 1968, 3rd ed.

PROBLEMS

Notice. Integral part of problem number refers to related chapter in text. In problems involving frictional forces, energy dissipated by such forces is to be considered negligible unless otherwise specified.

1.01. Given a block in the form of a rectangular parallelepiped of uniform density ϱ, base dimensions a and b, and height c. Block rocks back and forth without slipping on a horizontal plane so that first one edge of length b is in contact with the plane, then the other. Let the configuration be specified by the angle φ between the body vertical of the block (parallel to the edges of length c) and the true vertical. Motion of the block can be most easily demonstrated in the laboratory if c is considerably greater than a.

   (i) Express the potential energy in terms of φ. Two formulas will be needed, one for $\varphi > 0$ and one for $\varphi < 0$.
   (ii) Express the kinetic energy in terms of $\dot{\varphi}$.
   (iii) Find the period of oscillation in terms of the amplitude φ_m under the assumption that the latter is very small.

1.02. Given a one-particle system, an inertial Cartesian coordinate frame x_k, and a rotating Cartesian coordinate frame q_j. The origins of the two frames coincide and the q_3-axis coincides with the x_3-axis. Let the angle measured counterclockwise from the x_1-axis to the q_1-axis (with the x_3-axis pointed toward the observer) be ωt.

   (i) Write transformation equations (1.022).
   (ii) Write transformation equations (1.023).
   (iii) Derive $T(q_j, \dot{q}_j, t)$ by (1.030).

1.03. Given a horizontal track upon which two blocks of masses M_1 and M_2 are free to slide (without rotating) in one direction only. A spring of elastic constant K_1 is connected between a fixed point and M_1, a spring of elastic constant K_c is connected between M_1 and M_2, and a spring of elastic constant K_2 is connected between M_2 and another fixed point. All components are arranged in the common line of motion of the masses and, when the two are in their equilibrium positions, the springs are unstressed. Let q_1 and q_2 be the respective displacements of the two masses from their equilibrium positions. Find the equations of motion by the Lagrangian method.

1.04. Given a Cartesian coordinate frame with the y-axis vertical and a wire in the xy plane described by the function $y = f(x)$. A bead, considered to be a particle, slides on this wire. By the Lagrangian method, show that the equation of motion is:

$$\ddot{x}(1 + f'^2) + \dot{x}^2 f' f'' + gf' = 0.$$

1.05. A motor drives a vertical shaft S_1 in a prescribed manner. A second motor, whose frame is fixed to S_1, drives a horizontal shaft S_2 in a prescribed manner. To S_2 is attached a tube bent into the form of a large circle so that S_2 (produced) coincides with one of its diameters. Inside the tube a bead, regarded as a particle, is free to slide. How many degrees of freedom are there?

1.06. A straight tube of length a is fixed to the surface of a horizontal platform rotating about a vertical axis with angular velocity ω. One end of the tube is located at the axis of rotation and a bead, introduced into this end, is given an initial radial velocity v_0.

(i) Regarding the bead as a particle, find second-order equation of motion by the Lagrangian method and solve it under the initial conditions $r = 0$ and $\dot{r} = v_0$ at $t = 0$.

(ii) If $a = 15$ cm, $\omega = 0.100$ radian/sec, and $v_0 = 1.00$ cm/sec, find the time required for the bead to travel the length of the tube.

1.07. Given a horizontal platform with Cartesian coordinates q_1, q_2, affixed thereto. The platform rotates about a vertical axis through the origin at angular velocity ω. A particle of mass m is attached to a massless cord which passes through a small hole in the platform (at the origin) and connects to a massless spring situated beneath the platform and attached thereto. The spring has an elastic constant equal to $m\omega^2$ and is unstressed when the particle is at the origin, hence the force exerted by the cord on the particle exactly counterbalances the "centrifugal orce".

(i) Find the Lagrangian and the two second-order equations of motion in the q_j coordinates.

(ii) Obtain two first-order equations of motion by directly integrating the equations found in (i).

(iii) Show that $q_1 = A + R \cos(-2\omega t + \beta)$ and $q_2 = B + R \sin(-2\omega t + \beta)$ are solutions where A, B, R, and β are constants. Describe in words the motion involved, choosing a general case.

1.08. A harmonic oscillator consists of a block of mass M attached by a spring of elastic constant K to a platform B in the usual way so that a coordinate x describes the displacement of the block with respect to its normal equilibrium position denoted by a reference mark on B. Suppose that B is forced by a machine to move (in the same straight line as the direction of motion of M) so that its displacement with respect to an inertial reference frame is a prescribed function of time $f(t)$. Find the Lagrangian and the equation of motion in the coordinate x using such quantities as $f'(t)$ and $f''(t)$.

1.09. A charged particle of mass m and charge q moves in an electromagnetic field characterized by the potentials $\varphi = \Gamma z$, $A_x = -\frac{1}{2}\Lambda y$, $A_y = \frac{1}{2}\Lambda x$, [where Γ and Λ are constants.

   (i) Using (1.092), find the Lagrangian and the equations of motion.

   (ii) Evaluate the six components of the electromagnetic field in terms of Γ and Λ.

   (iii) What must be the relationship between the ω of Problem 1.07 and the B_z of this problem so that the equations of motion involving q_1 and q_2 on the one hand and x and y on the other shall be formally identical? (This comparison illustrates *Larmor's theorem*.)

1.10. Given a Cartesian coordinate frame with the y-axis vertical, A projectile of mass m is fired from the origin at $t = t_0 = 0$ with initial velocity $e_x A + e_y B + e_z 0$ and strikes the ground at $t = t_f = 2B/g$. Neglect air resistance.

   (i) Find the equations $x = x(t)$ and $y = y(t)$ for the true path.

   (ii) Find the definite action ΔS on the true path in $t_0 \leqslant t \leqslant t_f$ in terms of m, A, B, and g.

1.11. In the context of Problem 1.10, consider a varied path given by $x = At$ and $y = C \sin(\pi g t/2B)$.

   (i) Find the definite action ΔS_a on the varied path in terms of m, A, B, C, and g.

   (ii) Show that ΔS_a is minimum for $C = 32\pi^{-3} B^2/2g$ where $B^2/2g$ is the maximum height of the projectile on its true path. Also show that even with this value of C, $\Delta S_a > \Delta S$.

   (iii) Explain why minimizing ΔS_a by varying C does not yield a value of action equal to ΔS.

   (iv) As the varied path, take $x = At$ and $y = C_1 \sin(\pi g t/2B) + C_3 \sin(3\pi g t/2B)$. Again calculate ΔS_a using minimizing values of both C_1 and C_3; this ΔS_a should be closer to ΔS than the one found in (ii) above.

1.12. Suppose that the motion of the harmonic oscillator of Fig. 1.01 is given by $x = A \sin \Omega t$. Find a neat formula for the definite action on the true path in the interval $0 \leqslant t \leqslant t_f$ in terms of Ω, t_f, and E only where E is the total energy of the oscillator. *Hint:* $\cos^2 \theta - \sin^2 \theta = \cos 2\theta$.

2.01. Repeat Problem 1.06 using the Hamiltonian formulation.

   (i) Find $\mathcal{H}$ and use Hamilton's equations to check the relationship between $\dot{r}$ and p_r, also to find the second-order equation of motion.

   (ii) Is $\mathcal{H}$ a constant of the motion?

   (iii) Is $\mathcal{H}$ equal to the total energy?

   (iv) Invoke a constant of the motion to obtain a first order equation of motion and solve subject to the initial conditions $r = 0$ and $\dot{r} = v_0$ at $t = 0$. Compare result with solution obtained in Problem 1.06.

2.02. Given two particles of masses m_a and m_b connected by a massless inelastic cord. The cord passes through a small hole in the center of a horizontal table with m_a free to slide on the surface of the table and m_b hanging beneath it. The particle of mass m_b is constrained to move only vertically, e.g. by a vertical tube. Describe the position of m_a by the plane polar coordinates r and φ.

(i) Find the Hamiltonian.
(ii) Identify two constants of the motion and find the corresponding first-order equations of motion.

2.03. With reference to Problem 2.02, assume that the potential energy is zero when $r = 0$, then do the following:

(i) Find an expression for the effective potential energy $V'(r)$.
(ii) Sketch $V'(r)$ assuming that the angular momentum is not zero.
(iii) Using $V'(r)$, find the radius of the circular orbit in terms of angular momentum and other constants; check result by appeal to elementary concepts (e.g. centripetal force).

2.04. Assume that the Earth and the Moon describe circular orbits at angular velocity Ω around their common center of mass (which is inside the Earth). Let the mass of the Earth be M_1 and the radius of its orbit around the center of mass be R_1; let the mass of the Moon be M_2 and the radius of its orbit be R_2. The total distance from center of Earth to center of Moon is therefore $R_1 + R_2$. Consider a rotating Cartesian coordinate frame q_j with origin at the center of mass mentioned above, with the positive q_1-axis always passing through the center of the Moon and the $q_1 q_2$ plane coincident with the orbit plane. Let the system be an unpowered space ship whose position is given by the cylindrical coordinates r, φ, q_3, where $r = (q_1^2 + q_2^2)^{\frac{1}{2}}$ and $\varphi = $ arc tan q_2/q_1. Find the Hamiltonian using the r, φ, q_3 coordinates. *Hint:* the distance from the space ship to the center of the Earth is $(R_1^2 + r^2 + 2R_1 r \cos \varphi + q_3^2)^{\frac{1}{2}}$; a similar but not identical formula describes the distance from the space ship to the center of the Moon.

2.05. With reference to Problem 1.08, find the Hamiltonian.

3.01. Let $q_1 = x$, $q_2 = y$, $p_1 = p_x$, $p_2 = p_y$, $Q_1 = r$, $Q_2 = 0$, $P_1 = p_r$, $P_2 = p_\theta$. Show that the function $G_b = p_r(x^2 + y^2)^{\frac{1}{2}} + p_\theta$ arc tan y/x generates the transformation between rectangular and plane polar coordinates. Using (3.016) and (3.017), obtain all four of the forms (3.003), (3.004), (3.005), and (3.006). Because of the simplicity of the example, some of the steps will be trivial.

3.02. Given a system with one degree of freedom. Find all four of the forms (3.003), (3.004), (3.005) and (3.006) for the canonical transformation generated by $G_b = P \ln \sin q - P \ln P + P$. If the original Hamiltonian is equal to $Ap^2 + B \sin q$, where A and B are constants, find the new Hamiltonian.

3.03. With reference to the one-dimensional harmonic oscillator, show tha^l

the two constants $\zeta' = p_0$ and $\xi' = q_0$ are functions of the two constants $\zeta = \mathcal{H}$ and $\xi = t_e$ and that these functions do not involve x, p, or t.

4.01. Given a webbed string as in Section 4.02. It is determined experimentally that the wavelength is 2 meters at a frequency of 800 hertz and 1 meter at 1000 hertz.

    (i) Find the minimum frequency below which wave propagation is impossible.
    (ii) Find the phase velocity at infinite frequency.
    (iii) Find the wavelength, the phase velocity, and the group velocity for forward traveling waves at a frequency of 1500 hertz.

4.02. Consider a superposition of one-dimensional complex exponential waves or which:

$$A_k = \begin{cases} G & \text{if} \quad (k_0 - \delta) < k < (k_0 + \delta); \quad G \text{ is real.} \\ 0 & \text{otherwise.} \end{cases}$$

Work in the variable $k' = k - k_0$ and find $u(x, t)$ in terms of k_0, ω_0, and ω_0' as these quantities are defined in Section 4.03. Assume that the dispersion relation is adequately represented by its constant and first derivative terms.

4.03. Suppose that space is divided into three regions by the infinite imaginary planes $x = -a$ and $x = a$. In region 1, $x < -a$, the index of refraction is n_1; in region 2, $x > a$, the index of refraction is n_2; in the transition region, $-a < x < a$, the index of refraction is a smooth function of x, changing gradually from n_1 to n_2. The three regions are shown in Fig. P.01. A plane monochromatic electromagnetic wave with wave vector $k = k_1$ in region 1 is incident from the left. The wave vector also changes gradually within the transition region and, when the

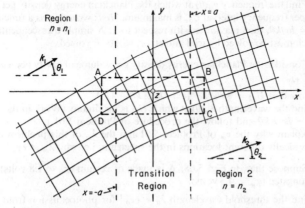

FIG. P.01. An electromagnetic wave crossing a transition region of increasing index of refraction. See Problem 4.03.

wave emerges into region 2, it once more becomes constant and is designated k_2. Both λ_1 and λ_2 are small compared with $2a$ and the wave in the transition region is quasi-plane. The wave vector has no z component anywhere. In this problem, it is recommended that the optical eikonal equation, $|\nabla \sigma|^2 - n^2 c^{-2}(\partial \sigma / \partial t)^2 = 0$, be used.

(i) Show that, for any closed contour, $\oint k \cdot dl = 0$.

(ii) Apply the result of (i) to the rectangular contour ABCD whose sides are parallel to the coordinate axes. Use information thus obtained to derive a relationship among k_1, θ_1, k_2, and θ_2 and, from this, another relationship among n_1, θ_1, n_2, and θ_2. The latter should be a familiar law of geometrical optics.

4.04. Verify that the spherical wave (4.056) is a solution to wave equation (4.035) provided that the condition mentioned in the text is satisfied by α and β.

5.01. Given an electromagnetic resonant cavity with dimensions $2 \times 2.5 \times 3$ cm.

(ii) Find the five lowest resonant frequencies.

(i) By actual count, find the total number of modes between $M = 0$ and $M = 1$ cm^{-1}.

(iii) By formula, find the total number of modes between $M = 1000$ cm^{-1}. and $M = 1001$ cm^{-1}.

(iv) To what frequency range does the interval of part (iii) correspond?

5.02. Draw several sets of three "boxes" symbolizing electromagnetic modes. Assume that there are four energy units, represented by dots. In the first set of three boxes, put four dots in the first box and none in the other two. Continue until all the different ways of distributing the four dots among the three boxes have been exhibited.

5.03. Find the frequency $\nu_m(\theta)$ at which the Planckian energy density per volume of cavity per frequency interval $U_{\tau\nu}$ is maximum. *Hint*: write $U_{\tau\nu}$ as a function of x where $x = h\nu/k\theta$, then maximize with respect to x. A simple transcendental equation, which must be solved by trial and error, will be involved.

5.04. Consider the Planckian energy density per volume of cavity per *wavelength* nterval, $U_{\tau\lambda}$.

(i) Show that $U_{\tau\lambda} = 8\pi hc \; \lambda^{-5} [\exp (hc/\lambda \, k\theta) - 1]^{-1}$.

(ii) Find the wavelength $\lambda_m(\theta)$ at which $U_{\tau\lambda}$ is maximum. (Work in the variable $y = hc/\lambda \, k\theta$ and follow the procedure of Problem 5.01.)

(iii) Explain why the ν_m of Problem 5.01 and the λ_m of this problem occur at physically different locations in the spectrum, i.e. why $\lambda_m \neq c\nu_m^{-1}$.

5.05. Suppose that, in Fig. 5.08, A is made of cesium ($\varphi_w = 1.8$ volts) and A' made of tungsten ($\varphi_w = 4.5$ volts).

(i) Find the threshold wavelength $\lambda_0 = c\nu_0^{-1}$ for photoemission from cesium.

(ii) Find the measured voltage φ_m at which the photoelectric current just stops if light of 5000 Å wavelength is used.

(iii) What is the shortest wavelength that can be used before photoemission from A' by scattered light begins to occur?

5.06. The ratio $\eta = (E_n - E_{n-1})/(E_n - E_1)$ compares the energy radiated in an atomic transition to an adjacent state with the total available energy, i.e. with the energy of a transition to the ground state. For the stationary states of hydrogen, do the following:

   (i) Derive an exact formula for η.

 (ii) Convert the formula obtained in (i) to an approximation suitable for large n.

(iii) Evaluate η for $n = 1000, 100, 10, 5$, and 3.

5.07. Draw a triangle diagram as in Fig. 5.10a for an electron with a de Broglie wavelength equal to its Compton wavelength.

   (i) What is v/c for such an electron?

 (ii) Is it a good rule to say that if the de Broglie wavelength is substantially larger than the Compton wavelength, the particle in question is non-relativistic but if substantially smaller, the particle is strongly relativistic?

5.08. Suppose that the geometry of Fig. 5.15 is kept unchanged in all details but the accelerating voltage is modified.

   (i) Find the two next higher voltages at which an electron maximum will be observed at the collector.

 (ii) Is there any voltage lower than 54 volts at which an electron maximum will be observed at the collector?

6.01. Consider the flow of substances of density ϱ in the following three cases:

   (i) $\mathbf{J} \neq 0$, $d\varrho/dt = 0$, $\partial\varrho/\partial t = 0$.

 (ii) $\mathbf{J} \neq 0$, $d\varrho/dt \neq 0$, $\partial\varrho/\partial t = 0$.

(iii) $\mathbf{J} \neq 0$, $d\varrho/dt \neq 0$, $\partial\varrho/\partial t \neq 0$.

Give an example and a brief description in words of a physical situation illustrating each case. *Notice:* $d\varrho/dt$ is the substantial derivative of ϱ; it is the rate of change of ϱ as experienced by an observer who moves with the substance.

6.02. Show that equation (6.036) is equivalent to equation (6.031).

6.03. Given a one-dimensional system in which:

$$\Psi(x, t) = \begin{cases} 0 & \text{in } -\infty < x < 0; \\ (2/a)^{\frac{1}{2}} \sin(\pi x/a) \exp - i\omega_0 t & \text{in } 0 < x < a; \\ 0 & \text{in } a < x < \infty. \end{cases}$$

   (i) Show that $\displaystyle\int_{-\infty}^{\infty} \Psi^*\Psi \, dx = 1$.

(ii) Find $\langle x \rangle$.

(iii) Find $\langle x^2 \rangle$.

(iv) Find $\langle \sin [(\pi x/a) - \omega t] \rangle$.

7.01. Given a gradual potential energy increase from region I to region II as in Fig. 7.02a. A plane wave of energy $E = 1.5\ V_0$ and amplitude A is incident in region I and a transmitted wave of amplitude B is formed in region II. Because of the gradual nature of the acclivity, no reflected wave is formed. Find B^*B/A^*A.

7.02. Space is divided into three regions by the infinite planes $x = 0$ and $x = a$. In region I, $x < 0$, $V < 0$; in region II, $0 < x < a$, $V = V_0$; in region III, $x > a$, $V = 1.5\ V_0$. A plane wave of energy $E = V_0$ and amplitude A is incident in region I. Solve problem completely and plot $\Psi^*\Psi/A^*A$ in $-\lambda_I < x < \lambda_I$ under the assumption that $a = 0.2\ \lambda_I$.

7.03. Given a rectangular barrier as in Fig. 7.06 and suppose that, in region I, the maximum value of $\Psi^*\Psi/A^*A$ is 3.24 and the minimum value is 0.04. Find the probability of transmission.

7.04. Given a simple potential energy step as in Fig. 7.02b. A particle is incident in the x direction in region I with a wave function in the form of a packet whose outer dimensions are those of a cube 1 mm on a side and whose inner structure is that of a plane wave. (In the context of quantum theory, this is a packet of enormous size.) $E = 1.5\ V_0$.

(i) Find A^*A, B^*B, and C^*C.

(ii) Find the outer dimensions of the reflected packet and of the transmitted packet, using packet speeds (group velocities) as the guiding consideration.

(iii) Show that the reduced length of the transmitted packet is consistent with the requirement of normalization.

7.05. Prove the following relationships for the three lowest-lying bound stationary states of the finite rectangular well displayed in Fig. 7.12:

(i) $\displaystyle\int_{-\infty}^{\infty} \Psi_0^*\Psi_1\, dx = 0.$

(ii) $\displaystyle\int_{-\infty}^{\infty} \Psi_1^*\Psi_2\, dx = 0.$

(iii) $\displaystyle\int_{-\infty}^{\infty} \Psi_0^*\Psi_2\, dx = 0.$

(This problem is to be done by the use of parity and known algebraic features of the wave functions; it is not to be done by dependence upon data from numerical solutions nor by appeal to a general theorem from a later chapter.)

7.06. Given a finite rectangular well as in Fig. 7.09 with $a = 1$ mm and $V_0 = 10$ electron volts. (This is a one-dimensionalized version of a typical problem involving an electron trapped inside a small metal crystal; for m, use the mass of the electron.) How many bound stationary states are there?

7.07. Locate the zeros and the other inflection points of the function $\psi_4(\xi)$ for the one-dimensional harmonic oscillator, then make a qualitative sketch of this function using an uncalibrated vertical scale.

8.01. Prove that: $[\hat{A}\hat{B}, \hat{C}] = \hat{A}[\hat{B}, \hat{C}] + [\hat{A}, \hat{C}]\hat{B}$.

8.02. Prove the *Jacobi identity*: $[[\hat{A}, \hat{B}], \hat{C}] + [[\hat{C}, \hat{A}], \hat{B}] + [[\hat{B}, \hat{C}], \hat{A}] = 0$.

8.03. Consider the following non-stationary superposition of the zeroth and the first stationary states of the infinite rectangular well:

$$\Psi(x, t) = \beta_0\Psi_0(x, t) + \beta_1\Psi_1(x, t).$$

  (i) Express the probability density as a function of x and t.
 (ii) What must be the value of $\beta_0^*\beta_0 + \beta_1^*\beta_1$?
(iii) Assume that $\beta_0 = \beta_1 = 2^{-\frac{1}{2}}$ and find $\langle x \rangle$ as a function of time.
 (iv) Assume that $\beta_0 = \beta_1 = 2^{-\frac{1}{2}}$ and find $\langle p_x \rangle$ as a function of time by using the momentum operator and integrating. Compare with $m\mathrm{d}\langle x \rangle/\mathrm{d}t$.

8.04. For the infinite rectangular well, $\Psi(x, t_0) = f(x)$, where:

$$f(x) = \begin{cases} (12/a^3)^{\frac{1}{2}}\,x + (3/a)^{\frac{1}{2}} \text{ in } -(a/2) < x < 0. \\ -(12/a^3)^{\frac{1}{2}}\,x + (3/a)^{\frac{1}{2}} \text{ in } 0 < x < (a/2). \end{cases}$$

  (i) Verify the normalization of $f(x)$.
 (ii) In the expansion $\Psi(x, t) = \sum_n \alpha_n\Psi_n(x, t)$, where the $\Psi_n(x, t)$ are the stationary states of the well, find the coefficient α_n as a function of n.
(iii) Verify, by actually adding up a few of the low-order terms, that $\sum_n \alpha_n^*\alpha_n = 1$.
 (iv) Evaluate $\langle E \rangle/E_0$, where E_0 is the energy level of the ground state.

9.01. Show that, for the stationary states of the harmonic oscillator:

$$\xi\psi_n = \left(\frac{n+1}{2}\right)^{\frac{1}{2}}\psi_{n+1} + \left(\frac{n}{2}\right)^{\frac{1}{2}}\psi_{n-1}$$

9.02. Using the results of Problem 9.01, prove (9.028).

10.01. With reference to Fig. P.02, find a formula for α_n as a function of n for each of the three examples shown.

10.02. What is the special significance of the coefficient α_0 in relation to a periodic function?

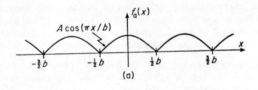

$$A \cos(\pi x / b)$$

(a)

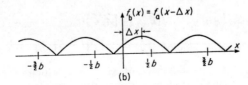

$$f_b(x) = f_a(x - \Delta x)$$

(b)

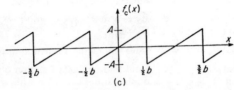

(c)

FIG. P.02. Periodic functions for Fourier analysis. See Problems 10.01 and 10.03.

10.03. With reference to Fig. P.02, find

$$\sum_{n=-\infty}^{\infty} \alpha_n^* \alpha_n$$

for each of the three examples shown.

10.04. Show that $\langle (x - x_0)^2 \rangle = \eta^2$ for the Gaussian function of (10.041).

10.05. Show that $(x - x_0) = \pm \eta$ are the inflection points of the Gaussian.

10.06. Consider the ground state $\Psi_0(x, t)$ of the infinite rectangular well, as given in (7.072), and the corresponding distribution function $\Phi_0(k_x, t)$ on momentum space.
  (i) Show that $\Phi_0(k_x, t) = \varphi_0(k_x) \exp - i\omega_0 t$.
  (ii) Find $\varphi_0(k_x)$.
  (iii) Plot $\varphi_0(k_x)$ with respect to the argument $k_x a$. Use a range long enough to show the essential features of the function.

10.07. Given a particle in the ground state $\Psi_0(x, t)$ of the infinite rectangular well. Suppose that, at $t = 0$, the infinitely hard containing walls at $x = \pm a/2$ are instantly removed. The particle then finds itself in free space ($V = 0$) with the wave function it had just prior to the removal of the walls. Write an integral expression for the future form of the wave function; do not attempt to evaluate.

11.01. If $a_1 = 2$, $a_2 = 4$, $a_3 = 7$, and $a_4 = 11$ and the probabilities are as given in the text, calculate $\langle A \rangle$ and ΔA for the distribution of Fig. 11.01.

11.02. Find Δx, Δp_x, and the numerical value of $\Delta x \, \Delta p_x / \hbar$ for:

(i) The ground state of the infinite rectangular well.
(ii) The first excited state of the one-dimensional harmonic oscillator.

12.01. Find $[\hat{L}_x, \hat{p}_z]$, $[\hat{L}_y, \hat{p}_z]$, and $[\hat{L}_z, \hat{p}_z]$.

12.02. Given the function $\psi = (3/4\pi)^{\frac{1}{2}} R(r) \sin \theta \cos \varphi$.

(i) Find $\hat{L}_x\psi$, $\hat{L}_y\psi$, $\hat{L}_z\psi$, and $\hat{L}^2\psi$.

(ii) Of which operator or operators in (i) is ψ an eigenfunction?

(iii) Find the α_l^m coefficients in the following expansion:

$$\psi = R(r) \sum_{l=0}^{\infty} \sum_{m=-l}^{l} \alpha \, Y^m(\theta, \varphi).$$

12.03. Prove that the maximum of $|F_{n,\,n-1}|^2$ occurs at n^2a/Z.

12.04. Consider the four Ψ_{2lm} states of the hydrogen $(Z = 1)$ atom.

(i) Write out the wave function in complete form for each state.
(ii) Write out $\Psi^*\Psi$ for each state.
(iii) Find the probability current density J for each state.

12.05. Show that the order of the degeneracy of the Ψ_{nlm} hydrogenic states is n^2,

13.01. Prove (13.029).

13.02. Notice that the C_{mn} in (9.026) and (9.028) are the matrix elements of (x) in the basis formed by the ψ_n functions of the one-dimensional harmonic oscillator.

(i) Find the matrix elements of $(\hat{p}_x)$ in the same basis.
(ii) Show that $[(\hat{x}), (\hat{p}_x)] = i\hbar(\hat{1})$ provided that the orders of $(\hat{x})$ and of $(\hat{p}_x)$ are recognized as infinite.

13.03. Find the eigenvalues and a set of orthonormal eigenvectors for the matrix:

$$\begin{bmatrix} 11 & -4i & -4i \\ 4i & 17 & 8 \\ 4i & 8 & 17 \end{bmatrix}.$$

13.04. Consider the 3×3 matrix $(\hat{L}_x)$ in the basis provided by Y_1^{-1}, Y_1^0, and Y_1^1.

    (i) Find the matrix elements of $(\hat{L}_x)$.

    (ii) Find the eigenvalues and eigenvectors of $(\hat{L}_x)$.

    (iii) Write out the complete wave function $\Psi(r, \theta, \varphi, t)$ for a hydrogenic atom with $n = 2$, $l = 1$, and $L_x = \hbar$.

13.05. Verify all the matrix elements of (13.112).

13.06. Given the well of Fig. 13.01 with V_0 and the χ_j functions as in Section 13.04, consider the second excited state of the well and find the energy E_2 and the function ψ_2 assuming that:

$$\psi_2 = \alpha_{02}\chi_0 + \alpha_{22}\chi_2 + \alpha_{42}\chi_4.$$

NAME INDEX

Albertson, J. S.   x

Balmer, J. J.   129
Bohm, D.   372
Bohr, N.   129–138, 168, 253, 281, 370
Born, M.   150, 156, 371
Borowitz, S.   372
Bragg, W. L.   147–149, 371

Compton, A. H.   138–142, 370
Condon, E. U.   371
Corben, H. C.   372
Crasemann, B.   372
Cropper, W. H.   372

Davisson, C. J.   143, 148–149, 370
de Broglie, L.   143–146, 370
Dicke, R. H.   372
Dirac, P. A. M.   150, 371

Ehrenfest, P.   246–254, 371
Einstein, A.   126–128, 132, 370
Emde, F.   371
Erdélyi, A.   371

Feynman, R. P.   277, 370–371
Foxworthy, J. E.   x
Franklin, J. L.   371
Friedrich, W.   371

Germer, L. H.   143, 148–149, 370
Goldstein, H.   ix, 372

Harnwell, G. P.   370
Heisenberg, W.   150, 283, 371

Ishiwara, J.   137, 370

Jacobi, C. G. J.   48
Jahnke, E.   371
Jammer, M.   371
Jeans, J. H.   112, 120, 370
Johnson, D. E.   371
Johnson, J. R.   371
Jordan, P.   150, 371

Kirchhoff, G. R.   110
Knipping, P.   371

Langenberg, D. N.   370
Leighton, R. B.   370–371
Livingood, J. J.   370
Lösch, F.   371
Ludford, G. S. S.   360, 371
Ludwig, G.   371
Lummer, O.   112
Lyman, T.   130

SUBJECT INDEX

OTHER TITLES IN THE SERIES IN NATURAL PHILOSOPHY